The GEOLOGY of AUSTRALIA

The Geology of Australia provides a vivid and informative account of the evolution of the Australian continent over the past 4400 million years. Starting with the Precambrian rocks which hold clues to the origins of life and the development of an oxygenated atmosphere, it then covers the warm seas, volcanism and multiple orogenies of the Palaeozoic which built the eastern third of the Australian continent. This illuminating history then details the breakup of Gondwana and development of climates and landscapes in modern Australia, and finally the development of the continental shelves and coastlines. Separate chapters cover the origin of the Great Barrier Reef, the basalts in Eastern Australia and the geology of the Solar System.

From Uluru to the Great Dividing Range, from sapphires to the stars, *The Geology of Australia* is a comprehensive exploration of the timeless forces that have shaped this continent and that continue to do so.

David Johnson holds an adjunct position as a Senior Principal Research Fellow in the School of Earth Sciences, James Cook University.

The GEOLOGY of AUSTRALIA

David Johnson

School of Earth Sciences
James Cook University

CAMBRIDGE UNIVERSITY PRESS

For my parents, Peter and Rua Johnson

PUBLISHED BY THE PRESS SYNDICATE OF THE UNIVERSITY OF CAMBRIDGE
The Pitt Building, Trumpington Street, Cambridge, United Kingdom

CAMBRIDGE UNIVERSITY PRESS
The Edinburgh Building, Cambridge CB2 2RU, UK
40 West 20th Street, New York, NY 10011–4211, USA
477 Williamstown Road, Port Melbourne, VIC 3207, Australia
Ruiz de Alarcón 13, 28014 Madrid, Spain
Dock House, The Waterfront, Cape Town 8001, South Africa

http://www.cambridge.org

© David Peter Johnson 2004

This book is in copyright. Subject to statutory exception and to the provisions of relevant collective licensing agreements, no reproduction of any part may take place without the written permission of Cambridge University Press.

First published by Cambridge University Press 2004
First paperback edition 2005

Printed in Australia by BPA Print Group

Typeface Sabon (Adobe) 11/13 pt. System QuarkXPress®

A catalogue record for this book is available from the British Library

National Library of Australia Cataloguing in Publication data

Johnson, David 1947 Nov. 5–.
    The geology of Australia.

    Bibliography.
    Includes index.
    ISBN 0 521 84121 6 (hbk).
    ISBN 0 521 60100 2 (pbk).

    1. Geology – Australia. I. Title.

551.700994

The illustrations in this book were partly sponsored
by Cannington Base Metals BHP Billiton

The author wishes to acknowledge the following institutions for permission to quote extracts of text from published work as follows: (pp. 103, 271) HarperCollins Publishers for permission to quote from pp. 87 and 240 of Laseron, C., 1954, *The Face of Australia*; (p. 270) Oxford University Press for Shelley, P.B., 'Ozymandias' from *The Oxford Book of Nineteenth-Century Verse*, 1965; (p. 103) the South Australian Department of Education and Industry Training for permission to quote from p. 411 of Howchin, W., 1918, *The Geology of South Australia*, Education Department; (p. 56) Penguin Group (Australia) for permission to quote from Carnegie, D.W., 1898, *Spinifex and Sand*, Penguin facsimilie edn, 1973. The sources for illustrations are given at the end of each chapter or under the illustration, and the author thanks those individuals and organisations for their assitance, especially the many colleagues who provided their pictures at not cost. All other photographs are from the author's personal collection. All diagrams have been completely redrawn, based on the author's original material or, in cases where an illustration has used material from other sources, these are acknowledged.

Contents

	Preface	ix
	Acknowledgements	xi
	Abbreviations and units	xii
1	**An Australian perspective**	1
	Australia: age, stability, climate, main features	1
	Box 1.1 Radiometric dating of rocks	3
	Box 1.2 What is geology?	13
	Australian geology	14
2	**The Earth: a geology primer**	19
	Model of the Earth	19
	Plate tectonics	21
	Box 2.1 Age-dating the rocks	22
	Minerals	30
	Types of rocks	32
	Box 2.2 Metamorphism	39
	Shaping of the landscape	41
	Box 2.3 The Australian regolith and soils	43
	Coastal and offshore areas	46
	Orogenic cycle	48
	Geological time scale	49
3	**Building the core of Precambrian rocks**	52
	The original Earth	52
	Archaean	55
	Proterozoic	60
	Box 3.1 Geology of Uluru and Kata-Tjuta	67
	Origin of life	68
	Box 3.2 Wilpena Pound and the Ediacaran fauna	71
	Supercontinents: Rodinia and Gondwana	73
4	**Warm times: tropical corals and arid lands**	79
	Part of Gondwana	79
	Explosive radiation of life	80
	Fossils	82
	Box 4.1 How are fossils preserved?	86
	Warm seas with arid plains, volcanic arcs and deep troughs	88
	Late Devonian upheaval	94
	Granites	95
	Box 4.2 Cooma – granite emplacement and metamorphism 435–433 Ma	96

5	**ICEHOUSE: CARBONIFEROUS AND PERMIAN GLACIATION**	99
	A glaciated continent	99
	The volcanic arc	106
	Development of the coal basins	108
	Box 5.1 *Glossopteris* and the vegetation of the cold-climate peatlands in Gondwana	110
	Box 5.2 Burning mountain: Mount Wingen	113
6	**MESOZOIC WARMING: THE GREAT INLAND PLAINS AND SEAS**	118
	Warm plains and then seas	118
	Box 6.1 The great extinction of life 251 Ma ago	119
	The great inland plains	120
	Box 6.2 The Sydney Basin	128
	Development of inland seas	131
7	**BIRTH OF MODERN AUSTRALIA: FLOWERING PLANTS, MAMMALS AND DESERTS**	139
	Australia emerges	139
	Box 7.1 Pollen data from brown coal and other Tertiary deposits	149
	The last 15 million years: cooling and growth of the ice-caps	151
	Box 7.2 Evidence for climate change	154
	Australia's arid interior	155
8	**EASTERN HIGHLANDS AND VOLCANOES BARELY EXTINCT**	164
	Volcanic provinces	164
	Box 8.1 Basalts as a source of gemstones	173
	Seamount chain offshore	176
	Origins of the volcanics and the Great Divide	177
9	**BUILDING THE CONTINENTAL SHELF AND COASTLINES**	182
	Origin of the outline	182
	Box 9.1 Australia's Exclusive Economic Zone	186
	Sea levels	188
	Types of coasts	190
	Box 9.2 Tsunamis	194
	Box 9.3 Coastal erosion problems	197
	The Australian coastline	200
	Box 9.4 Comparison of Sydney Harbour and Port Phillip Bay	207

10	Great Barrier Reef	213
	Introduction to reefs	213
	Box 10.1 Effects of cyclones on the Great Barrier Reef	214
	Reef types	217
	Reef deposits	221
	Formation of the Great Barrier Reef	223
	Box 10.2 Extent of terrigenous sediment in the Great Barrier Reef	227
	Continental slope and trough seaward of the Great Barrier Reef	229
11	Planets, moons, meteorites and impact craters	232
	Earth in context	232
	The Sun	232
	The planets	233
	Earth's Moon	236
	Meteorites	240
	Impact craters	241
	Box 11.1 Large meteorite impacts: Eltanin and Chicxulub	246
	Past and future of Earth in the Solar System	248
12	Cycles in a continental journey	252
	Global wandering	252
	Cycles of deformation	255
	Cycles of climates	259
	Evolution and extinctions	267
	Epilogue – lessons of geological perspective	269
	Index	273

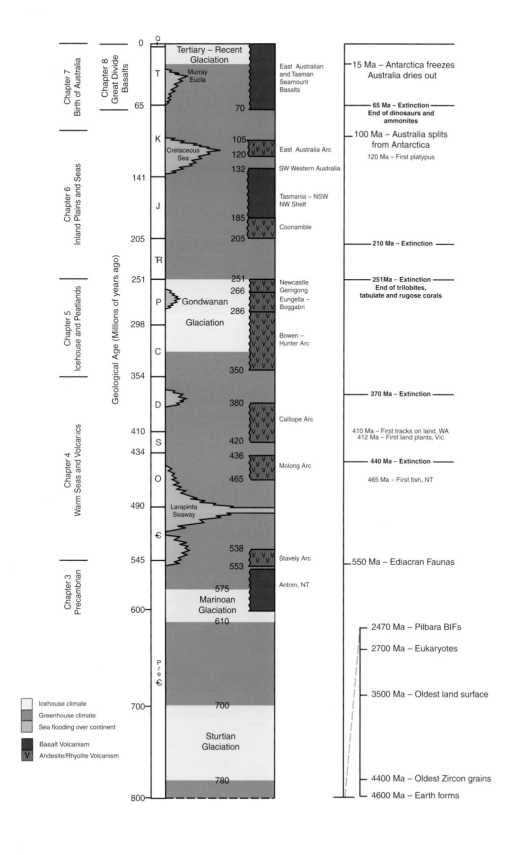

Preface

Most of the general books on Australian geology were written in the 1800s and early 1900s, and the popular classics by Charles Laseron, *The Face of Australia and Ancient Australia*, were published in the 1950s. A list of general books on the geology, soils and fossils of Australia is given at the end of Chapter 1.

It is time for a new summary of Australian geology, since so many new understandings have been generated in the last 50 years. This book is written basically in the order in which Australia formed, starting with the oldest rocks and working towards the most recent events. In this way we build Australia block by block, episode by episode, and also trace the development of the Earth's climate and life. The diagram at left shows the major events in Australian geological history. Geological time is written as 'millions of years' for general statements and as 'Ma' (mega-anna) when measured accurate dates are given for particular events.

This book uses a minimum of scientific jargon, though it is impossible to bypass all technical words. Indeed, in coming to terms with the scientific basis for many of the decisions we make about managing the Australian environment and commercial development we all need a smattering of technical knowledge. I have kept it to a minimum. Each technical term is explained in a geology primer (Chapter 2). Instead of a glossary, which merely defines the word using other technical terms, Chapter 2 briefly sets each in context. With respect to the sources of the data and theories summarised in this book, I have included the principal sources and other useful references at the end of each chapter, with a short list of websites.

Finally, this book is about the development of the Australian continent and the evolution of its major components and of the landscape in particular. Many of the localities mentioned are listed in the index, so those travelling can understand and appreciate the underlying geology. The map (p. x) shows many of the main localities referred to in this book.

However, it has not been possible to include details of the origin of our many world-class ores, coal and petroleum deposits, and of the economic geology which underpins so much of our quality of life – that will have to wait for another day.

Main localities mentioned

Acknowledgements

My particular thanks to my parents and especially my late father for fostering my interest in geology as a youngster and for the many outcrops and meat pies we sampled together. My thanks also to my wife and children who tolerated my absences from our family life on field trips.

I have been fortunate to study and work with a wide range of very knowledgeable geologists and other scientists during my time in industry and at the three universities with which I was associated as a student and staff member. It is impossible to single out every contribution made by these people – whether they were those who lectured me, my colleagues or my own students – but certainly all have contributed to my own knowledge and continued interest in Australian earth sciences. Because this is a general book I trust those scientists not cited individually will accept my apologies for I could not clog the text with every source as is normal in a scientific paper.

Many people have provided advice on source material and data, access to illustrations, or critical comments on sections of text, and the book has been much improved by their advice. I hope I have included everyone in this list and apologise if anyone has been omitted: Ross Andrew, Mark Barley, Al Bashford, Peter Betts, Alex Bevan, Ted Bryant, Gavin Birch, Ray Cas, Allan Chivas, Jonathan Claoue-Long, Lindsay Collins, Jim Colwell, Keith Crook, Mohinudeen Faiz, Michael Gagan, David Gillieson, Vic Gostin, Iain Groves, Bob Henderson, Chris Herbert, Robert Hill, Richard Jenkins, Barry Kohn, Bill Laing, Mark Leonard, Bernd Lottermoser, David Lowry, Ian MacDougall, Ken McNamara, Nick Oliver, Ken Page, Alex Ritchie, Mick Roche, Peter Roy, Mike Rubenach, Peter Schouten, Jeffrey Stillwell, Caroline Strong, Lin Sutherland, Fons VandenBerg, Peter Whitehead, Simon Wilde, Stephen Wroe, Ann Young, Bob Young.

My wife Patricia, Topsy and David Evans, Susan Allison, Alan Gillanders and other family members were a great help in proofing drafts of the text, and pointing where my explanations were unclear.

Stuart Johnson drafted most of the diagrams, and I am very grateful to him for his help, attention to detail and artistic balance of the visual material.

My special thanks to the anonymous referees, to Bob Henderson, and especially to the publisher, Jill Henry, and the editorial and production staff at Cambridge University Press, and editor David Meagher, for their help in bringing this volume to production. The accuracy and ideas are of course my responsibility.

DAVID JOHNSON
Herberton, April 2004

Abbreviations and units

Units follow the International System of Units (SI).

Length, thickness
- kilometres — km
- metres — m
- millimetres — mm
- micrometres — µm
- nanometres — nm

Slope
- metres per kilometre — m/km

Area
- square kilometres — km²
- hectares — ha

Volume
- cubic metres — m³
- cubic kilometres — km³

Density
- kilograms per cubic metre — kg/m³

Pressure
- kilobars — kbar

Atmospheric pressure
- hectopascals — hPa

Temperature
- degrees Celsius — °C
- Kelvin — K

Time
- year — yr
- second — s
- million years (10^6) — Ma
- billion years (10^9) — Ga

Speed
- kilometres per second — km/s
- metres per second — m/s
- millimetres per year — mm/yr

Vertical height
- metres above sea level — m ASL

CHAPTER 1 *Australia: age, stability, climate, main features*

AN AUSTRALIAN PERSPECTIVE

Australia – a continent of great age, stability, aridity and flatness.

Radiometric dating is one of the keys to reading the story in the stones.

How has deep sea and space exploration changed our perspective on the Earth and Australia?

Some call Australia the world's largest island, others the smallest continent. I want to describe Australia as a continent, one of the primary building blocks of planet Earth, and one which spans almost the entire record of Earth history.

Traditional stories of the Aboriginal people tell of the creation of this continent, and of many of its landscape features. These stories are a parallel to this book because they emphasise the close spiritual relationship between all of us and the land that supports us.

This book tells how the rocks formed, and how the present climate developed over millions of years. As the Reverend J. Milne Curran wrote in one of the earliest explanations of Australian geology, in 1898:

> Australia has a history far more ancient than any written by men – to read this history is one of the objects of geology – records preserved in the great stone-book of nature.

There is evidence that the Chinese knew of a land in the south, from perhaps as early as the Sui dynasty (AD 589–618), and that Chinese ships had mapped part of Australia in 1422, even landing to search for metalliferous ores.

The first documented European discovery of this continent was by the Dutch commander Willem Jansz, who sighted the 'Great South Land' in 1606. He had sailed the *Duyfken* around the southern coast of New Guinea early in 1606, and veered southward to 13 degrees 45 minutes latitude. Here a party landed, on the eastern side of the Gulf of Carpentaria, before heading back to Java. He named the point *Cabo Keerweer* (Cape Keerweer), which means: 'Cape Turnabout' in Dutch. It lies at the mouth of the Kirke River, about 70 km south of Aurukun. Thus a small, remote sandspit on a flat coastline was the first European name given to this continent.

Later that year the Spanish captain Luis Vaez de Torres sailed from the New Hebrides (Vanuatu) through the gap south of New Guinea now called Torres Strait, on his way to the Philippines. Torres did not realise what lay on the southern side of his vessel.

The early Dutch explorers named this continent New Holland, but they decided there was not much of value in spices or precious metals to warrant further exploration. Then came the main English and French explorations in the late 1700s. James Cook charted the eastern coast of Australia, and later the French explored much of the western coast. To Cook this was still 'New Holland'.

The Latin term for 'southern land' is *terra australis*. Although this term had been used as well as New Holland, it was the man who first circumnavigated and mapped the coastline, Matthew Flinders, who gave real impetus to the present name of Australia. He published an account of his discoveries in his great work *A Voyage to Terra Australis*, a printed copy of which was placed in his unconscious hands on 18 July 1814, the day before he died.

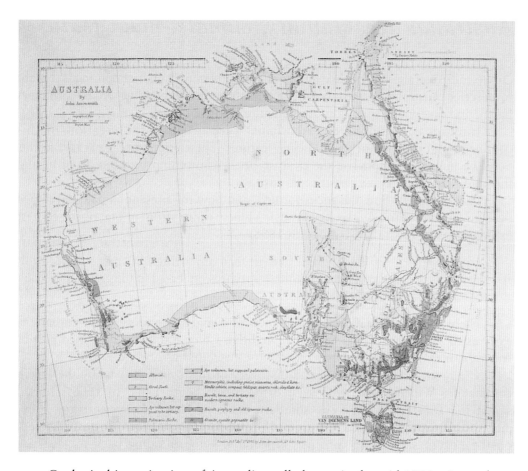

Figure 1.1 *Early map of the geology of Australia by J. Beete Jukes, published about 1850, showing the geology mapped along the coast but a dearth of information on the inland. Tasmania is labelled as Van Diemen's Land. There is detail on the Great Barrier Reef, since this was also investigated by Jukes during his time in Australia.*
SOURCE: THE MITCHELL LIBRARY, STATE LIBRARY OF NEW SOUTH WALES.

Geological investigation of Australia really began in the mid 1800s. So we have about 150 years of information to draw upon. Perhaps the earliest geological map of Australia was that prepared for publication in 1850 by J. Beete Jukes, who had been the naturalist on the survey vessel HMS *Fly*. A map of the whole continent shows old rocks around the edge, and the centre completely blank.

For many people in those days Australia was so different to Europe that extraordinary theories were proposed to explain the differences. The Reverend W. B. Clarke, in an address to the Philosophical Society of New South Wales on 20 November 1861, took time to refute the following notions:

> It was once maintained by a philosopher of some eminence that New Holland, being so singular and anomalous a region, must have originated in a corner of the sun knocked off by a comet, and that, tumbling into the Pacific Ocean, it soon became the suitable abode of those bizarre marsupials...which are only an imperfect development of life upon the more recently raised lands, such as Galapagos and Australia.

All these early explorers, and the naturalists who sailed with them, had noted the strange animals and plants. Australia did not fall out of the Sun, but it is still a very individual place with unusual animals and plants inhabiting an extraordinarily diverse landscape.

Age

To early theorists, Australia was not only very young but also populated by totally different life to that known from the natural worlds of Europe, America and Africa. They thought they had discovered a completely new existence, something which must have formed *after* their established order. However this theoretical stance was not supported by the rocks. In the 1800s many geologists spent time confirming that there were fossils and rocks in Australia that were demonstrably of the same age as those in Europe. Prominent early geologists in New South Wales were the Reverend W. B. Clarke, the Reverend J. Tenison-Woods, the Reverend J. Milne Curran and C. S. Wilkinson, in Queensland Robert Logan Jack, in Victoria Alfred Selwyn, and in Western Australia Dr W. G. Woolnough.

We know now that Australia is very old. Continental crust in Western Australia has been dated at older than 3500 million years, and metamorphosed sedimentary rocks in the Yilgarn region contain zircons that are even older, dated at 4404 million years. This implies that a crust of rocks was formed before 4400 million years ago and then eroded, with the zircons subsequently deposited in sediments. It is the oldest material known on Earth, and is only about 200 million years younger than our present estimates for the age of the planet. The youngest materials are still forming in the river systems, beaches and reefs around our coastline.

It is also clear that much of the fossil life has links to the other continents. In particular, the similarity of rocks around 250–300 million years old in such distant places as Australia, India, Africa and South America suggested these continents were once joined, and this observation led to the theory of continental drift. Scientists in the late 1800s and early 1900s proposed that a huge continent, which they called Gondwana, broke up, and that the fragments then drifted apart. But at that time there was no known mechanism to effect such massive changes, and the theory was not strongly supported.

In the 1960s new evidence showed how the break-up might have happened, and the theory of plate tectonics was developed. We now know Gondwana was not the only supercontinent: there was an earlier one, which we call Rodinia. The rocks in Australia have an extraordinary story to tell, starting before the formation of any other rocks still preserved on the planet.

Box 1.1 Radiometric dating of rocks

In the 1800s geologists could only interpret the *relative* ages of rocks. Guesses of the actual ages were made, but there was no way to do absolute dating. It was the discovery of radioactive decay in 1896 by the French physicist Henri Becquerel, and later work by the Curies, which provided the means. The key fact is that radioactive elements decay to daughter products in a special way. The daughter products can also be radioactive and lead to another decay cycle, though most commonly they are stable and not radioactive. For each radioactive pair (parent and daughter) there is a period of time – the half-life – after which exactly half the original number of radioactive parent atoms remain, half having been transformed into daughter atoms.

For instance, radioactive uranium atoms decay to stable lead atoms. In 1907 Boltwood had found that the ratio of the decay product (lead) to the parent radioactive uranium increased with geological age. This discovery confirmed the potential of radiometric dating to provide accurate measurements of the absolute age of rocks.

The radioactive decay process is random, and it is not possible to predict when an individual atom will disintegrate. However, given a large number of atoms, it is possible to predict when half will have decayed. It is like watching a heated pan of popcorn. You cannot tell which kernel will pop next, but you can predict how long it will take to finish the batch.

For the decay of uranium-235 (U-235) to lead-207 (Pb-207) the half-life is 704 million years. That is, over 704 million years, or one half-life:

1000 atoms of U-235 → 500 atoms of U-235 + 500 atoms of Pb-207

After a further half-life, or a total of 1408 million years:

the remaining 500 atoms of U-235 → 250 U-235 + 250 Pb-207

which, with the 500 Pb-207 already formed, will mean there would be 250 U-235 plus 750 Pb-207. Note that this implies that the Earth was far more radioactive earlier than it is now.

We can determine very accurately both the amounts and the decay rates of radioactive elements, so we can then calculate the age of the specimen. Samples for dating are selected very carefully to avoid rocks that are altered, contaminated, or disturbed by later heating or chemical changes. Multiple samples are taken to provide cross-checking.

Different radioactive isotopes are used for different age ranges, depending on the half-life of the decay process. The following pairs have the stated half-lives:

rubidium-87 to strontium-87	48800 million years
uranium-238 to lead-206	4470 million years
potassium-40 to argon-40	1250 million years
uranium-235 to lead-207	704 million years

A common method for measuring the amounts of each isotope is thermal ionisation mass spectrometry (TIMS). The isotopes are extracted from the rocks and vaporised over a hot filament, and the gas is then analysed in a mass spectrometer.

The latest technique involves using a high-power electron microscope – the superprobe – to analyse minute amounts of radioactive materials in mineral grains such as monazite.

These elements are useful for dating rocks up to hundreds of millions of years old. But what about younger materials? In 1951 Walter Libby discovered that minute

Figure 1.2 *Radioactive decay of uranium-238 and uranium-235. (a) Parent and daughter atoms and half-lives. (b) Decay curves over 6000 million years, showing the more rapid decay of U-235, which has a shorter half-life (704 million years) than that of U-238 (4470 million years). Note that when the Earth formed there would have been about twice as much U-238 and 32 times as much U-235 as at present. The diagram also emphasises the enormous time span of the Precambrian era, which spans most of the history of the planet.*

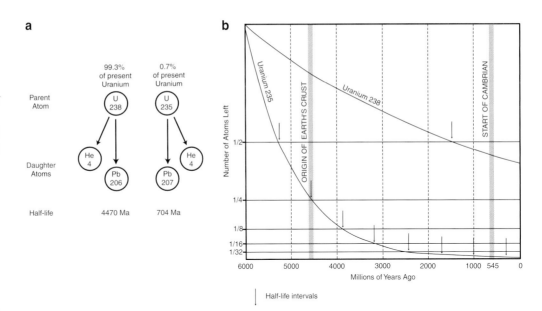

Figure 1.3 *Electron microscope image, taken in back-scatter mode, of microcraters formed in a cut and polished zircon grain during SHRIMP analysis. The grain is 150 micrometres wide and the microcraters are the eight, slightly darker, rounded blotches in the grain. This is the oldest grain dated on Earth – 4404 million years old.*
IMAGE: JOHN VALLEY AND SIMON WILDE

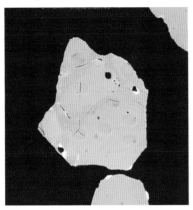

amounts of a radioactive isotope of carbon – carbon-14 (C-14) – exist in air, natural waters, plants and organisms. Radioactive C-14 is produced continually by cosmic ray bombardment of nitrogen atoms in the atmosphere. The C-14 decays to nitrogen-14, with a half-life of 5568 years. Radiocarbon dates are normally reported as 'before present' (BP), with reference to the year 1950. So if carbon from prehistoric wood is found to have only half the amount of C-14 present when compared to 1950 wood, the estimated age for the prehistoric wood is 5568 years. The age would normally be expressed as 5570 ± 150 years.

The natural production of C-14 has not been constant over time. Ages must therefore be calibrated, for instance by reference to C-14 dating of trees whose true ages have been measured by counting the tree rings.

SHRIMP

Australian geologists have made a speciality of dating ancient rocks, and some of the technology was developed at the Australian National University (ANU). Sensitive high-resolution ion microprobe (SHRIMP) involves very detailed uranium–lead age dating of individual grains, typically zircon, within a rock. The very fine microbeam can also probe and date individual growth zones within a single grain.

Previous radiometric dating was limited to analyses of whole rocks or of groups of grains extracted from the rock, and almost always for igneous rocks. The SHRIMP capability enables absolute dating of individual grains.

The SHRIMP date records the age of the magmatism (or rarely, metamorphism in high-grade rocks) which formed the zircons, and hence the age of the original rock. If the rocks are uplifted and eroded the zircons retain their age signature, even when recycled from igneous or metamorphic rocks into sedimentary rocks, and weathered again into loose sand. Thus even the age of grains in modern beach sands can be determined.

Stability

Australia is tectonically stable and has no active mountain building or major fault systems. We have no mountains as young as those in New Zealand or the Himalaya, nor faults like the San Andreas Fault in California. We have no active chains of volcanoes like those in New Zealand, Japan or Indonesia. Active tectonism, involving earthquakes and volcanism, occurs mainly at the boundaries of the large plates that make up the outer part of the Earth. Australia is situated well within a very large plate.

Most earthquakes occur in the outer crust of the Earth, where rocks are cold enough to be brittle. At greater depths, where the temperature is high enough for the rocks to deform by plastic flow, there is rarely sudden fracturing to trigger an earthquake. In eastern Australia earthquakes occur at depths of up to about 20 km. In very active seismic zones, like Indonesia or Fiji, earthquakes occur from close to the surface to depths of 700 km. Shallow earthquakes can cause much damage, but deep earthquakes rarely do.

Australia is just one landmass sticking above the waves, riding on a much larger plate on the surface of the Earth. This plate is the Indo-Australian plate which contains parts of New Zealand and Fiji in the east, and India in the northwest. We are all travelling northeastwards towards Asia at about 70 mm/yr.

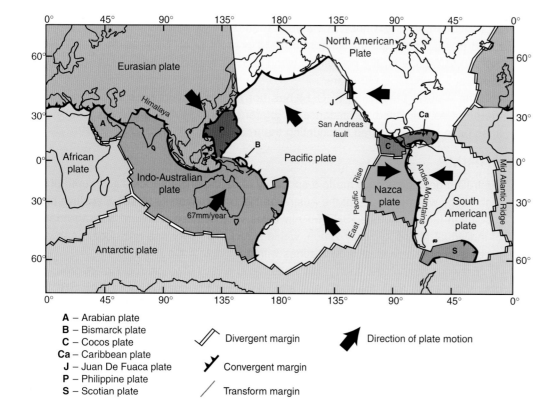

Figure 1.4 *The Earth's surface consists of interlocking plates. Most plates comprise both continents and ocean floors. The original concept of continental drift emphasised continents because that was where the rocks were mapped and the connections drawn. However, we now use the term 'plate tectonics' because it is the plates that are moving, and the continents are being carried upon them. The main plates are named in the figure. The minor plates are noted.*

The three types of plate margin are shown: divergent (such as the mid-ocean ridges), convergent and transform. The teeth on the convergent margins point from the downgoing plate into the over-riding plate. Two major, modern mountain belts have formed at convergent margins: the Himalaya and the Andes.

The Indo-Australian plate carries two continental masses (Australia and India) and three smaller masses (Papua New Guinea and the north island and the western side of the south island of New Zealand). To the south of Australia the ocean floor is spreading as the Indo-Australian plate moves away from Antarctica. To the north, the plate is converging with Asia.

The plate motion shown is as measured at Darwin: 67 mm/yr in the direction 35° east of north.

REPRODUCED AND MODIFIED WITH PERMISSION OF THE MCGRAW-HILL COMPANIES

To put this in perspective, consider that in 216 years of White settlement the Australian land mass has moved about 15 m closer to Asia. It is hard to notice this when there is no permanent marker by which to judge the movement. However, imagine if this movement was occurring in the street outside your front door, so that one side of the street was stable and the other was moving sideways at 70 mm/yr. A house built directly across the street by members of the First Fleet would now be down the road, and the block next to it would be opposite you. In 50 000 years – a minimal estimate for the time Aboriginal people have been here – the same block of land would have moved three and a half kilometres.

Despite the stability of the plate there are small earthquakes within Australia (intra-plate tectonism). The reasons for these earthquakes are not obvious in comparison to those along major plate boundaries. Some earthquakes are caused by new movements on old faults, and some may be caused by heating effects in the deep sedimentary basins.

Analyses within drillholes and mines do show that most of the continent is being gently compressed towards a northeast–southwest axis, which lies parallel to the plate movement. Some compression is to be expected because the plate is starting to be forced against the edge of the Eurasian and Pacific plates to the north. The southeastern corner shows more east–west compression, perhaps reflecting the nearby boundary at New Zealand, which is transmitting some pressure inwards towards Australia. More details on plate tectonics are given in Chapter 2.

Low-intensity earthquakes occur in Australia almost daily. Early White settlers noted slight earthquake shocks; the first was felt by Governor Phillip on 22 June 1788. However, most tremors are never felt by people even though they are detected by sensitive instruments. Fortunately Australia is sparsely populated so that most significant earthquakes cause little damage.

The most common method of describing the size of an earthquake is the Richter magnitude scale. The ground displacement is measured with a seismograph, and a correction for the distance from the earthquake epicentre to the seismograph is applied. Each unit increase in magnitude represents a tenfold increase in ground displacement and about a thirtyfold increase in the seismic energy released.

Another measure of earthquake size is the moment magnitude, derived from mathematical modelling of the surface wave generated by the earthquake. Moment magnitudes are better measures of the energy released, especially for earthquakes above Richter magnitude 7. The 1960 Chile earthquake, which measured 8.5 on the Richter scale, had a moment magnitude of 9.5.

A significant earthquake is followed by smaller events, called aftershocks, and is sometimes preceded by foreshocks. Worldwide, for every magnitude 5 earthquake, 10 magnitude 4 earthquakes are expected, 100 magnitude 3, 1000 magnitude 2, and so on.

A magnitude 1.0 earthquake (commonly written as M1) releases a similar amount of seismic energy as a typical quarry blast. A magnitude 5.0 earthquake releases about the same seismic energy as the explosion of 10 000 tonnes of TNT – about the energy released by the atomic bomb dropped on Hiroshima. Magnitude 9 earthquakes, such as those in Chile in 1960 or Alaska in 1964, are extremely damaging.

Australia has had large earthquakes, but they occur very infrequently, and on average an earthquake exceeding magnitude 7 occurs somewhere in Australia every 100 years or so. Such an earthquake would affect only a small area. Any particular place in Australia would only expect to be within 50 km of a magnitude 7 earthquake about once every 100 000 years. Seismically active areas such as Japan, New Guinea, the Philippines and California experience magnitude 7 earthquakes every few years.

Significant earthquakes in Australia

There have been several major earthquakes recorded in Australia over the past 160 years. The largest was magnitude 7.1 at Meeberrie, west of Meekatharra in Western Australia, on 29 April 1941. Damage was limited because of the isolated location, but included cracked walls in farm houses and burst water tanks.

The largest earthquake measured in South Australia was the Beachport earthquake on 10 May 1897. At magnitude 6.5 it caused serious damage at Kingston, Robe and Beachport, and caused minor damage even in Adelaide. It was felt as far away as Port Augusta and Melbourne. It is thought that the epicentre was offshore.

In New South Wales the Newcastle earthquake on 28 December 1989 resulted in 13 deaths, over 120 injuries, and building damage exceeding AU$1.5 billion. Although of only magnitude 5.6 its location in a major city meant there was serious damage to life and property.

The Tennant Creek series in the Northern Territory involved three tremors between magnitudes 6.3 and 6.7 on 22 January 1988. They were felt in Darwin, and the largest was felt as far away as Cairns in northern Queensland and in high rise buildings in Perth and Adelaide. In Tennant Creek, walls were cracked in well-constructed buildings, objects fell from shelves, and furniture was shifted. There were no serious injuries, but the strong shaking from the main events and the prolonged aftershocks badly frightened many residents. The epicentres were in the desert about 40 km southwest of the town, so total damage was limited to about AU$1.2 million (1990 values) – mainly damage to a high-pressure gas pipeline between Alice Springs and Darwin.

Some earthquakes produce small fault scarps where they displace the ground surface. The Tennant Creek earthquakes produced a surface rupture about 35 km long trending east–west, with the southern block thrust over the northern block by up to one metre vertically and about 2 m horizontally. The Meckering earthquake in Western Australia on 14 October 1968 also produced a surface fault rupture, 32 km long and trending north–south. The land east of the fault was lifted vertically by up to 1.5 m, and moved westwards by up to 2 m. Thus the eastern block was thrust over the western block, indicating strong compression of the crust.

Southwestern Australia has experienced several earthquakes: Meckering (1968), Cadoux (1979) and the Burakin series, which started in July 2000 but had

Figure 1.5 *Earthquake occurrences around and onshore in Australia, 1841–2000. Note the dense concentration of high magnitude earthquakes in Indonesia and Papua New Guinea associated with the convergent plate margin.*
Figure courtesy of Geoscience Australia, Canberra. Crown copyright ©. All rights reserved. www.ga.gov.au

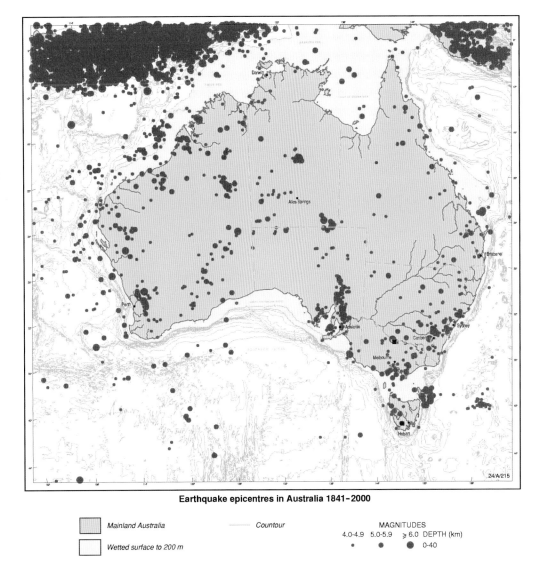

its most significant events on 28 September 2000 (M5.0), 5 March 2001 (M5.1), 23 March 2002 (M5.1) and 30 March 2002 (M5.2). The Meckering quake (M6.7) destroyed a bank, hotel, shire hall, three churches and 60 of about 75 houses, with total damage about A$29 million (1990 values). There were no fatalities and only minor injuries, largely because the earthquake occurred in mid-morning.

Despite the traumas at Newcastle, earthquake intensity on the Australian land mass is relatively low. We do not experience the frequent strong quakes associated with seismically active regions such as Indonesia, Japan, New Guinea and New Zealand. On a global scale this is a stable continent. But as the national population grows and more buildings and infrastructure are developed in Australia, even low-intensity quakes may cause damage and even loss of life.

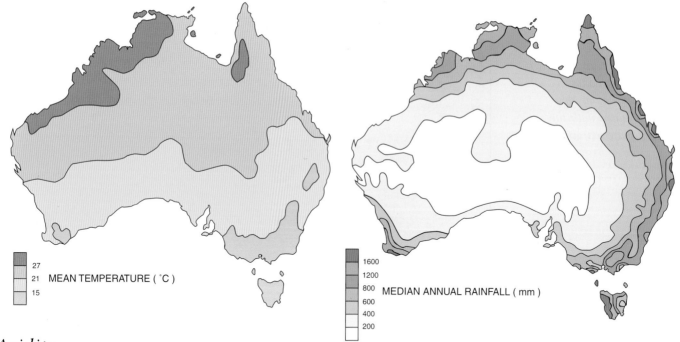

Figure 1.6 *Annual rainfall and temperature patterns in Australia. The patterns are not the same. Temperature decreases in a north–south direction, while the rainfall pattern is concentric, decreasing with distance inland.*
Figure courtesy of Geoscience Australia, Canberra.
Crown copyright ©. All rights reserved. www.ga.gov.au

Aridity

The present Australian climate is overwhelmingly arid: one third of the continent is desert and another third is semi-arid. The temperature gradient is primarily a diagonal trend across Australia, with the hottest regions in the north and especially northwestern Australia, and the coolest in the south and especially in the southeast. Rainfall zones are more or less concentric, with most of Australia receiving less than 400 mm annually.

The country has a wide climatic range, from wet tropical in northeastern and northwestern Australia, to arid deserts in the Centre, to humid temperate regions across most of the coastal southern continent, with seasonal freezing and snow on the southeastern highlands and in Tasmania.

The extremes are evident in the facts. The highest recorded temperature is 53°C at Cloncurry in Queensland in 1889. Marble Bar in northern Western Australia recorded the longest hot spell – 160 days over 37.5°C in a year. The lowest recorded temperature is –23°C at Charlotte Pass in New South Wales in 1994. In terms of rainfall, the driest area is Lake Eyre with an average annual rainfall of 125 mm, and the wettest is Mount Bellenden Ker in northern Queensland with a median annual rainfall of 4048 mm. In 1979 Bellenden Ker had the highest recorded annual rainfall: 11 251 mm, or over 11 metres.

Yet the climate has not always been so. Much of Australia's geology, and its present landscape and soils, formed in vastly different climates. There were times when almost the whole of Australia was glacial.

We tend to think of climate mainly in terms of human comfort and the requirements of farmers and graziers. However it is also important in understanding the geology. Climate determines the types of sedimentary rocks that form, and especially the accumulation of peat, which later changes to coal. Climate is a principal factor in the weathering and erosion of rocks, and hence the change from rocks into soils, and the chemistry of groundwaters. It also affects the development of landscapes and the liberation of sedimentary materials into the river systems. In many places climate controls the development of valuable ores concentrated in the soil zones; one example is bauxite, which is the main source of aluminium.

Flatness

From a long-distance perspective Australia is one of the lowest and flattest land areas on Earth. The average height is 330 m. The highest point, Mount Kosciusko, is only 2228 m above sea level, and Lake Eyre is the lowest point, 16 m below sea level. The total range of topographic relief on the onshore land is less than 2250 m. Offshore, around the edge of the continental shelf, there are steep undersea cliffs where the continental mass drops into 3000–5000 m of water in the ocean basins.

The Great Divide forms a prominent spine along much of eastern Australia, and there are ranges in central Australia and elsewhere (Figure 1.7), but the continent is dominated by a series of broad plains. The flat terrain means drainage patterns are also poorly defined for much of the continent. There are four drainage groups (Figure 1.8).

The first is a series of relatively short river systems that drain towards the coast and form a ring around all but the southern margins of the continent. The second is the major active river system of the Murray–Darling, which drains most of inland New South Wales and part of southern Queensland towards Adelaide. The third is a group of well-defined internal drainage systems, especially the Lake Eyre system but also including the Amadeus, Bulloo, Frome and Torrens systems, which end in large salt lakes. The fourth is an immense area within western and central parts of the continent where the ground is so flat that water flows as sheets down local slopes, ending in a poorly connected series of salt lakes that mark the paths of very old drainage systems.

The Murray–Darling, the Lake Eyre system, and these salt lake chains represent drainage patterns at three stages of landscape development. The Murray–Darling system is still young and very active; its headwaters are sculpturing the ranges, and the water not used for crops or stock eventually reaches the sea through Lake Alexandrina in South Australia. The Lake Eyre and associated systems are less active except after major inland rains, and flow more slowly, ending up in the large lakes.

The western systems represent the remnants of major river systems that were active millions of years ago, when the hills were higher and the valleys well defined. The landscape now is too dry and flat to channel flows, and the watercourses are rarely well defined. However, as we will discover, these are just the senile remnants of systems that were very vigorous rivers millions of years ago. Unless there is an uplift of new mountain ranges, eventually all of the continent will be reduced to this state.

Figure 1.7 *Oblique three-dimensional image of Australia and surrounding ocean regions, showing land masses in brown, shallow seas in yellow (< 2000 m water depth), and deeper sea floors in blue (2000–4000 m) and green (> 4000 m).*

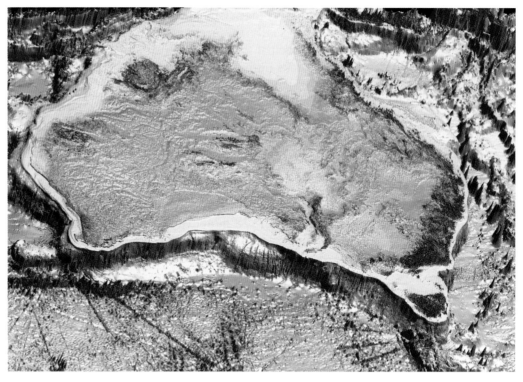

The overall flat terrain is obvious. The Great Divide marks a spine of higher country along eastern Australia. Remnants of old mountain ranges, now much reduced in size by erosion, are evident in the Mount Lofty Ranges of South Australia, the MacDonnell, Musgrave and Petermann Ranges in Central Australia, the Hamersley Range in the Pilbara region and the King Leopold Ranges in the Kimberley of Western Australia.

Geologically the continent extends to the edge of the continental shelf, with a very steep drop to the surrounding deep-sea floor. Note the extensive continental shelves that connect the Australian mainland to Papua New Guinea in the north and to Tasmania in the south. The shallow area on the eastern side of the image is the northern part of the Lord Howe Rise, which broke away from the Australian mainland during the opening of the Tasman Sea, between 84 and 48 million years ago (see Chapter 7). The continental shelves rise steeply from the deep ocean floors and are incised along their edges by numerous canyons. During the lower sea-levels of the ice ages these continental shelves were largely dry, and there was easy access to and from Papua New Guinea.

Adjoining plateaus – pieces of the Australian continental mass which were rifted apart but did not drift far during breakup of Gondwana (see Chapter 9) – are also obvious: the Queensland and Marion Plateaus, the Exmouth Plateau off the North West Shelf, the Naturaliste Plateau off the southwestern corner, and the South Tasman Rise, which is tenuously linked south of Tasmania.

East of Australia lie two north–south seamount chains: the Tasman chain, and further east the Lord Howe chain. These mark successive volcanoes developed over a mantle plume hotspot as the Indo-Australian plate drifted northwards (see Chapter 8).

(Note: The lines, like scratches, in the seafloor south of Australia are artifacts of the data collection: they represent concentrations of data along ships' tracks.)

FIGURE COURTESY OF GEOSCIENCE AUSTRALIA, CANBERRA. CROWN COPYRIGHT ©. ALL RIGHTS RESERVED. WWW.GA.GOV.AU

Figure 1.8 *Major drainage divisions, showing coastal drainage, the Murray–Darling system, the internal Lakes Eyre–Bulloo–Frome–Torrens and desert drainage, upland areas of irregular local drainage, and the salt-lake palaeo-drainage system of Western Australia – South Australia – Northern Territory.*

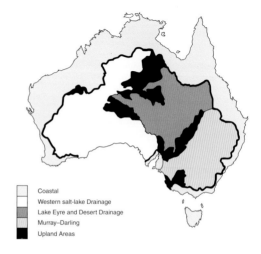

Coastal
Western salt-lake Drainage
Lake Eyre and Desert Drainage
Murray–Darling
Upland Areas

Figure 1.9 *Students on a geology field trip examining Devonian limestones with fossil corals on the Burdekin River near Charters Towers, north Queensland.*
PHOTO: DAVID JOHNSON

Box 1.2 What is geology?

Geology is the science that deals with the materials that constitute our Earth, and with the dynamic changes that have affected the Earth from its origin about 4600 million years ago to the present time. Geology also encompasses the study of the surfaces and interiors of other bodies in the Solar System, to place Earth in its planetary context.

Geologists and other earth scientists interpret the geological history of a region from the features preserved in the rocks exposed at the Earth's surface. They also use subsurface information obtained from drillholes, quarries, cuttings, caves, mines and geophysical surveys.

Geology provides a time perspective for other natural sciences that are more concerned with the present. It also investigates former ecosystems and traces the development of past fauna and flora from their very beginnings through to the present. Geologists can read past climates from the rock sequences, to help understand the climatic changes that the Earth is now experiencing.

Palaeontology is the study of fossils and the evolution of life. By comparing fossils with modern animals and plants, palaeontologists can interpret past conditions for life on the planet.

Geophysics is the study of the Earth by the measurement of the effects of physical processes. Examples include recording of earthquake travel times and seismic waves to understand the internal structure of the Earth, or small changes in electric and magnetic fields to determine the nature of subsurface rocks. Geophysics also involves the mathematical modelling of internal Earth processes, such as the flow of solid materials, to understand deformation of rocks, movements of the crust and continents, and circulation in the core which is the dynamo for the Earth's magnetic field.

Geomorphology is the particular study of surface landforms and the processes that created them under the wide variety of climatic and other natural conditions that have prevailed on Earth.

Geochemistry involves specialised studies of the chemistry of rocks and associated fluids. The studies involve both inorganic and organic chemistry. Inorganic studies include the analysis of igneous lavas and rocks to determine their origins and crystallisation histories, the analysis of sediments to unravel climate changes, and the analysis of metalliferous ores to determine their origins. Organic studies include probing the burial, heating history and formation of coals and petroleum, and searching for signs of life in meteorites from outer space.

Geologists are primarily people who have a strong interest in the natural world and want a professional career that provides outdoor activity as well as scientific challenge. Geologists are employed by exploration and mining companies, by government organisations, as consultants, by financial firms, and as researchers and teachers by universities and schools.

Contrary to a popular belief that geologists are only interested in ripping up the ground for mining, most geologists are very conscious of the delicate balance of the world's environment on the land, in the sea and in the atmosphere. That is one of the messages in rocks. Many geologists are intimately involved in ensuring that the extraction of ores and energy is done with minimal impact on the environment.

Australian geology

Why we know more than we did 50 years ago

There are two main reasons why our understanding of Australian geology is far different to that in the mid 1900s. Firstly there has been an unprecedented level of exploration of the Earth at an international level due to advances in technology, especially electronics and imaging from space after World War II. Secondly, our knowledge of the Australian land mass has increased with the expanding population and development of the country. At the international level we can consider three examples.

The development of sonar during the World War II started as a relatively simple exercise using sound pulses to detect the position and distance of an underwater object such as a submarine. This has now been extended to sensing the nature of the seabed to form images analogous to air photos, and also to using much higher powered seismic energy to penetrate the Earth and decipher its structure. Paralleling this work was the electronic acquisition of vast amounts of data on the magnetic and gravitational properties of the ocean floor and the continental margins. With the capabilities of modern computers we can process these data to reveal much about the origin of the rocks.

Also the most comprehensive, international co-operative scientific program ever mounted began with the start of the Deep-Sea Drilling Program (DSDP) in the 1960s. The DSDP program has evolved into the Ocean Drilling Program (ODP), of which Australia was a member. The next phase, the International Ocean Drilling Program (IODP), started in 2003.

The original DSDP was designed to allow scientists to recover cores of the deep-

Figure 1.10 *The ODP drill ship* Joides Resolution, *which can drill holes in the seabed in water 5 km deep.*

PHOTO: JOHN BECK, OCEAN DRILLING PROGRAM

sea bed, in water depths up to 5000 m, and to 1000 m below the seabed. The core is 50–100 mm in diameter. You can imagine the scene when the drill is extracted to recover one length of core, and the hole is to be re-entered for the next coring run. The job is to re-enter a small hole in 5000 m of water with the ship buffeted sometimes by gale-force winds, and currents moving at up to 2 metres per second. The process has been likened to taking a pen refill on the end of a piece of thread and lowering it into the pen tube wedged in a crack on the road pavement many storeys below, while sitting on a rocking horse on the top of a tall city building.

Why the effort? Geologists thought that the seafloor was spreading open from the mid-ocean ridges. If this were so, the ocean floor should be younger near the mid-ocean ridges and progressively older farther away. How could we sample and then date this seafloor in 5000 m of water? The DSDP was born to recover rocks for dating from the deep ocean floor.

The results were worth the effort, for they proved beyond doubt that the age of the ocean floors increases symmetrically out from the mid-ocean ridges towards the continents. It proved that the ocean floors are moving. New ocean floor is generated by basalt erupting at the ridges and spreading outwards. For the first time we know how fast the crust is moving, and how fast the inner parts might be churning around. Recent work using the Global Positioning System of satellites (GPS) has allowed us to measure these movements on a scale of decades.

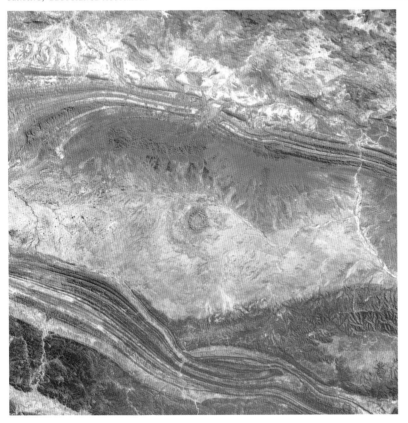

Figure 1.11 *Satellite image, approximately 100 km across, of the MacDonnell and James Ranges west of Alice Springs. The meteorite crater of Gosse Bluff is in the centre of the image. The Finke River runs southwards in the eastern part of the image. The image is falsely coloured to accentuate the trends of the rock ridges.*
IMAGE: COPYRIGHT © COMMONWEALTH OF AUSTRALIA (2000). LANDSAT ETM SATELLITE IMAGE ACQUIRED BY THE AUSTRALIAN CENTRE FOR REMOTE SENSING, GEOSCIENCE AUSTRALIA

Finally, the ability to obtain images of Earth and the other planets and moons from vehicles in space has increased our limited perspective of this planet. Before the advent of high-altitude aircraft and satellites we could only see small areas at one time, perhaps a few kilometres across on a single photo. Now, individual satellite images provide coverage of areas 60–70 km across. The images allow us to see how rock masses relate to each other on a broad scale, show us the large fracture systems that cut through the Earth's crust, and help us understand how it all came together.

We know now that the Earth is just one example of a rocky planet in the Solar System, positioned at a delicate distance from the Sun where it is not too close for the atmosphere to be blown away by the heat and the solar wind, and not too distant to freeze water. We know that the Moon and probably Mercury have rocks similar to those on Earth. Each of these other bodies records the early history of the Solar System, which has been wiped out by erosion on Earth.

After World War II the Australian Bureau of Mineral Resources was given the task of preparing a series of 'four-mile maps', at 1:250 000 scale of the whole of Australia to provide a basis for mineral exploration and proper planning. This task was essentially finished in the 1970s. Now the Bureau's successor, Geoscience Australia, is preparing the second generation of maps and also undertaking fundamental exploration of the offshore areas. Soon Australia's Exclusive Economic Zone (EEZ) will be proclaimed under an international treaty, the United Nations Convention on the Law of the Sea. Following the finalisation of an agreement with the United Nations concerning Australia's areas of extended continental shelf under the terms of the United Nations Convention on the Law of the Sea, the Australian marine jurisdiction is likely to be in the order of 14 million square km compared to the 7.7 million square km area of onshore Australia.

Australia has been comprehensively mapped, explored and drilled by companies looking for minerals, coal and petroleum. Almost all the country has at one time or other been subject to exploration, so samples have been recovered by drill holes from areas where the rocks are hidden by desert sands or river floodplains. Thousands of analyses have been made, giving us details of oil contents, coal types, gold and other metal values, geological ages, and compositions of groundwaters.

One of the consequences of these advances is that we do not see Australian geology as a land-based study stopping at the seashore. For example, most of our oil and gas fields are offshore. The nature of the subsurface geology of the seabed around Australia is intimately linked to the continental shelves and coastal regions. It clear that many onshore rocks formed in conditions similar to those offshore today, so by studying the offshore we can understand more about the onshore geology.

New discoveries are being made monthly. This is a wonderfully exciting time to be an earth scientist, with major new discoveries being made and big challenges ahead as we try to develop the resources to provide for society without destroying the purity of air and water, and the vast natural world which we value as Australian.

Summary

Europeans originally thought Australia was very young, compared to their well-known 'old' Europe. We now know that Australia contains some of the oldest rocks known on Earth. Most of the central and western parts have been stable for many millions of years and preserve an extraordinary record of Earth history. The long stability combined with erosion means it is one of the lowest and flattest land areas on Earth.

Modern scientific advances on land, under the sea, and from space, combined with strenuous exploration for minerals and energy in the last 50 years, mean we have a much greater understanding of this continent than before.

SOURCES AND REFERENCES

Sources and references for chapter 1

Buick, R., Thornett, J.R., McNaughton, N.J., Smith, J.B., Barley, M.E. & Savage, M., 1995. 'Record of emergent continental crust ~ 3.5 billion years ago in the Pilbara craton of Australia'. *Nature*, 375: 574–577.

Clarke, Rev. W.B., 1861. *Recent Geological Discoveries in Australasia*. Second Edition. Joseph Cook & Co.

Feeken, E.H.J., Feeken, G.E.E. & Spate, O.H.K., 1970. *The Discovery and Exploration of Australia*. Thomas Nelson. 318 pp.

Flannery, T., 2001. *Terra Australis*. Text Publishing, Melbourne. 268 pp.

Jukes, J. Beete, 1850. *A Sketch of the Physical Structure of Australia, So Far As It Is Present Known*. T & W Boone, London. 95 pp.

Menzies, Gavin, 2003. 1421. *The Year China Discovered The World*. Bantam Books. 650 pp.

Milne Curran, J., 1898. *Geology of Sydney and the Blue Mountains*. Angus & Robertson. 391 pp.

Petkovic, P. & Buchanan, C., 2002. Australian bathymetry and topography grid (January 2002), Data package, Geoscience Australia, Canberra.

Plimer, I., 2001. *A Short History of Planet Earth*. ABC Books. 250 pp.

Sigmond, J.P. & Zuiderbaan, L.H., 1976. *Dutch Discoveries of Australia*. Rigby Limited. 176 pp.

Wilde, S.A., Valley, J.W., Peck, W.H. & Graham, C.M., 2001. 'Evidence from detrital zircons for the existence of continental crust and oceans on the Earth 4.4 Gyr ago'. *Nature*, 409: 175–178.

General books on Australian geology, soils and palaeontology

Archer, M. & Clayton, G., 1984. *Vertebrate Zoogeography and Evolution in Australasia*. Hesperian Press. 1203 pp.

Archer, M., Hand, S. & Godthelp, H., 1991. *Australia's Lost World – Riversleigh, World Heritage Site*. Reed New Holland. 264 pp.

AUSGEO. A quarterly publication on current topics published free by Geoscience Australia. Subscribe online at www.ga.gov.au/about/corporate/ausgeo_news.jsp

David, T.W.E. & Browne, W.R., 1950. *The Geology of the Commonwealth of Australia*. 3 volumes. Edward Arnold, London. [*A classic compilation, out of date, but of great historical interest*]

Gostin, V.A. (ed.) 2001. *Gondwana to Greenhouse*. Geological Society of Australia Special Publication No. 21. 356 pp. [*A compilation of short scientific papers on environmental issues from a geoscience perspective*]

Hilliss, R.R. & Muller, R.D., 2003. *Evolution and Dynamics of the Australian Plate*, Geological Society of Australia Special Publication 22 and Geological Society of America Special Paper 372. 432 pp. [*Available from www.gsa.org.au*]

Laseron, C.F., 1954. *Ancient Australia*. Angus & Robertson. 210 pp. [*Third edition revised in 1984 by Rudolph Brunnschweiler.*]

Laseron, C.F., 1954. *The Face of Australia*. Angus & Robertson. Second Edition. 244 pp.

Mackness, B., 1987. *Prehistoric Australia*. Golden Press. 191 pp.

Paton, T.R., Humphreys, G.S., Mitchell, P.B., 1995. *Soils: A New Global View*. UCL Press, London. 212 pp.

Sutherland, L. & Webb, G., 2000. *Nature Guide to the Gemstones and Minerals of Australia*. New Holland. 128 pp.

Twidale, C.R. & Campbell, E.M., 1993. *Australian Landforms: Structure, Process, Time*. Glen Eagles Publishing, Glen Osmond, SA. 560 pp.

Veevers, J.J. (ed.) 1984. *Phanerozoic Earth History of Australia*. Oxford Monographs on Geology and Geophysics No.2. Clarendon Press. 418 pp.

Veevers, J.J. (ed.) 2000. *Billion-year earth history of Australia and neighbours in Gondwanaland*. Gemoc Press, Sydney. 388 pp. [*Both books with specialist detail, though difficult for general readers*]

Vickers-Rich, P. & Hewitt-Rich, T., 1999. *Wildlife of Gondwana*. Indiana University Press. 304 pp.

White, M.E., 1986. *The Greening of Gondwana*. Reed. 256 pp.

Wright, A.J., Talent, J.A., Young, G.C. & Laurie, J.R., 2000. 'Palaeobiogeography of Australasian Faunas and Floras'. *Memoirs of the Australasian Association of Palaeontologists 23*. 515 pp.

Young, A. & Young, R., 2001. *Soils in the Australian Landscape*. Oxford University Press, Melbourne. 210 pp.

Geology of individual states and territories

New South Wales

Branagan, D.F. & Packham, G.H., 2000. *Field Geology of New South Wales*. Third edition. NSW Department of Mineral Resources. 418 pp.

Scheibner, E. & Basden, H. (eds) 1996. *Geology of New South Wales. Volume 1: Structural Framework*. Geological Survey of New South Wales Memoir 13. 295 pp.

Scheibner, E. & Basden, H. (eds) 1996. *Geology of New South Wales. Volume 2: Geological Evolution*. Geological Survey of New South Wales Memoir 13. 666 pp.

Northern Territory

No current book, but a map is available.

Queensland

Day, R.W. and others, 1975. *Queensland Geology*. Geological Survey of Queensland Publication 383. 194 pp.

Bain, J.H.C. & Draper, J.J., 1997. *Atlas of North Queensland Geology*. Australian Geological Survey Organisation and Geological Survey of Queensland. 95 pp.

Henderson, R.A. & Stephenson, P.J. (eds) 1980. *The Geology and Geophysics of Northeastern Australia*. Geological Society of Australia Inc. 468 pp.

South Australia

Drexel, J.F. & Preiss, W.V., 1995. *The Geology of South Australia. Volume 2: The Phanerozoic.* South Australian Geological Survey Bulletin 54. 347 pp.

Drexel, J.F., Preiss, W.V. & Parker, A.J., 1993. *The Geology of South Australia. Volume 1: The Precambrian.* South Australian Geological Survey Bulletin 54. 242 pp.

Tasmania

Burrett, C.F. & Martin, E.L. (eds), 1989. *Geology and Mineral Resources of Tasmania.* Geological Society of Australia, Special Publication 15. 574 pp.

Victoria

Birch, W.D. (ed.), 2003. *Geology of Victoria.* Geological Society of Australia Special Publication No. 23. 842 pp.

Ferguson, J.G. (ed.), 1988. *The Geology of Victoria.* Geological Society of Australia Inc. 663 pp.

Western Australia

Geological Survey of Western Australia, 1990. *Geology and Mineral Resources of Western Australia.* Western Australia Geological Survey Memoir 3. 827 pp.

Myers, J.S. & Hocking, R.M., 1998. *Geological Map of Western Australia, 1: 2 500 000* (13th edition). Western Australia Geological Survey.

Websites

Geological Society of Australia Inc.
www.gsa.org.au

Geoscience Australia
(*formerly the Australian Geological Survey Organisation, AGSO*)
www.ga.gov.au

General interest and Australian fossils
www.austmus.gov.au/exhib/dinosaur.htm
www.amonline.net.au/palaeontology/faqs/fossil.htm

Glossary of terms
www.agso.gov.au/education/glossary.html

Satellite imagery of Australia
www.auslig.gov.au/acres

Radiometric dating
www.wrgis.wr.usgs.gov/docs/parks/gtime/

Radiocarbon dating
www.c14dating.com

Radiocarbon calibration
www.units.ox.ac.uk/departments/rlaha/orau/01_04.htm

The age of the Earth
www.wrgis.wr.usgs.gov/docs/parks/gtime/

Earthquakes

Earthquakes in Australia
www.ga.gov.au/urban/projects/1011573567-3899.jsp

Seismology Research Centre, La Trobe University
www.seis.com.au/

Earthquakes in South Australia
www.pir.sa.gov.au/pages/minerals/earthquakes/

Earthquakes in the USA and the San Andreas fault system
www.earthquake.usgs.gov/

Plate tectonics
www.socal.wr.usgs.gov/office/ganderson/es10/lectures/lecture06/lecture06.html
citt.marin.cc.ca.us/ring/rplates.html

Maps

Geological Map of Australia
This (1:5 000 000 scale) map can be ordered online from Geoscience Australia at www.ga.gov.au

Online mapping facility
www.ga/gov/au/map

Environment and water maps of Australia
www.earthsystems.com.au

Illustrations

Figs 1.2a, b: modified and redrawn with permission from Taylor and Francis Publishers after figs 254, 255, pp. 356, 357 in Holmes, A., 1965. *Principles of Physical Geology.* Thomas Nelson & Sons. 1288 pp.

Fig. 1.4: redrawn with permission from The McGraw Hill Companies after fig. 19.1 from Plummer, C.C. & McGeary, D., 1996. *Physical Geology.* 7th edn. Wm C. Brown. 538 pp.; information for the plate motion (for Darwin) from fig. 58 in Veevers, J.J. (ed.) 2001. *Billion Year Earth History of Australia Atlas.* GEMOC Press, Sydney.

Fig 1.5: digital image from Geoscience Australia, http://www.ga.gov.au/archive/earthquakes/GA077; Crown copyright ©. All rights reserved. www.ga.gov.au

Fig. 1.6: redrawn from an image on the website www.auslig.gov.au/facts/climxtre.htm

Fig. 1.7: digital image from Petkovic, P. & Buchanan, C., 2002. Australian bathymetry and topography grid (January 2002), Data package, Geoscience Australia, Canberra. Crown copyright © All rights reserved. www.ga.gov.au

Fig. 1.11: satellite image provided by the Australian Centre for Remote Sensing (ACRES), Geoscience Australia. Crown copyright © All rights reserved. www.ga.gov.au

CHAPTER 2 *Model of the Earth*

THE EARTH: A GEOLOGY PRIMER

This chapter introduces the main ideas of geology, and explains briefly the technical terms used in this book. We deal with the overall structure of the Earth, the theory of plate tectonics, the main mineral groups and rock types (igneous, sedimentary and metamorphic), the deformation of rocks into folds and faults, the processes of erosion and formation of the landscape, and the geological time scale.

The Earth is a generally spherical planet, slightly flattened at the poles. Inside the Earth are three components (see Figure 2.2): a central, hot **core** with a 3470 km radius, composed mainly of iron and nickel with a density of 10 000–13 000 kg/m^3 (10–13 t/m^3), solid in the centre and liquid in the outer core; a **mantle** 2900 km thick, composed of dense (3–6 t/m^3) rocky material in a hot plastic state; and an outer, cool, solid **crust** with a density of 2–3 t/m^3. The core and mantle make up most of the Earth – 83% by volume – while the crust is a relatively thin skin.

The core–mantle boundary is a zone of fundamental importance. Circulation in the liquid outer core is the dynamo that produces the Earth's magnetic field. A combination of thermal circulation and gravitation, as heavier constituents settle towards the centre, fuels this engine of the Earth's magnetism. This boundary has thin zones that slow seismic waves by 10–30%, but no-one knows why these thin zones occur. Recent work suggests that these ultra-low-velocity zones are caused by irregular patches of liquid material at the base of the mantle that contain impurities driven out as the core continues to cool and crystallise.

Circulation in the overlying mantle is the main mechanism for heat transfer from the centre to the outer regions of the Earth, and is the primary engine driving the movements of the continents and the associated volcanism and earthquakes.

We know that the Earth's magnetic field has reversed (see Box 2.1), which means that there has probably been a major reversal in core circulation. Unfortunately we have much less information on the changes in the palaeointensity of the magnetic field, and this is hampering accurate modelling of the past events. The causes of short-term changes (< 1 million years) are probably different to the causes of long-term changes (> 20 million years), such as the reversals. Considering the long time scales of these reversals (millions of years) compared to the shorter circulation times modelled for the outer core (thousands of years) it may be that variations in the lower mantle exert a control on outer core circulation.

The shallower boundary between the mantle and the crust was discovered in 1909, and is called the **Mohorovičić discontinuity** after its discoverer, or the **Moho** for short. It is a zone between one and several kilometres thick where the velocity of seismic compression waves (P-waves) increases suddenly from about 6.8–7.2 km/s in the crust to 8.1–8.2 km/s in the mantle. This velocity change is attributed to a change in chemical composition of the rocks: crustal rocks are richer in silicon but poorer in iron and magnesium than those in the mantle. The Moho lies at 30–40 km depth under inland western Australia, and 37–47 km under central New South Wales.

This thin crust is of two types: continental and oceanic. **Continental crust** is heterogeneous and of relatively low density (2.0–2.8 t/m^3), composed mainly of granites and sedimentary rocks. **Oceanic crust** is basaltic and has a density of 3.0–3.1 t/m^3. Both continental and oceanic crusts are floating on the denser mantle. As a result of its lower density, continental crust floats on the mantle at a higher elevation, forming the land masses and mountains, and varies from 30 to 70 km thick, averaging 45 km. In contrast, denser oceanic crust floats at a lower elevation, forming the ocean basins, and is typically 8 km thick.

This three-part model of core, mantle and crust explains compositional variations but not how the rocks behave. Temperature increases towards the centre of the Earth, tending to make rocks molten, but higher pressures tend to keep them solid.

There is a zone, typically about 100–200 km down, far below the base of the crust, where the rock behaviour changes, although the chemical composition appears to remain essentially the same. But in this zone the rocks are most easily deformed. They have softened because of temperature increase but have not become immobilised by confining pressures. The temperature at the top of this zone is estimated to be about 1300°C. Because temperature increase is greater below the oceanic crust, the top of this zone is shallower (100 km) there than below the cooler continental crust (200 km).

This zone marks the base of the **lithosphere**, and it is very important because it marks the level at which the major plates of the Earth's surface move. Below this level the mantle deforms more easily, and we think it contains major convection cells that release heat from the Earth's interior. This lower part of the mantle is called the **asthenosphere**. The lithosphere acts as a series of coherent plates floating on the more fluid asthenosphere, like large icebergs on the sea. So the moving lithospheric plates are composed of both the crust and part of the upper mantle.

Outside the solid part of the Earth lies a **hydrosphere** comprising the oceans, with a density of 1.09 kg/m^3 and an average thickness of 4 km, and an **atmosphere**, with an average density of 0.00012 kg/m^3. The atmosphere has 75% of its mass in

Figure 2.1 *The Earth from space, taken by Galileo spacecraft on 11 December 1990 about 1.5 million miles from Earth. This photograph emphasises the watery nature of our planet, floating in the black vacuum of space. Australia is at the centre right. The blue oceans cover 70% of Earth, and the white clouds show the water present in the atmosphere.*
PHOTO: NASA

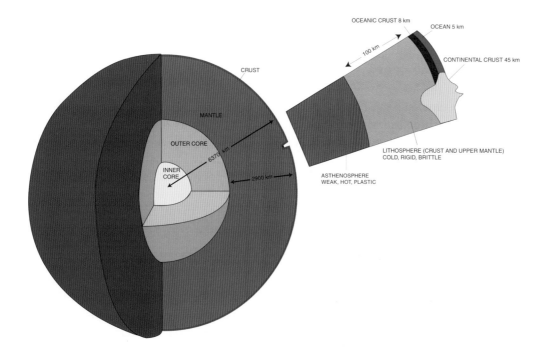

Figure 2.2 *Cross-section of the Earth showing the core and mantle, and an expanded wedge to show the thin crust. The crust is of two types: thin oceanic crust and thicker continental crust. The mobile plates consist of lithosphere, made of the crust and part of the upper mantle, and these plates move on top of a convecting asthenosphere.*
REPRODUCED AND MODIFIED WITH PERMISSION OF THE MCGRAW-HILL COMPANIES AFTER FIGURE 17.5 FROM PLUMMER, C.C. & MCGEARY, D., 1996, *Physical Geology*, 7TH EDN, WM C. BROWN, AND ADDITIONAL DATA FROM SCHEIBNER, E., 1999, *The Geological evolution of NSW – A brief review*, GEOLOGICAL SURVEY OF NSW, SYDNEY

the first 10 km and becomes less dense farther from the Earth's surface, merging into outer space. The main components of the atmosphere are nitrogen (78%), oxygen (20%), argon (1%) and carbon dioxide (0.03%), and there are traces of other gases. The amount of water vapour is variable and can be as high as 4% near the surface.

These thicknesses confirm that the hydrosphere and atmosphere, together about 15 km thick, are only a very thin skin on a rocky planet some 6370 km in radius at the equator.

Plate tectonics

Tectonics is the study of the forces and movements in the Earth, especially in the outer parts. It is aptly named, because the ancient Greek word *tekton* means carpenter or builder. In 1912 the German meteorologist Alfred Wegener summarised the linear pattern of mountain belts on the surface of the Earth and explained these in terms of large rock masses moving horizontally – continental drift – with the resulting compression pushing rocks upwards to form the mountains. The theory of continental drift was taken up by a few enthusiasts, notably Alex du Toit in South Africa and later Sam Carey at the University of Tasmania. Most of the scientific community rejected the ideas. After all, what mechanism could be responsible for such movement?

The answer came in the 1960s with detailed analyses of the magnetism recorded in rocks. Although direct measurements of the Earth's magnetic field have been made only in about the last 400 years, the rock record preserves evidence of the Earth's palaeomagnetism back to at least 3.5 billion years ago.

The discovery of patterns of 'magnetic stripes' in rocks on the ocean floors showed that the Earth has experienced **magnetic reversals**. The stripes formed because the Earth's magnetic field has reversed periodically – that is, the north magnetic pole switches position to the south geographic pole, and vice versa. We call the present situation 'normal polarity'; and when the north pole is in Antarctica it is 'reversed polarity' (see Box 2.1). Obviously, during periods of reversed magnetism the compass needle would point southwards!

As far as we can tell from analysing the change in polarity through a thick pile of basalts, the Earth's magnetic field does not flip instantly. The magnetic poles move, often in a convoluted fashion, across the globe and can take 10 000 years or more to reposition themselves.

By dating the rocks we can put ages on these periods of normal and reversed polarity, as shown in Figure 2.3. The oscillation pattern is irregular. We do not know what is driving these changes, except that we think they are due to changes in the circulation of the Earth's molten, iron-rich core. These reversals have been observed as far back as the Precambrian, though the more recent events are best known and most accurately dated.

The magnetic fields preserved in the rocks of the ocean floors display striped zones of normal and reversed polarity. These stripes occur symmetrically on both sides of the mid-oceanic ridges. Furthermore, the rocks are younger near the ridge and older farther away. This symmetry of age and magnetic signature confirms that rocks rise in a molten state at the mid-oceanic ridges and then spread sideways in both directions.

As the continental plates spread apart, some minerals within the rocks incorporate the magnetic polarity at that time. During periods of reversed polarity, some millions of years later, the minerals incorporate the reversed polarity. It is as if two identical recording tapes are being slowly pulled out of the crack at the mid-ocean ridge, one to either side, and each magnetic reversal is printed onto each tape.

Here was firm evidence for the movement envisaged by Wegener, du Toit and Carey. The theory of plate tectonics was born. But what is driving this movement?

It is thought that convection cells in the mantle drive heated material upwards, expanding and elevating oceanic lithosphere along the spreading zones. That is why the spreading zones are marked by mid-ocean ridges – huge submarine mountain chains built by the sea-floor basaltic volcanoes. In some places these volcanoes poke above sea level, for example at Iceland and in the Azores in the mid Atlantic Ocean. Then as the lithosphere spreads laterally it cools, contracts and subsides, forming the broad abyssal plains of the ocean basins, typically at 4000–5000 m water depth. However, it is not clear that the rising mid-ocean ridge just pushes the oceanic parts of the plates away to the sides. Mantle convection is driven by rising heat but also by the density difference between the hot mantle and sinking slabs of cold lithosphere. Much of this plate movement is probably achieved as the slab of cold lithosphere sinks down in the subduction zones along the margins of the ocean basins. In this scenario the slab-pull keeps opening the crack along the mid-ocean ridge, which is filled by rising basaltic magma.

There is also lively debate about whether the mantle convection involves cells through the whole depth of the mantle, or whether the mantle is divided into two layers at a prominent seismic discontinuity at 660 km depth, with the plates being driven by cells in the upper part.

Box 2.1 Age-dating the rocks

In Chapter 1 (Box 1.1) radiometric dating was described as the principal method used to determine the absolute age of rocks. Geologists commonly use four other methods to date rocks: magnetic signature, thermoluminescence, fossils, and fission track dating.

MAGNETIC SIGNATURE

The magnetic signature of a rock is influenced by the orientation of the Earth's magnetic field at the time it is formed, and we know the magnetic field varies in direction and intensity. For instance, the magnetic poles move relative to the geographic poles or spin axis of the Earth. The South Magnetic Pole, first visited by Australian geologists Edgeworth David and Douglas Mawson and medical doctor Forbes Mackay in 1908, has now moved to a position in the sea.

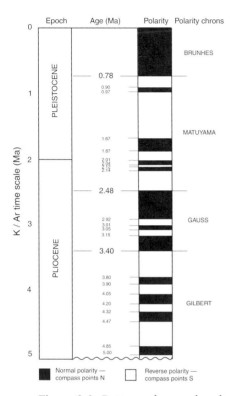

Figure 2.3 *Pattern of normal and reversed polarity of the Earth's magnetic field over the last 5 million years. The last major reversal, the Brunhes–Matuyama, was 780 000 years ago.*

Redrawn and reproduced with permission of Springer Verlag GmbH & Co. and Mebus Geyh from figure 7.2 of Geyh, M.A. & Schleicher, H., 1990, *Absolute Age Determination*, Springer Verlag, 503 pp., which also used data from Harland, W.B., et al., 1982, *A Geologic Time Scale*, Cambridge University Press; Craig, L.G., Smith, A.G. & Armstrong, R.L., 1986, 'A geologic time scale', *Terra Cognita*, 6:141, and Obradovitch, H.M. Sutter, J.F. & Kunk, M.J., 1986, 'Magnetic polarity chron tie points for the Cretaceous and early Tertiary', *Terra Cognita*, 6: 140

Variations in the Earth's magnetic field can be recorded in both igneous rocks and sediments. In igneous rocks, the magnetism is recorded instantly as magnetic minerals crystallise from the magma. As magma cools, mineral grains pass though the **Curie point**, which for magnetite is about 575°C. Pierre Curie discovered in 1895 that all magnetic materials lose their magnetism when heated above a certain temperature. Consider the process in reverse when a magma cools and its minerals crystallise. As a mineral cools below the Curie point it becomes magnetic, and the magnetism it acquires has the same orientation as the local orientation of the Earth's magnetic field. Because of this phenomenon, scientists can decipher the orientation of the magnetic field in the past by measuring the remnant magnetism in carefully oriented samples of rock containing magnetite.

If the magnetite is heated above 575°C it will lose the magnetism, and on cooling it will gain a new magnetism that is aligned with the magnetic field at that time. Other minerals can also preserve magnetism, and have different Curie point temperatures, but magnetite is the most magnetic and hence most useful mineral.

In fine-grained sediments, detrital magnetic grains settle parallel to the geomagnetic field. Careful measurements along sediment cores show variations in magnetic signature that can be matched between cores, and also to patterns established in known sequences. While sediments are poorer magnetic recorders than igneous rocks, they were laid down continuously and hence have potentially better time coverage. However, the magnetism is weaker because of varying oxidation of the magnetic particles, and may be acquired over a longer time period and hence be time-averaged.

Most spectacularly, the rocks show that the Earth's magnetic field has completely reversed in the geological past. These oscillations can be measured on oriented rock samples in the laboratory, along sediment cores, and by towing a magnetometer at sea to measure the variations across the ocean floors. The most recent polarity reversals are numbered in sequence to number 33, which occurred 79 million years ago. A very important time marker is the Brunhes–Matuyama reversal, which occurred 780 000 years ago.

Four major polarity **chrons** (time zones with a duration of around 1 million years) are recognised over the last 5 million years: the present Brunhes and earlier Gauss chrons are dominantly normal polarity, and the Matuyama and Gilbert chrons are dominantly reversed polarity. Within each chron are short periods when the polarity is temporarily changed. There are also **superchrons** – very long periods with no reversals – such as the Cretaceous normal superchron from 118–83 Ma ago, and the late Palaeozoic reversed superchron 317–262 Ma ago. To convert a palaeomagnetic signature to an absolute age, marker beds have to be dated by radiometric dating. For the magnetic reversal scale this has been done many times to establish the ages of each reversal.

Having established this pattern we can use it to date rocks. For instance, many weathered rocks in Australia, the deeper deposits in salt lakes, and some of the glacial deposits in Tasmania pre-date the Brunhes–Matuyama transition 780 000 years ago.

The palaeomagnetic record in rocks shows that there is a range of variations in the Earth's magnetic field, from secular variation as the magnetic pole rotates, through excursions where there are major changes in the field (but of the same polarity), to full polarity reversals.

A comparison of magnetic data recorded by satellites in 1979–1980 and 2000 shows that the Earth's magnetic field has partly collapsed. The reversed magnetic flux is evident in two places: beneath southern Africa and the North Pole. Such a situation may be just a temporary excursion, but it may also be the state to which the Earth's dynamo moves before a full magnetic reversal.

THERMOLUMINESCENCE DATING

Grains of quartz and feldspar on the surface of the Earth are exposed to external ionising radiation, which is preserved as charges trapped in defects in the atomic lattice. When the sediment grain is buried this external ionising radiation is stopped. This change provides a basis for dating the time of burial of the sediments.

The trapped charges can be liberated by heating the crystalline grain, typically to 270–400°C, and a proportion of the trapped charge recombines, causing an emission of light that can be measured. This thermoluminescence (TL) is recorded digitally on a sensitive instrument as a glow curve. The trapped charge can also be released by irradiating the grain with a particular wavelength of light (e.g. an argon–ion laser with a wavelength of 514 nm) and measuring the emission at a different wavelength. This is called optically stimulated luminescence (OSL). All grains contain traces of radioactive uranium and thorium, which can also cause light to be emitted. The age is finally determined by measuring the glow curve and then taking account of any glow caused by the inherent radioactivity.

TL and OSL dating are useful for dating material up to 100 000 years old, and in some cases to 500 000 years – well beyond the age range of radiocarbon dating. Moreover, they enable geologists to date younger quartzose sediments such as old beaches, dune systems and river channel sands that are devoid of the organic materials required for radiocarbon dating.

FOSSILS

The evolutionary sequence of fossils through the geological record is well established. We know there are large parts of the record missing because organisms were not preserved or because the sedimentary sequences were eroded. However, on a large scale we know that, for example, trilobites (see Chapter 4) existed only from the Cambrian through the Permian (545–251 Ma). Within this span, trilobite age ranges have been determined for individual genera and species. So if we find rocks containing particular trilobites we can place them in the age sequence. Many other fossils, such as shells, corals, plants, foraminifera and graptolites, can be used to determine the geological age of sedimentary rocks.

We call this method **biostratigraphic dating**. In some cases the fossils can provide ages only to the nearest 10 million years. In places where detailed studies have been made and the rocks through the sequence have been dated by absolute methods, the age of individual fossils can be resolved to a range of less than 1 million years.

FISSION-TRACK DATING

Minute amounts of uranium (U-238) occur in some minerals. When present the uranium undergoes radioactive decay (fission), and the emitted particles damage the atomic lattice in the host mineral. These defects can be etched with acid to make them visible as small cylindrical holes (tracks) on polished grain surfaces under a microscope. Initially formed tracks are round 16 µm long and 1–2 µm in diameter. The number of defects depends on the amount of uranium within the grain and the age of the sample. Greater uranium content and longer time will produce more fission tracks.

The amount of uranium content is measured using a slice of a mineral free of uranium, typically muscovite, which is placed on top of the sample. Together they are exposed to radioactivity in a nuclear reactor, causing some of the uranium to fission and create defects in the muscovite slice. The number of tracks recorded in the muscovite is a measure of the uranium content in the underlying mineral. The original tracks in the grain (caused by natural, spontaneous fission) are compared with the number in the muscovite detector (caused by induced fission) to derive an age.

At higher temperatures the lattice naturally heals itself (anneals), resetting the fission track clock to zero. The two minerals commonly used for fission track dating – apatite and zircon – have annealing temperatures of 100–120°C (apatite) and 210°C (zircon). Heating is required over a 1 million to 10 million year timescale to anneal the tracks. When apatite is heated above 60°C the fission track length shortens, until over 120°C it is totally healed.

So the fission-track age is an apparent age that records the time since the rock cooled through the annealing temperature. It does not record the original age of the rock formation. Although increased heat flow as a result of nearby magmatism or movement of hydrothermal fluids is a possible cause, the primary cause of regional annealing is crustal heating. While the rocks are deeply buried the fission tracks are being continually annealed. Tectonic uplift and elevation of rocks out of the deeper, hotter regions allows fission tracks to be preserved. That is, the fission track ages can provide information on the age of continental uplift, and through this the rates of erosion, or denudation, of the landscape.

Regional studies using fission-track dating show that the long-term denudation rate across Australia has been around 10 m per million years. Higher rates of 15–20 m per million years occurred in southwestern Western Australia and the Kimberley region between 250 and 200 Ma ago, and very much higher rates of up to 40 m per million years in southeastern Australia around 50 Ma ago.

Other uses of fission-track dating are:

1. Dating tektites, the siliceous glassy buttons which form during meteorite impacts (see Chapter 11). Fission tracks record when this material solidified and fell back to Earth.

2   Dating fused glasses around impact structures. The rocks were melted by the impact, so the fission-track age is the age of impact.
3   Dating uplift of mountain belts and sedimentary basins, where the fission-track age denotes the time the rocks were elevated to cooler levels in the crust. By establishing fission-track ages across a continent we can tell when various parts were uplifted relative to one another.

Plate margins

Three types of plate margin are recognised: divergent, convergent and transform.

Divergent plate margins form where upwelling of the mantle along fractures thousands of kilometres long causes the plate to break apart. This upwelling can underlie continents or oceans. Above lines of rising mantle the plates are subject to tension, and commonly crack like a piece of plaster board that has been flexed. In these areas of extension, lithospheric plates can be broken apart, leaving deep rifts. Examples are the Great Rift Valley, which runs down central eastern Africa, and the Red Sea, a rift now fallen below sea level. Generally there is volcanism along the cracks in the rift where hot magmas rise towards the surface, and along fractures thousands of kilometres long beside the rifts. If extension continues, the fragmented plate splits apart and the two parts drift away from each other. The gap is filled by basalt, forming a new ocean floor. Thus there are two phases of break-up – rift and drift.

Divergent plate margins have formed the major ocean basins, in which upwelling basaltic magma is injected between the diverging plates to form new crust at spreading centres. The active spreading centres form long **mid-oceanic ridges (MORs)**, which are undersea volcanic mountain chains.

Water and other fluids emanate from vents along the spreading ridges. The vent fluids are mainly seawater that has circulated through the upper parts of the ocean crust, becoming hotter and enriched with sulphur and commonly with iron and other metals such as copper and zinc. Extraordinary invertebrates live around these vents, functioning on a metabolic sulphur cycle. The water can be at temperatures up to 350°C, though it does not boil because of the high water pressure.

It is clear that basalts from the mantle can be channelled through **hotspots** where massive amounts of magma are erupted. Unlike the long plate boundary fractures, hotspots are only 50–100 km wide. Large hotspots pour out massive amounts of basalt, estimated at up to tens of cubic kilometres over several million years. These result in the **large igneous provinces** (LIPs) such as the Deccan basalts of India, the Siberian LIP, and the Kerguelen LIP which formed over some 50 million years during and after the split of India and Australia from Antarctica.

Hotspots have been attributed to **mantle plumes**, which are tall columns of mantle material rising from the base of the mantle and concentrating extra heat at one spot. However, there is also debate whether the plume concept is correct. If plumes rise from the base of the mantle there would have to be a 3000 km vertical continuity, and such a scenario does not hold up in models. It may be that the

Figure 2.4 *Basalt lava in Hawaii, showing the glassy surface formed by chilling.*
Photo: David Johnson

locations of hotspots and large igneous provinces are determined more by weaknesses in the lithosphere, perhaps as cracks that propagate during plate movement.

Smaller hotspots form a single but perhaps multi-vented volcano. As a plate drifts over a hotspot, a series of volcanoes are formed in a line parallel to the direction of movement. A well known example of this is the Hawaiian chain, where the present active volcanoes are only the most recent in a long chain that goes back millions of years. Older volcanic islands to the northwest are now extinct because they have passed over the hotspot and are now removed from the source of hot basalt. The Tasman seamounts east of Australia (see Chapter 8) are another example.

Convergent plate margins form where the convection cells of the mantle are moving downwards, forcing plates together.

A **subduction zone** is formed where one plate sinks beneath another. In ocean areas, the down-going plate can form a deep **trench** in the ocean floor along the boundary, such as the Marianas Trench near the Philippines or the Peru Trench off South America. Such trenches can be 10 000–11 000 m deep – some 6000 m deeper than the floors of the main ocean basins.

The denser and cooler converging plate sinks back into the mantle, and this sinking is typically marked by a zone of earthquakes at increasing depth, called the **Benioff zone**. The down-going plate becomes hotter and the mantle above it starts to melt. The resulting magma rises through the overlying plate to form a line of volcanoes parallel to the convergent margin, about 150 km from the edge of the upper plate. Typically these volcanoes form curved zones on maps, and are called **volcanic arcs**. Subduction that occurs in ocean crust, e.g. around Tonga, is called an **island arc**. Subduction under the edge of a continent, e.g. along the Andes or northwestern USA, is called a **continental arc**.

A recent analysis of earthquake travel times showed that zones of higher seismic velocity extend from the subduction zones deep into the mantle. The higher velocities imply denser rock masses. These higher velocity zones cross the 660 km thermal boundary into the lower mantle, though there is evidence of deflection and kinking of the plate between 500 and 1000 km depth. These zones are interpreted as remnant slabs of cold crust that are sinking deep into the mantle, persisting to around 1800 km depth, and then disintegrating across a zone at 1800–2300 km depth.

All this evidence points towards whole mantle convection with small, irregular chemical and thermal variations in the upper mantle, especially under the continental lithosphere. It also suggests that the very deep interval between 2000 km and the core–mantle boundary contains important clues to understanding the dynamics, composition and evolution of the Earth's interior.

A **collision zone** is formed where subduction ceases and the convergence of two continents forces the rock masses vertically upwards, forming huge mountain chains; for example, at the boundary between the Indo-Australian and Eurasian plates where India is colliding with China, forming the Himalaya. These collisions deform the rocks and can push slabs of rock hundreds of kilometres horizontally on thrust faults (see Figures 2.6, 3.10b).

Transform plate margins occur where two plates are sliding past each other, as in the San Andreas Fault in California. The Great Alpine Fault, which extends down the south island of New Zealand and has separated rock masses by 300 km over the

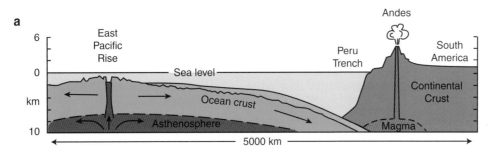

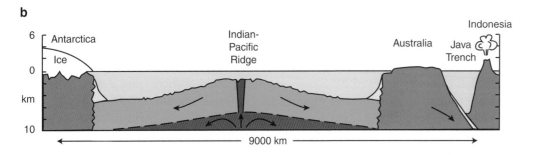

Figure 2.5 *Oceanic extension and subduction zones. a. New ocean crust generated at the East Pacific Rise is spreading on the eastern side, to be subducted in the Peru Trench along the coast of South America. The resultant volcanism is evident in the continental arc of the Andes. b. Seafloor spreading in the Southern Ocean is increasing the separation of Australia from Antarctica.*

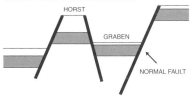

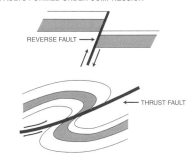

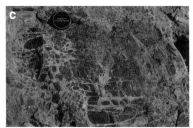

Figure 2.6 *a. Faults and folds. b. Folded quartz vein. c. Breccia.*
PHOTOS: DAVID JOHNSON

past few million years, is another example. Earthquakes are very common in these regions, but volcanism is absent.

Transform faults are very common in the ocean crust, at right angles to the spreading axis. The spreading rate is not uniform along the length of the axis, so one section of the ocean crust is moving faster than the one next to it. This differential movement creates a transform fault.

Plate tectonics has been developed as a complete theory only over the past 35 years, and has explained many features that had puzzled geologists. For example, vertical movements were thought to be due primarily to **isostasy** – the changes in buoyancy experienced by the crust floating on the mantle as mountains grew or were eroded. An analogy was made with wooden blocks of different sizes floating on water. A large block sits higher above the surface and also extends lower beneath the water. According to this idea, as mountains grow a deep mass of rock or mountain root forms under mountain belts to compensate for the increased weight. During erosion the mass rises to compensate for the loss of weight. While this effect is undoubtedly true, it does not explain why mountains form in the first place or why deep basins subside.

Plate tectonics accounts not only for the lateral movements of plates over the globe but also for the vertical movements known as uplift and subsidence. **Uplift** is caused by the expansion and rising of hot crust, heated by hot mantle from underneath, or by the buckling as two plates collide. **Subsidence** is generally caused by the contraction and sinking of colder crust, or by sag and cracking as a plate is pulled apart.

The convergence of plates is the major process causing rocks to be compressed into **folds**, just like the rucks or wrinkles that form in a tablecloth when it is pushed across a table top. In folds the rocks are essentially being bent like plastic. There is some recrystallisation and small-scale slippage, but on a gross scale the rocks are deformed as if they were soft. The process occurs very slowly, on a scale of millions of years.

Some folds are relatively broad and open, though with greater compression the limbs of the fold can be squeezed parallel, forming isoclinal folds. Isoclinal means 'same inclination' because the limbs are now at the same angle. Lateral compression can also push the fold over so that the limbs are then horizontal, forming a recumbent fold.

It is interesting to note that folds show the amount of shortening experienced within the Earth's crust. By taking a piece of rope and tracing it over the folds, then drawing it out straight, the amount of shortening can be measured. Measurements of 50% shortening are not uncommon (see Figure 2.6).

Rocks can also suffer brittle deformation, breaking and moving past each other along flat surfaces called **faults**. Faults range from small movements of less than 1 m to major displacements of more than 10 km, as along parts of the Darling Fault in Western Australia. **Normal faults** form during tension when one block slides down against the other. **Reverse faults** form during compression when one block moves steeply over the next block. **Thrust faults** form when one rock mass moves laterally over an essentially horizontal fault plane by lateral compression of the Earth's crust. **Strike slip faults** develop when one rock mass is moved horizontally.

THE EARTH: A GEOLOGY PRIMER   29

Brittle deformation on a smaller scale forms rocks that look as if they have been shattered, forming an accumulation of angular, fractured pieces known as **breccia**.

Large-scale tectonism occurs over millions of years, and multiple deformation of the rocks is common. Major deformed rocks, such as those in the Willyama Complex at Broken Hill, the Cooma Complex in New South Wales, and the Mount Isa region, typically show one set of folds overprinted by another, with perhaps four separate generations of deformation.

Minerals

Minerals are naturally occurring crystals (see Figure 2.7). Minerals have individual crystal structures and can be identified by their chemical composition, their crystal structure (determined by X-ray analysis), their hardness and their tendency to break along preferred directions (or **cleavages**) which correspond to planes of weakness in the atomic lattice structure. Hardness has proved a useful test; on a standard hardness scale called **Mohs' scale,** known to all geologists, 1 is the softest mineral and 10 is the hardest:

Table 2.1 Mohs' scale of increasing mineral hardness.

1	Talc
2	Gypsum
3	Calcite
4	Fluorite
5	Apatite
6	Orthoclase (feldspar)
7	Quartz
8	Topaz
9	Corundum
10	Diamond

Some minerals consist of just one chemical element such as a sulphur crystal, or pure carbon as **diamond**. Most minerals contain two or many elements.

Oxides are simple minerals consisting of oxygen and one or two other elements, such as the iron oxide **haematite** (Fe_2O_3), which colours rocks and soils red or brown. **Magnetite** (Fe_3O_4) is a particular iron oxide which is magnetic. Other examples are **rutile** (TiO_2) and **ilmenite** ($FeTiO_3$), which has iron substituted in some of the titanium sites of the lattice.

Another, more valuable oxide is pure aluminium oxide or **corundum** (Al_2O_3), which occurs as ruby and sapphire. Spinel group minerals are oxides of two metals, such as spinel ($MgAlO_4$) and jacobsite ($ZnFe_2O_4$).

The most common oxide is **quartz** (silica, SiO_2), which occurs in many forms, including clear quartz crystals, smoky quartz, gems such as citrine and rose quartz, and in a less crystalline form as **opal**. Quartz is very hard and does not break along cleavages, so it is very resistant to abrasion. This is why it is such a common

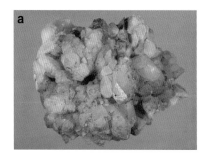

Figure 2.7 *a. Quartz, discoloured by iron oxides. b. Galena (cubic crystals). c. Fluorite. d. Sphalerite. e. Feldspar (left) and mica (the silvery flakes are due to the single cleavage) f. Garnet crystals in a schist. g. Pyrite.*
PHOTOS: DAVID JOHNSON

sediment grain, remaining after all the others have been worn away or decomposed.

Sulphides are widespread as the brassy crystals of **pyrite** (FeS_2) and the copper ore **chalcopyrite** ($CuFeS_2$). **Sphalerite** (ZnS) and **galena** (PbS) are ores of zinc and lead respectively.

However, the most abundant minerals on Earth are the **silicates**. These minerals all have the same basic building block: a silicon atom surrounded by four oxygens, forming a tightly bonded triangular pyramid, or tetrahedron. These SiO_4 tetrahedra are then linked in various structures to form the different groups of silicate minerals. Other elements may be within the silicate structures or linked to them (see Figure 2.8).

The simplest silicate is **olivine** $(Mg,Fe)_2SiO_4$, a green mineral of which clear crystals are known as the gem peridot. The magnesium and iron ions form tight links between individual tetrahedra. The formula indicates that olivine minerals can range from pure magnesium silicate to pure iron silicate.

Zircon ($ZrSiO_4$) also has a tight structure, with each zirconium atom linked between silica tetahedra. Zircon crystallises early from igneous melts and incorporates high proportions of radioactive elements such as uranium or thorium. Zircon is also very hard and resistant, and generally survives weathering to be incorporated in sedimentary rocks.

The **garnet** group comprises silicates in which the tetrahedra are still linked by other elements but in a slightly different structure with a wider range of other elements, usually including aluminium, and with magnesium, iron, calcium, manganese or chromium. Most garnets are red. Spessartine ($Mn_3Al_2Si_3O_{12}$) is a black, red or orange garnet, the colour depending on the exact composition. Uvarovite garnet is green because of its chromium content.

Pyroxenes and **amphiboles** are chain silicates; the pyroxenes have single chains and the amphiboles have a double chain. The crystals tend to be elongate, with two intersecting cleavages that form needle-like fragments.

Pyroxenes have the general formula $X_2(Si,Al)_2O_6$, where X represents mainly iron and magnesium, although a very wide range of elements may be substituted, including manganese, calcium, sodium, and titanium. The amphiboles have the formula $X_7Si_8O_{22}$ and have a similar variation in X, but also space for fluoride or hydroxyl (OH) groups in the hexagonal spaces between the linked chains. Pyroxenes and amphiboles are dark greenish or black minerals.

Sheet silicates consist of hexagonal networks resembling layers of wire netting.

They include green **chlorite** and the **micas**, such as black **biotite** and silver **muscovite**. The micas have a single cleavage and split easily into flakes. **Talc** and **serpentine**, and clay minerals such as **kaolinite** and **montmorillonite** are also sheet silicates. Many clays have the ability to swell by accommodating water molecules between the lattice sheets when wet, and then shrinking as the clay dries out. Such behaviour can cause severe problems for building foundations and road bases.

Feldspars are the most abundant minerals in the Earth's crust, and are typically white, pale grey or pink in colour. They have a continuous framework of SiO_4 and AlO_4 tetrahedra linked in three dimensions. Potassium, sodium, calcium and rarely barium occupy appropriate spaces in this three-dimensional lattice. Three members contain only potassium, sodium or calcium:

orthoclase $KAlSi_3O_8$
albite $NaAlSi_3O_8$
anorthite $CaAl_2Si_2O_8$

Members of the series from albite to anorthite are called **plagioclase** feldspars. However, in nature feldspars occur with intermediate compositions which may be anywhere in a continuous series, from 100% Na / 0% Ca to 100% Ca / 0% Na content.

There are also non-silicate minerals such as the carbonates **calcite** $CaCO_3$, and **dolomite** $(Ca,Mg)CO_3$ and **siderite** ($FeCO_3$). Sulphates include **gypsum** ($CaSO_4.2H_2O$), which is used in agriculture to make soils more crumbly, and **barite** ($BaSO_4$), which is added to drilling mud in petroleum exploration to make the mud heavier and so help stabilise subsurface gas discharge.

Phosphates include **apatite** $Ca_5(PO_4)_3(OH,F,Cl)$ and **monazite** $(Ce,La,Th)PO_4$, which is mined in beach sands as a source of cerium and lanthanum.

Our teeth and bones are made primarily of apatite, and in the lattice we can substitute more fluoride for the less stable hydroxyl (OH) and chloride groups by ingesting fluoride. This makes the chemical structure of our teeth stronger and hence less prone to decay.

There are also naturally occurring halides such as **fluorite** (CaF_2), which can occur as beautiful green or violet crystals. Common salt is the mineral **halite** (NaCl).

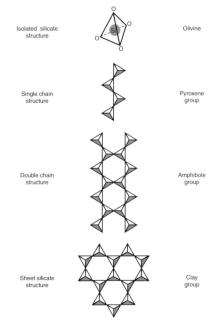

Figure 2.8 *Four main types of silicate structures: isolated, single chain, double chain and sheet.*

Types of rocks

Rocks are made of minerals. Some rocks are composed entirely of one mineral – quartzite, for example, is made of quartz grains – but most comprise many minerals. Three types of rocks are recognised: igneous, sedimentary and metamorphic.

Igneous rocks

The word igneous comes from the Latin *igneus*, which means 'on fire' or 'burning'. Igneous rocks are those that crystallise from hot, molten magmas.

Figure 2.9 *A volcanic ash field.*
Photo: David Johnson

Magma is liquid rock that has been melted by heating within the Earth. The magma rises because of over-pressuring of magma within or at the base of the crust, because of buoyancy, or because of pressure release, particularly as volatiles are released under lower pressure near the surface. Where magma reaches the surface it can be erupted from volcanoes as **lava**, forming **extrusive** igneous rocks. Some magmas, such as **basalt**, are very fluid and can flow for more than a hundred kilometres. Basalt has abundant dark minerals such as pyroxene and olivine, and no quartz, so it is a dark grey to black rock. It is derived primarily from the Earth's mantle. Some terranes have abundant basaltic rocks that, when altered, have a greenish black colour and are colloquially named **greenstones**.

Volcanoes such as those on Hawaii, Iceland, and Vanuatu are basaltic. Basalt lava issues forth in large streams, flowing down into hollows and along valleys. The lava can cool and solidify on the base against the ground, and on the surface against the air, while still flowing red-hot internally. If this lava continues to drain away when the eruption ceases, a **lava tube** is left behind.

Other lavas, such as **rhyolite**, are derived from melting of continental crust, and so are much richer in silicon. They are pale and very viscous, forming domes of limited extent. The eruptions can be extremely violent when the pent-up pressure is released. The volcano that erupted on Martinique (West Indies) in 1902 produced a fiery ash cloud that annihilated the town of St Pierre in 10 minutes. A spine of stiff rhyolite was then squeezed 300 m vertically into the air over a period of seven months. Rhyolites often contain altered patches where the glassy sections have broken down to soft clays. Groundwater can permeate these patches and voids,

Figure 2.10 *Intrusion of magma forming a dyke, near Mt Garnet, Queensland.*
Photo: David Johnson

precipitating concentric layers of coloured quartz that form balls of agate known as 'thunder eggs'.

Some lavas are intermediate in composition between basalts and rhyolites. These lavas are known as **andesite** (or **trachyte**, depending on the alumina content), after many of the volcanoes which erupt such lavas and ashes in the Andes mountains of South America. Most andesites are formed by melting of mantle above the descending plate, which is caused by the addition of fluids from the dehydrating sediments and wet oceanic crust going down into the subduction zone.

Both rhyolitic and andesitic volcanoes tend to be far more explosive and dangerous than basaltic volcanoes. On many occasions the magma does not erupt as a continuous fluid but explodes as showers of volcanic ash, forming cinder cones or blankets of **tuff**. **Scoria** is the term used for deposits of basaltic ash.

Explosive eruptions of rhyolitic or andesitic volcanoes commonly form huge columns of ash which can rise over 25 km into the atmosphere and be spread by winds over thousands of square kilometres. The ash showers from Mount St Helens, Mount Pinatubo and the El Chichon eruptions are examples. The volumes of material erupted are quite stupendous – anything from 1 to 10 cubic kilometres may be ejected within days. These volcanoes also cause devastating pyroclastic flows of hot ash and gas (also known as nuées ardentes), or debris avalanches accompanying the explosive blast. Such flows and avalanches flatten and burn everything in their path.

Magmas reach the Earth's surface through deep cracks that end as fissures from which the lava flows, or through pipes that form circular vents. In some cases erosion or quarrying will expose these feeder systems. **Diatremes** are pipes formed by gas explosions. The pipes may subsequently be filled partly by crystallised magma and partly by wall rocks and sediments that have collapsed back into the vent.

A **caldera** forms where the magma chamber under the volcano collapses. The final lavas and ashes sink back down, accompanied by collapse of the central part of the volcano along circular fractures. This leaves a very deep depression, commonly 2–5 km across, at the original apex of the volcano – the caldera.

When a volcano is eroded, any solidified lava in the central vent may remain, and is known as a volcanic **plug.**

Not all magma reaches the surface. Magmas can intrude rocks underground, pushing apart rock units or filling fractures. These are called **intrusions**. An intrusion parallel to the bedding planes of sedimentary rocks is called a **sill**, while one that fills a vertical fracture cutting through rocks is called a **dyke**. Since both sills and dykes form only in rock that has fractured and moved apart, they occur close to the surface where the confining pressure is low enough for this to happen.

The crystal size of a magma is related largely to the cooling history, although the numbers of seed crystals can also affect the final grain size. When lava issues onto the surface it can cool quickly so that crystals do not have time to grow, and the rock is fine grained. Where the lava flows into water and is quenched, or even just cooled rapidly in cold air, the surface of the lava will solidify into **volcanic glass**. Dark rhyolitic glass is known as **obsidian**.

As it crystallises, lava can also preserve gas bubbles rising through the magma. The final rock then contains round holes, called **vesicles**. Vesicles that have been filled by minerals, forming little pea-shaped or almond-shaped lumps within the rock, are called **amygdules**.

In contrast to rapid cooling on the surface, crystallisation underground by slower cooling allows the growth of larger crystals, forming coarser-grained rocks. Basaltic magmas form **dolerite** or **gabbro** underground, and rhyolitic magmas form **granite**.

There are also much larger intrusions in which granitic igneous rocks, in particular, have replaced the overlying rocks over areas of thousands of square kilometres, forming a **batholith**. Batholiths are believed to form in the crust, typically at least 5 km below the surface.

The minerals present in a rock depend mainly on the chemical composition of the parent magma. However there is a general sequence of crystallisation of igneous rocks.

Basaltic and doleritic magmas are the hottest magmas, typically erupted at around 1100–1300°C. They are rich in iron and magnesium, and have low silicon contents. The first minerals to crystallise as the magma cools are olivines and pyroxenes with calcium-rich plagioclase. The abundant pyroxenes and fine magnetite give basalt its dark colour. As the magma cools the plagioclase becomes progressively richer in sodium and poorer in calcium. Quartz, and sodium-rich and potassium-rich feldspars, amphiboles and micas have not yet crystallised at these temperatures.

Rhyolitic and granitic magmas contain very little iron and magnesium. They are typically around 900–1000°C when they start to crystallise. Feldspars rich in sodium and potassium constitute over 50% of the rock, and with more than 25% quartz, they are pale coloured. Minor quantities of mica and amphiboles may be present. The final material to crystallise is normally very rich in quartz and forms thin veins of **aplite** cutting across the granite.

Sedimentary rocks

Sedimentary rocks are composed of sediments – particulate materials such as river or beach sand. Sediments accumulate in areas where the crust is subsiding, known as a **sedimentary basin**.

Because sediments are laid down upon a surface they can form in layers, and this is known as **bedding**. Very fine layers, less than 10 mm thick, are called **laminae**. Where the sediment was later mixed or was laid down in a single event there may be no layering, and the unit is then said to be **massive**.

Geologists differentiate sediments in terms of size and composition. The size scale classifies **gravel** as coarser than 2 mm, **sand** as 0.063–2.00 mm in size, and **mud** as less than 0.063 mm (63 µm or microns) in size. Mud can be divided into **silt** (coarser than 4 µm) and **clay** (finer than 4 µm).

Sediments can be composed of individual mineral grains such as quartz, feldspar or mica, or of rock fragments, all of which are derived from the erosion of older rocks. Another source is ash blown out of volcanoes and washed down river systems. Alternatively, sediments can be made of skeletons of animals such as coral or shells, or of plant debris. There is also an accumulation of very fine sediment from the microscopic shells of organisms floating in the ocean waters, called **pelagic** sediment.

Figure 2.11 *Sedimentary rocks. a. Bedded sandstone with conglomeratic layers filling a scour eroded in the older sediments below it. b. Ripple marks on a sandstone surface, formed by currents flowing from left to right. Width 200 mm. c. Cross-laminae formed by ripples migrating across a Permian tidal flat, with mud layers deposited during slack water. Rock 60 mm thick.*
PHOTOS: DAVID JOHNSON

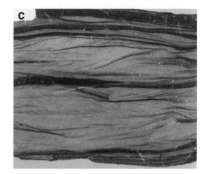

Sandstone is formed when loose sand is buried and turned into rock, **mudstone** is formed from consolidated mud, and **conglomerate** is formed from gravel. Sedimentary rocks commonly preserve structures such as ripple marks, mud cracks and animal burrows, which enable us to interpret the original environment of deposition.

The sediments can be bound together in two ways. Finer sediment such as mud can be deposited as a **matrix** between coarser grains in the original environment, and then harden to hold the grains together. Alternatively, waters percolating through the sediments during burial can precipitate new minerals such as quartz, calcite or clays to form a **cement**.

Limestone is formed by the accumulation and cementation of the hard parts of animals, such as coral or shell debris. The mineral in limestone is mainly **calcite**, a crystalline form of calcium carbonate. Some limestones form in reefs, and others by the accumulation of carbonate sediments in tidal flats, bays or on continental shelves where there is little influx of sediment from the land. **Dolomite** contains equal quantities of magnesium and calcium in the carbonate, and forms by the alteration of limestone.

Many sedimentary rocks, especially limestones, contain **fossils** such as the shells or skeletons of animals that lived at the time the sediments were being deposited, perhaps millions of years ago. Fossils range in size from the large bones of dinosaurs to corals, plant leaves, shark teeth, the earbones of whales (cetotoliths), down to microscopic items such as insect legs. Sometimes we can relate the fossils to animals or plants still living today, but in most cases the fossil organisms are now extinct.

Because life has evolved over millions of years, many fossils can be linked to particular periods in Earth history and can be used to help date the rocks in which they occur.

Figure 2.12 *Grey fossiliferous limestones, Wee Jasper, near Yass in New South Wales. The beds were deposited horizontally, and have been folded and then uplifted and eroded, so these remnants dip at a high angle.*
PHOTO: DAVID JOHNSON

Cherts are very fine-grained rocks composed of quartz. Many cherts formed as sediments, especially as accumulations of siliceous microfossils in the deep sea, which were later recrystallised to form chert. Some cherts formed by the alteration of fine siliceous volcanic ash or by precipitation from hot fluids in the crust.

Sedimentary rocks contain the world's resources of fossil fuels – coal and petroleum. **Coal** forms from peat when plant material accumulates in wet areas where it is buried before it is dried out or burnt. The transition passes from peat to brown coal to black coal with increasing temperature, pressure and time. In contrast, **petroleum** forms from algal and plant materials dispersed through sediments.

The formation of both coal and petroleum requires burial. As sedimentary rocks accumulate and subside into hotter parts of the crust, the algal and plant materials are subjected to higher temperatures. This turns the peat into coal and liberates oil and gas from the algal and plant material. Whether petroleum is oil or gas depends mainly on the original source of the organic materials and also the burial history. Excessive heating within the Earth can turn oil to gas, leaving behind a carbonaceous residue.

Evaporites are sediments formed by the evaporation of water, which leaves layers of salt and gypsum, as in the large salt lakes of arid inland Australia.

Figure 2.13 *Black coal. The pale laminae are layers of mud or volcanic ash washed or blown into the original peat swamp. Specimen is 150 mm wide.*
PHOTO: DAVID JOHNSON

Metamorphic rocks

Metamorphic rocks form through the alteration of other rocks under high temperatures and pressures. Any existing igneous, sedimentary or metamorphic rock can be subject to metamorphism. The new minerals developed in the rock depend on the original chemical composition of the rock, the temperature and pressure conditions, and the composition of fluids circulating through the rock.

Rocks can metamorphose even at shallow depths, though at first the original structures and grains are preserved. Such changes at low temperature, up to 200°C, are called **burial metamorphism**. Above this temperature significant changes begin to occur, which obliterate original structures and form new minerals, and if the temperature is high enough there will be a transition to igneous rocks with melting. Granite begins to melt at 625–650°C and basalt at 850–900°C, and such melting typically starts at crustal depths of 7–10 km.

Sometimes this metamorphism (**contact metamorphism**) is caused by the heat generated by a large igneous intrusion that forms a contact aureole around the intrusion. The rocks altered by this heat are known as **hornfels** and are typically fine grained and very hard.

Alternatively, rocks may be metamorphosed by the regional heating that accompanies a major deformation of the crust (**regional metamorphism**). The rocks are folded and faulted, minerals are transformed and new structures are imposed, so that in many cases the nature of the original rock is completely lost. By analysing the mineral assemblage and chemistry of the minerals, geologists can determine the actual temperatures and pressures of metamorphism. For instance, different assemblages

Figure 2.14 *Folded rock with vertical cleavage.*
Photo: Michael Rubenach

develop in high temperature–low pressure metamorphism (low grade), compared to those formed in high temperature–high pressure (high grade) zones (see Box 2.2).

Metamorphism commonly segregates minerals, forming a planar fabric or **foliation** through the rock. In rocks with abundant mica minerals, these minerals crystallise parallel to each other and at right angles to the principal stress, like the gradual shuffling of a mass of cards into a neat and parallel pack. The plane of weakness along the aligned micas is called a **cleavage**. Fine-grained metamorphic rock with a well-developed cleavage (which makes the rock easy to split) is called **slate**. One with a coarser grain size and less parallel cleavage is called **schist**.

High-grade metamorphism forms a coarse-grained, layered rock called **gneiss**. Typically there are lenticular patches of quartz and feldspar held within the weaving micaceous layers. At high temperatures some minerals reach their melting points, and **partial melting** of the rock occurs. Quartz and sometimes feldspar are 'sweated out' from the rock and form veins, cutting across or partly enmeshed with deformation structures of the rock.

Marble is formed by the metamorphism of limestone. During this process organic material is burnt off (so a grey limestone becomes a white marble) and the grain size commonly increases, resulting in a coarse sugary texture.

Box 2.2 Metamorphism

Metamorphism is the change in mineral content of a rock, and was once considered to be a solid state change in which ions diffusing through the rock mass allowed new minerals to grow. However, much of the work on Australian metamorphic rocks has emphasised the role of hot fluids in moving materials through the rocks. Metamorphism is commonly associated with structural deformation and folding, with foliations imposed on the rocks.

Temperature and pressure are the two main variables that drive metamorphism. The range of temperatures and pressures experienced during metamorphism is shown in Figure 2.15a. Rising temperature can be caused by local sources such as an intruded magma, by accumulated radiogenic heat or by regional heating above a rising mantle. Temperatures of 200–800°C are common.

Pressure is commonly quoted in bars. One bar is close to the normal air pressure – the pressure exerted by the atmosphere at the Earth's surface. One bar is also the pressure under a 10 m column of water. So the pressure at the seabed in the deep oceans (5000 m water) is 500 bars. Ocean crust 10 km below the surface experiences a pressure of around 3 kbar (3000 bars), continental crust 35 km deep experiences a pressure of 10 kbar, and the deep crustal root below the Himalaya at 70 km experiences a pressure of 20 kbar.

Pressure is caused largely by the weight of overburden rock, called the lithostatic load. Directed pressure caused by rock masses being forced together by tectonic processes is less than 100 bars, and relatively unimportant. Heated rocks can lose large amounts of water and gases such as carbon dioxide, which build up a fluid pressure in the surrounding rocks. While the partial pressures of differing components in the fluids have a major effect on the ensuing chemical reactions, the physical pressure effect is minimal, apart from some hydraulic fracturing.

METAMORPHIC GRADE

Increasing temperature and pressure cause a well-known series of changes in the mineral assemblage in the rocks. These are called changes in **metamorphic grade**. The original work on metamorphic rocks in the Scottish Highlands identified a series of indicator minerals for the increasing metamorphic grade:

chlorite → biotite → garnet → staurolite → kyanite → sillimanite

This sequence is recognised elsewhere, though the appearance of any one index mineral is not enough to define the pressure–temperature conditions: the whole assemblage must be examined. Furthermore, the minerals present are especially dependent on the composition of the original rocks; for example, metamorphosed basalts and metamorphosed shales produce very different minerals. Detailed studies have been done in Australia on such metamorphic transitions in the Broken Hill, Cooma and Mount Isa–Cloncurry regions.

The sequence of metamorphism depends on the tectonic situation, and the consequent relative rates and degrees of heating and of lithostatic load pressure.

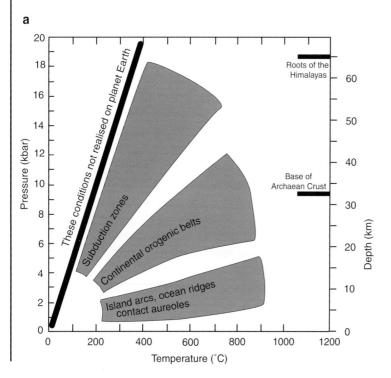

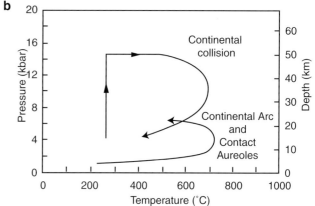

Figure 2.15 *a. Temperature and pressure conditions in different tectonic situations. The bases of the Archaean and modern Himalayan crust are noted. Subduction zones, where slices of the crust are stacked on top of successive thrust faults, reach very high pressures though without high temperatures. Island arcs, ocean ridges and contact aureoles experience low pressure but may reach high temperatures. Continental orogenic belts are usually subject to intermediate pressure– temperature (PT) conditions.*

Pressures can range up to 18 kbar under the 70 km thick continental crust of the Himalaya, and temperatures can exceed 800°C. At temperatures below 200°C the original features of the rocks are clearly visible, but increasingly above this temperature and above pressures of 3 kbar their nature changes. The rocks become deformed and the mineralogy changes – they are metamorphosed.

b. The trend of PT change in two contrasting tectonic settings. Continental collision along convergent margins creates an anticlockwise PT path, whereas continental arcs, where there is major magma injection without major pressure increases, produce clockwise PT paths.

REPRODUCED BY PERMISSION OF FRANK S. SPEAR, MODIFIED FROM FIGURES 1.1, 3.10 IN SPEAR, F.S., *Metamorphic Phase Equilibria and Pressure-Temperature-Time Paths*, MINERALOGICAL SOCIETY OF AMERICA MONOGRAPH

CONTINENTAL COLLISIONS

The development of greatly thickened crust, such as under the Himalaya, causes very high pressures. The depression of this crust lowers rocks into high-temperature regions. When the rock mass reaches maximum depth it also reaches maximum pressure, followed by maximum temperature conditions. Uplift and continuing erosion of the mountains will lead to reduced overburden pressure. This uplift removes the rocks from high temperature regimes, and gradual cooling leads to decreasing temperature. Such a course can be plotted as a clockwise curve on a pressure–temperature diagram, as shown in Figure 2.15b.

CONTINENTAL ARCS AND CONTACT AUREOLES

Most preserved, convergent margin sequences result from continental rather than island arcs. Convergent margins can generate substantial magma intrusions within moderate crustal thicknesses, and so lead to rapid increases in temperature without large pressure increases. Continued subduction can move the sequences upwards. This uplift and the associated erosion, especially as the pulse of hot magma intrusion passes, lead to lower temperatures. So in continental arcs the rocks may never reach the high pressure environments experienced in the continental collisions. Such a course plots as an anticlockwise path lower on the PT diagram, as shown in Figure 2.15b. Similar low pressure, high temperature metamorphic conditions affect country rocks in the contact aureole around plutonic intrusions such as granites.

Australian metamorphic rocks are most commonly of the low pressure, anticlockwise path.

Shaping of the landscape

Weathering is the process that converts solid rocks into particulate soils, partly by physical breakdown of rocks to smaller pieces and partly by chemical change. Chemical changes can form new minerals, such as clays and various iron oxides, that were not present in the bedrock. This weathering forms a gradation from fresh bedrock to completely altered soil. A **soil profile** consists of layers or zones representing these gradual changes.

During weathering of basalt all the component minerals break down easily, forming clays stained by iron oxides and generating rich, red volcanic soils. Weathering of granite, in contrast, releases clays from the breakdown of feldspars, but the quartz remains to form the pale, sandy, poorer soils typical of granite country.

Weathering of bedrock proceeds more quickly along fractures, which help the penetration of plant roots and groundwater. One result is that decomposition of the bedrock moves inward from these fractures, and can leave **corestones** of relatively fresh rock sitting in the soil mass.

A wide range of soil types has developed in Australia, in response to the wide range of rock types and climatic regimes over millions of years (see Box 2.3). Many soils are soft and easily worked. In contrast there are three soils which form cemented materials: laterite, silcrete and calcrete.

Ferricrete, commonly also called **laterite,** is characterised by a hard, red-brown, iron-rich cemented crust, commonly nodular, and underlain by mottled or bleached white clays. Laterites are of surprisingly similar composition regardless of the original material that has been lateritised.

Landscapes with flat-lying, hard units such as a laterite cap or a resistant rock like sandstone or basalt, commonly form plateaus. A **plateau** is an extended area of higher topography with a flat or gently undulating surface. A **mesa** is a flat-topped hill (see Figure 2.16). Erosion of a plateau eventually leaves some isolated mesas.

Silcrete is similar to ferricrete but the crust is composed of very hard pale grey or white rock cemented by silica. **Billy** is a colloquial term for silcrete crusts or boulders.

Calcrete is a layered or nodular rock formed of calcium carbonate in the soil zone. Typically the nodules have other soil particles mixed with the precipitated calcium carbonate and are stained red or brown by iron oxides. Calcretes also form in the shallower parts of cave deposits.

As a group these cemented soils are called **duricrusts** because of their hard durable nature. Each may form on bedrock or on eroded materials, and lateritised river deposits are common.

Figure 2.16 *The Mount Mulligan plateau, north Queensland, is capped by Triassic sandstones.*
Photo: Richard Rudd

Box 2.3 The Australian regolith and soils

Northern hemisphere land masses extend to the polar regions so that during the Pleistocene, ice sheets and glaciers extended into Europe and central north America. These ice masses scraped off the old soils, exposing fresh bedrock. Modern soils in Europe and North America are only 5000 years old, generally thin (1–2 m), and commonly rich in organic matter. This is not so in Australia, where glaciers were limited to highlands in Tasmania, Victoria and New South Wales. Accordingly, Australia retains soils formed over millions of years. The very deep weathering profiles and their great lateral extent across the Australian landscape are caused primarily by the accumulated effects of around 200 million years of weathering.

Given the dry, oxidising environment of much of Australia, it is little wonder so much of the landscape is now red from the remnant iron oxides accumulated after millions of years of weathering.

Regolith is the weathered and transported blanket of material that covers fresh bedrock. The deep and widespread regolith provides a special challenge to mineral explorers because it generally hides features that elsewhere in the world are used to locate ore bodies.

Regolith includes soils developed in place, the weathered bedrock, materials moved down hill slopes, transported alluvium and windblown sands with any subsequent soil development thereupon, and also salt lake sediments.

The nature of a soil depends on many factors: the composition of the bedrock or alluvium on which the soil is forming, the climate, the effects of plants, animals and human activity, the topographic relief and groundwater movement, and time.

Australian soils have the following features:

- They are very deep and retain features of past climates because they are very old. Two contrasting climatic regimes have influenced the development of the Australian regolith: (1) Seasonally humid, tropical to subtropical conditions from the Jurassic to Middle Miocene, and (2) Arid conditions since the Miocene, with several phases of extremely windy conditions.
- They are low in organic matter because of the prevailing dry climates, except in Tasmania and in some wet depressions.
- They commonly contain duricrusts resulting from cementation by iron oxides, silica, or calcium carbonate.
- They have a subsoil that is more clayey than surface soils, because of the filtering down of clays weathered in shallow horizons.
- They commonly have high salt levels in the subsoil, and this salt can be flushed to the surface by a rising water table.
- They commonly show the effects of wind action, either by the deflation of a surface or by the addition of fine sand and silt to soils hundreds of kilometres from their source. For instance, quartz silt forms up to 25% of soils over a wide area of New South Wales, and calcareous silt blown from westerly sources has been added to soils in South Australia and western New South Wales.

Consequently, Australian soils can fall anywhere between two end members. Firstly there are very old soils with well zoned profiles, most of which formed in the moister climatic regimes. These soils develop from the top down, though they may be multi-cyclic and the zonation can extend through alluvial units as well as bedrock.

Secondly there are soils forming over younger transported materials, much of which is still mobile. These soils have a very weak zonation and cover 50% of the continent – most of the western half of the continent and on the eastern highlands. In the west these occur as vast sandplains and dune systems, composed mainly of sediments derived from erosion of older duricrust profiles.

The broad alluvial plains especially in the wetter tropical areas have massive grey, yellow or brown soils, which are commonly mottled.

It is common for a landscape in Australia to be multicyclic, and road cuttings often expose several layers representing different soil-forming and climatic events.

Erosion is the removal downslope of rock and soil by gravity, water, wind or ice. It is erosion that gradually shapes the landscape into the hills and valleys and plains we see around us.

Initial downslope movement occurs as **soil creep,** where the surface soil layers gradually move downhill due to gravity. On steeper slopes there can be rockfalls and avalanches. When the soil becomes saturated with water after heavy rain, material can become mobilised as a mass flow. In this case a whole part of the hillslope fails at once, and moves rapidly downhill as a lobate mass, generally leaving a small fault scarp cut into the hill where it started. These flows are called **debris flows** or **mud flows,** depending on their composition and water content.

Sediment then reaches the watercourses. Narrow creeks and small streams converge, eventually forming larger rivers that flow down wide valleys. The area containing a major river with all its tributaries is called a **drainage basin,** and the line along the ridges separating one drainage basin from the adjacent basin is called the **watershed.**

Water flow through the ground can transport soluble materials, and in limestone country can dissolve out large amounts of the rock to form underground caves and tunnels, such as those on the Nullarbor Plain, at Bungonia and Jenolan in New South Wales, and Chillagoe in northern Queensland. Such a landscape is called **karst.** Karst forms about 4% of Australian landscapes. Within the caves, evaporating water can concentrate the dissolved materials until the calcium carbonate is reprecipitated as **stalagmites** or **stalactites,** forming spectacular displays such as the Cathedral Cave at Wellington in New South Wales. Stalagmites grow from the ground upwards, and stalactites from the roof downwards.

Ground collapses over caves and the long-term flow of water into the underground openings form **sinkholes,** where creeks disappear underground. Commonly in areas of limestone terrain the creeks do not form large drainage nets but flow into these sinkholes, leaving isolated tower-like hills rather than the elongate ridges typical of normal stream systems.

Figure 2.17 *Rivers deposit coarse sediment in channels, fine sands and silts in levees and muds on floodplains. This photograph shows part of a meander belt of the Murray River. Note the abandoned channel forming an ox-bow lake in the central foreground.*
PHOTO: RICHARD RUDD

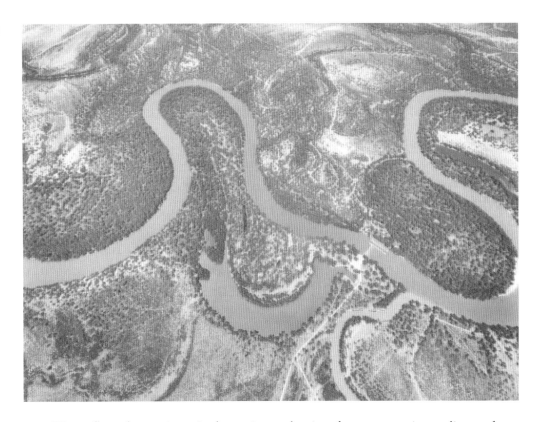

Water-flow down rivers is the main mechanism for transporting sediment from the hills to the sea. In doing so the rivers themselves develop distinctive patterns on the landscape. A typical river consists of an incised **channel** with raised **levees** along both sides. These levees consist of sediments deposited from suspension as the water velocity drops when floodwaters overtop the bank. Stretching for vast distances either side of the channel and levees are the **floodplains**. The floodplains are vegetated in wet areas, but are very bare, with salt lakes and even desert sands, in arid areas.

Some rivers are confined to one main channel (e.g. Gascoyne River in Western Australia, Burdekin River in Queensland) while others, especially those in flat inland areas, form a belt of anastomosing channels up to a hundred kilometres across, such as the Diamantina and Thomson Rivers in the Channel Country in western Queensland.

Wind can transport sand and finer sediments forming **aeolian** deposits such as sand dunes. Because wind is much less dense than water it can only transport fine sediment, and typically it transports a narrow size range of sediment. That is why sand dunes contain such well graded sands, which geologists term **well sorted**. Poorly sorted sediments contain a wide range of grain sizes.

In the continental interior, wind transport leaves behind the heavier stones, forming the gibber or **stony deserts**, which commonly overlie rock pavements. The sand is blown away forming the dunes of **sandy deserts**. These dunes are of two

  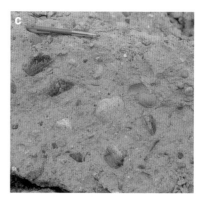

main types – **barchan dunes** which have a curved shape, and longitudinal dunes which can extend as long ridges for a hundred kilometres or more. The wind blows across the curve of a barchan dune, and parallel to the ridges of **longitudinal dunes**.

Wind transport is also important near the coast where onshore winds blow beach sands inland, forming coastal sand plains.

Sediments can also be transported by ice. **Glaciers** are rivers of ice, formed when the temperature is below freezing. Snow accumulating on highland areas becomes compacted into ice, which then starts to flow downhill. As the ice flows it scours the underlying rocks, plucking out boulders and grinding some of them to fine rockflour. Sand grains in the ice polish the rock surfaces like sandpaper, while larger rock fragments gouge out striations and grooves parallel to the direction of ice flow. When the ice melts this material is deposited as a very poorly sorted mass of boulders, sand and silt in a deposit on land called **moraine**. Geologists call the material deposited in moraines **till**; when consolidated into rock it is called **tillite**.

Some glaciers extend out to sea or onto an ice shelf. Part of the ice can melt allowing sediment to fall to the seabed, or large blocks of ice can break off and form **icebergs**. Icebergs can drift with the currents for hundreds of kilometres, gradually discharging their cargo of sediment as they melt. Both on land and under the sea, ice can leave an **erratic**, which is an exceptionally large boulder sitting in finer sediments. The boulder is composed of rock totally different to nearby bedrock, but has been derived from a source maybe hundreds of kilometres away and transported by ice. Flowing water simply could not move such a large boulder that distance along a river bed.

Glaciers do not occur in Australia now, though they have been regular features during the geological past (see Chapters 3, 5).

Coastal and offshore areas

When rivers reach the coast, sediment is deposited. The triangular piece of land at the river mouth where sediment is deposited by the spreading channels of a river is called a **delta**.

Figure 2.18 *Glacial features. a. Deep U-shaped valley cut by glacier beside Mount Garibaldi (Athabasca), Canada. b. Glacial striations on limestones, Ontario, Canada. c. Poorly sorted till, Canada.*

PHOTOS: DAVID JOHNSON

The heavier gravel and sand is dumped at the river mouth when the velocity slows as the river reaches sea level. Some of this sand stays to form the sand bars in the river openings and is often dredged to allow the safe passage of ships, while the rest of the sand is washed along shore, principally by waves.

Thus there is a continual supply of sediment to the coast. At the mouth there is transport of sand along the beach and surf zone. Mud is commonly carried offshore as a turbid, freshwater plume floating on the denser, salty ocean water. Waves and currents eventually mix the two waters and the mud settles to the seabed offshore, or is carried by the tides into protected lagoons and tidal flats.

Continental shelf and deep sea

The Australian continent is surrounded offshore by a relatively flat **continental shelf** with a slope of about 1–5 m/km, extending up to 150 km offshore to water depths of 100–200 m. Deep seismic profiles show that the shelf is made of sediments underlain by fault blocks of the bedrock similar to the main Australian land mass. The continental shelf represents the broken edge of the continental mass left after break-up due to seafloor spreading. The upper surface of the shelf is formed of sediments eroded from the onshore land mass.

The break-up also left isolated masses of the former land mass stranded offshore as **plateaus**, separated from the present Australia by new seafloor or by a narrow **trough**. An example of the latter is the Queensland Plateau off northeastern Australia, which is separated from the continental shelf and Great Barrier Reef by the Queensland Trough.

Beyond this shelf, the seafloor drops more steeply, forming a **continental slope** that is incised by submarine canyons, similar to the valleys and gullies that cut into hills on land. Material may be transported down these canyons and deposited at depths of 1000–4000 metres, forming lobes of sediment called **deep-sea fans**. These fans are formed by **turbidity currents** – masses of highly turbulent water, heavy with suspended sediment, flowing along the seabed. These turbidity currents start on the shelf edge or slope and are generated by pounding storm waves or earthquake-induced slumps. Sediment is mixed with water and, the suspension being denser than seawater, flows downhill on the seabed. The coarser sediment is dropped near the base of the slope, forming the deep-sea fans, and the finer

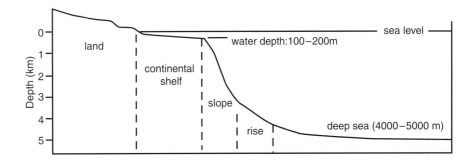

Figure 2.19 *The relationship between a continent and the continental shelf, slope, rise and deep sea. Water depth on the continental shelf is deep by human standards – 100–200 m – but shallow when compared to the depths of the main ocean basins – 4000–5000 m.*

sediment stays in suspension as the current flows further to the deep sea. The resulting deposits, called **turbidites**, are characteristically graded from coarse at the base to finer at the top of the bed. The coarser, heavier grains settled from the suspension first, with finer material successively on top. Turbidites represent the transport of huge amounts of sediment and can extend 3000 km across the deep ocean floor.

Beyond the fans, the deep seabed of the main oceans consists of flat **abyssal plains,** which are typically 4000–5000 m deep. Sedimentation consists of extensive turbidites and also layers of clays and pelagic ooze. The plankton are mainly algae and foraminifera, which have tiny skeletons of calcium carbonate, and accumulate to form a pale grey or brown calcareous ooze. In the deepest parts are siliceous oozes, which are composed of the opaline skeletons of diatoms and radiolaria. The reason these very deep water sediments are siliceous is that the carbonate dissolves in the deep cold waters, thereby concentrating the siliceous materials.

Orogenic cycle

Plate tectonics explains what geologists have recognised for a long time: that the Earth is in continuous change. Rocks contain evidence of repeated cycles of erosion and sedimentation, followed by burial and metamorphism, followed in turn by granite emplacement with uplift to form mountain chains. Then the cycle of erosion and sedimentation starts again. **Orogenesis** is the old term for mountain building, and the sequence has been called the **orogenic cycle.**

The cycle starts when mountain regions and hills are eroded, and the resulting sediment accumulates in a sedimentary basin. Some of this basin may be onshore so that the sediments form in river or glacial deposits, depending on the climate. Some of the basin may be offshore so that marine sediments are formed along the coast, on the continental shelf or in the deep sea. Volcanism is also common at the start of the cycle, when there is active tectonism and the crust is fractured, allowing magma to rise to the surface. Consequently, lavas commonly occur at the bases of the sedimentary basins, and volcanic debris is a prominent contributor to the sediments filling the early part of the basin.

A typical sedimentary basin forms by subsidence caused by the extension or downwarping of the crust. The basin thus formed can accommodate the eroded sediments. During the initial rapid subsidence, oceans can flood the basin, forming marine sediments. Then subsidence slows and the basin fills above sea level, the shoreline recedes, and onshore sediments are formed.

Continued subsidence lowers the sediments into hotter parts of the Earth's crust, especially when the sedimentary basin is carried on a moving plate into a subduction zone. The sediments and volcanics become faulted, deformed into folds and metamorphosed by the heat and pressure. Compression in the subduction zone can force the rocks upwards into mountain ranges. Finally, granites are formed by the melting of these rocks, and the hot granitic magma rises into the overlying crust.

Erosion of these mountains and deposition of sediments in a nearby subsiding basin starts the orogenic cycle all over again. The point where newly deposited sediments or volcanics overlie an eroded older terrain is called an **unconformity**. Commonly the older rocks below the unconformity have been tilted or folded, so that there is an angular relationship because horizontal sediments overlie dipping rocks below the unconformity. The key point about an unconformity is that it represents a time gap: the period of time during which uplift and erosion occurred is not represented in the rocks.

The continental cycle can be in the order of 250–650 million years. In the oceans the cycle is shorter, and the oldest known oceanic crust is 180 million years old. The average oceanic cycle has been estimated to be closer to 60 million years. Both of these cycles seem very long, but they have happened many times to Australia.

But as with most rules, there is one exception. **Cratons** are parts of the continental crust that have not been affected by tectonic processes for over one billion years, and some for over 3 billion years. Considering that most of the Earth's crust has been recycled – buried, deformed and uplifted repeatedly – these cratons are special because they preserve evidence of Earth's most ancient history, which has been obliterated elsewhere by tectonic processes. Cratons have a thickened, cold crust with a relatively low surface heat flow, and high seismic velocities up to 250 km depth. Australia has three of these cratons (see Chapter 3).

One result of these orogenic processes is that rocks on a continent can commonly be mapped into regions which have a similar geological history and range of rock types, and which differ from adjacent regions. Geologists call these regions **terranes**. In some places, such as California, plate movements have shuffled widely differing terranes against each other.

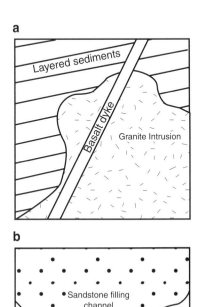

Figure 2.20 *The relative age of rocks can be determined from their relationships. a. A basalt dyke is younger than the granite into which it intrudes, and the granite is younger than the sedimentary rocks into which it has intruded. Compare with the photo in Figure 2.10.*
b. Sandstones that infill channels are younger than the rocks in which the channel has been cut.

Geological time scale

The geological time scale recognises that some rocks formed later than others. For instance, sediments must be younger than the layers they overlie, and igneous dykes are younger than the rocks they intrude. These observations allow us to assign relative ages – that this rock is younger than that one. However, they do not allow us to assign an absolute age in terms of thousands or millions of years.

We are used to the history of world events, measured in spans of hundreds or thousands of years. The geological time scale measures the history of the Earth on a span of millions of years.

Radiometric dating has determined an accurate age for many rocks, and by combining all the data we can derive a geologic time scale. The scale is divided into major periods that represent the rocks formed during each time interval. The scale is set out in Figure 2.21, with the boundaries noted in millions of years.

The span of geological time is almost too much to comprehend. Present estimates place the age of the Earth at 4.6 Ga – that is, 4600 million years. Our record is far more detailed the closer we get to the present. For instance, we can separate the stages in the waning of the last ice ages, with an accuracy of the order of 10 000 years. However further back in time, we can resolve changes only on a time scale of around 1 million years.

Life on Earth as we know it is surprisingly young: even the oldest complex life forms preserved as fossils are less than 570 million years old. That is only the last one-eighth of geological time.

One way to grasp the magnitude of geological time is to compare it to the events in a year, to compress the entire 4.6 billion years into 365 days. On that scale the oldest rocks we can date formed about mid-March, and we can find living creatures such as algae and jellyfish in May. Plants and then animals did not emerge on land until the end of November, and the valuable coal basins of eastern Australia were formed from vast peatlands over a period of 19 hours on 10 December. Dinosaurs roamed central Queensland from shortly after this until 26 December, when Australia separated from Antarctica and started its drift northwards towards Asia. The Great Barrier Reef probably formed mainly after 11.03 pm on 31 December. Aboriginal people arrived perhaps around 11.54 pm, and the last of the giant kangaroos died out by 11:59:35. Twenty-five seconds to go. Rome ruled the Western world for 5 seconds from 11:59:45 to 11:59:50. James Cook arrived on the Australian coast at one second to midnight. And, as they say, the rest is history.

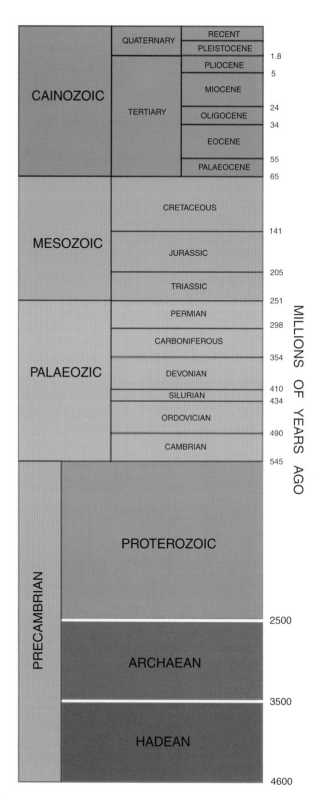

Figure 2.21 *The geological time scale.*

TIMESCALE IS BASED ON THE GEOSCIENCE AUSTRALIA TIMESCALE BOOKMARK, MODIFIED AND REPRODUCED COURTESY OF GEOSCIENCE AUSTRALIA, CANBERRA. CROWN COPYRIGHT ©. ALL RIGHTS RESERVED. WWW.GA.GOV.AU

SOURCES AND REFERENCES

Anand, R. & Paine, M., 2002. 'Regolith Geology of the Yilgarn Craton, Western Australia'. *Australian Journal of Earth Sciences*, 49: 3–162.

Betts, P.G., Giles, D., Lister, G.S. & Frick, L.R., 2002. 'Evolution of the Australian lithosphere'. *Australian Journal of Earth Sciences*, 49: 661–695.

Binns, R.A., 1964, 'Zones of progressive regional metamorphism in the Willyama Complex, Broken Hill District, NSW'. *Journal of the Geological Society of Australia*, 11: 283–330.

Clark, I.F. & Cook, B.J., 1983. *Perspectives of the Earth*. Australian Academy of Science, Canberra. 651 pp.

Conrad, C.P. & others, 2004. 'Great earthquakes and slab pull: interaction between seismic coupling and plate-slab coupling'. *Earth and Planetary Science Letters*, 218: 109–122.

Deer, W.A., Howie, R.A. & Zussman, J., 1966. *An Introduction to the Rock Forming Minerals*. Longmans Green & Co., London. 528 pp.

Dunay, R.E. & Hailwood, E.A., 1995. *Non-biostratigraphical Methods of Dating and Correlation*. Geological Society Special Publication No. 89. The Geological Society, London. 265 pp.

Geyh, M.A. & Schleicher, H., 1990. *Absolute Age Determination*. Springer-Verlag. 503 pp.

Gurnis, M., Wysession, M.E., Knittle, E. & Buffett, B.A. (eds) 1998. *The Core-Mantle Boundary Region*. Geodynamics Series, Volume 28. American Geophysical Union. 334 pp.

Heller, R., Merrill, R.T. & McFadden, P.L., 2002. 'The variation of intensity of the earth's magnetic field with time'. *Physics of the Earth and Planetary Interiors*, 131: 237–249.

Holmes, A., 1970. *Principles of Physical Geology*. 2nd Edition. Thomas Nelson & Sons. 1288 pp.

Hulot, G., *et al.*, 2002. 'Small-scale structure of the geodynamo inferred from Oersted and Magsat satellite data'. *Nature*, 416: 620–623.

Kohn, B.P., Gleadow, A.J.W., Brown, R.W., Gallagher, K., O'Sullivan, P.B. & Foster, D.A., 2002. 'Shaping the Australian crust over the last 300 million years: insights from fission track thermotectonic imaging and denudation studies of key terranes'. *Australian Journal of Earth Sciences*, 49: 697–717.

Marjoribanks, R.W., Rutland, R.W.R., Glen, R.A. & Laing, W.P., 1980. 'The structure and tectonic evolution of the Broken Hill region, Australia'. *Precambrian Research*, 13: 209–240.

Osborne, R., Tarling, D. & Gould, S.J., 1996. *The Historical Atlas of the Earth*. Henry Holt & Co. 192 pp.

Pillans, B., 2003. 'Subdividing the Pleistocene using the Matuyama-Brunhes boundary (MBB): an Australian perspective'. *Quaternary Science Reviews*, 22: 1569–1577.

Rubenach, M.J., 1992. 'Proterozoic low pressure/high temperature metamorphism and anti-clockwise P-T-t path for the Hazeldene area, Mount Isa Inlier, Queensland'. *Journal of Metamorphic Geology*, 10: 333–346.

Spear, F.S., 1993. *Metamorphic Phase Equilibria and Pressure-Temperature-Time Paths*. Mineralogical Society of America, Monograph. 799 pp.

Sutherland, F.L., 1995. *The Volcanic Earth*. University of New South Wales Press. 248 pp.

Sutherland, L. & Webb, G., 2000. *Nature Guide to the Gemstones and Minerals of Australia*. New Holland. 128 pp.

Skinner, B.J. & Porter, S.C., 1995. *The Dynamic Earth*. 3rd edition. John Wiley. 567 pp.

'The Dynamic Earth'. *Scientific American*, 249(3) September 1983.

Vernon, R.H., 2000. *Beneath Our Feet. The Rocks of Planet Earth*. Cambridge University Press. 216 pp.

Wagner, G. & van den Haute, P., 1992. *Fission-Track Dating*. Kluwer. 285 pp.

Young, A. & Young, R., 2001. *Soils in the Australian Landscape*. Oxford University Press. 210 pp.

Young, G.C. & Laurie, J.R. (eds) 1996. *An Australian Phanerozoic Timescale*. Oxford University Press. 279 pp.

WEBSITES

General geology
Geoscience Australia
www.ga.gov.au
General basic geology
www.kidsvista.com/Sciences/geology.html
United States Geological Survey
www.usgs.gov
Maps with tectonic features anywhere on Earth
www.aquarius.geomar.de/omc/omc_intro.html
Seismic Wave simulations
www.epsc.wustl.edu/~saadia/page2.htnl

Volcanoes
How volcanoes work
www.geology.sdsu.edu/how_volcanoes_work/
Global Volcanism Network
www.nmnh.si.edu/gvp/index.htm
Volcano World
www.volcano.und.nodak.edu/vw.html
Hawaiian Volcano Observatory
www.hvo.wr.usgs.gov/
US volcanism and including joint US-Russian work in Kamchatka region
vwww.olcanoes.usgs.gov/
Latin American volcanoes from Mexico to Chile with photos, activity reports and webcams
www.tabitha.open.ac.uk/williamg/LAVolc.htm

Glaciers, ice erosion and moraines
www.tvl1.geo.uc.edu/ice/image/imageref.html

The original Earth

Formation of the Earth

There are some times when I wonder if we have any idea at all of how the Earth was formed. The most accepted scientific theory is that a huge, rotating shell of dust and gas, spread out across billions of kilometres of space, was left behind after the contraction of the matter that formed the Sun. This shell gradually collapsed due to gravity, the dust coalescing into a series of variously sized lumps, or planetismals. The planetismals, which ranged in size from a few metres to Mars-sized objects, then accreted by gravity over a period estimated to have been 29–100 million years. They formed spinning balls of rock – the terrestrial planets – and Earth is one of them.

The early Earth and other planetismals were bombarded with meteorites and comets, probably the debris left from formation of the Solar System. The craters formed are visible on our Moon through binoculars. It is quite possible that fragments of the early Earth were blasted out at supersonic velocities and reached the Moon, where they remain waiting to be discovered, probably buried by the debris from later impacts.

Few of the craters left by these impacts remain on Earth, for most have been eroded by running water. However, the craters are preserved on other rocky planets in the Solar System that are dry (see Chapter 11). These rocky planets have no visible water left, nor does our Moon, although Mars has river valleys, which implies there was running water there in the past. The water may have been evaporated from the planetary surface or may now be frozen into the ground.

What about the oceans on Earth? There are respectable theories that all the water on Earth came from comets, or icy masses that accreted early in the formation of the Solar System, or from water blown away from the inner rocky planets close to the Sun.

What about the atmosphere? The general feeling is that most of the atmosphere derives from the degassing of the early Earth, and there is continuing escape of deep gases along major breaks in the crust, such as the mid-ocean ridges. The composition of the atmosphere is also thought to have changed from one richer in methane in the Precambrian to one with very little methane now. And the earliest atmosphere may have contained no oxygen gas at all.

But we have so little evidence and so little real appreciation of these galactic processes that our estimates could be way off. What about the facts?

The age of the Earth

The Earth originated around 4600 million years ago. Sometimes this is written as 4.6 billion years or 4.6 giga-years (Ga). This estimate is based on calculations of the residual amount of lead left from the radioactive decay in the Earth's crust. Very old rocks are known from the NW Territories of Canada, dated at 4030 Ma, and from Greenland at over 3800 million years.

CHAPTER 3

BUILDING THE CORE OF PRECAMBRIAN ROCKS

What do we know of the origin and age of the Earth? Was the Earth hotter or colder than now? What do we know of the origin of the Earth and of life?

The oldest rocks tell us there was a time when there were no complex life-forms. Then there was a time when life was evident, and there were the first glaciations.

Precambrian plate tectonics involved two supercontinents – Rodinia and Gondwana – of which only fragments now remain.

Figure 3.1 *Archaean metamorphosed sedimentary rocks in the Jack Hills area, northern Yilgarn craton, Western Australia. These rocks have yielded the oldest mineral grain, a zircon, yet dated on the planet. The actual zircon dated is shown in figure 1.3.*
PHOTO: SIMON WILDE

We can also date individual mineral grains in rocks using radioactive decay rates of elements such as uranium, thorium, lead and potassium. SHRIMP dating of zircons in metamorphosed conglomerate from the Mount Narryer and Jack Hills areas of Western Australia has provided evidence of even older materials. Detrital zircons – that is, grains that have been eroded from a granite and redeposited – have been individually dated as old as 4276 Ma, and most recently 4404 Ma, meaning there was Earth crust ready to be eroded before 4400 Ma ago. Allowing for the fact that these are probably not the first rocks *ever* formed, we place the age of the Earth about 4600–4500 Ma, very close to the ball-park estimates made using the decay rates of radiogenic lead.

Incidentally, these ages are very close to, but slightly younger than, those measured on rocks brought back from the Moon (see Chapter 11), which are the oldest rocks known. Finally, meteorites have been dated to ages in the range 4500–4600 million years, indicating this time was a major phase of consolidation of rocky materials in the Solar System.

The oldest known extensive erosion surface is in the Pilbara region of Western Australia, where geologists have found cherts interbedded with volcanic basalts, at almost 3500 million years. The cherts are made of silica and are probably chemical precipitates, formed in quiet periods between outpourings of the basalt. These rocks overlie an eroded older surface. Hence we have evidence of a land surface on the Earth probably more than 3500 million years ago.

When do we see the first sedimentary rocks, which clearly indicate the transport and deposition of sand and mud by water? Again in the Pilbara, between 2500 and 3500 Ma ago. These sediment particles must have been eroded from older rocks, and soon after 3000 Ma there were rivers and lakes or seas in which to deposit them.

It was a different world then. The Moon was closer to Earth, leading to bigger tidal ranges. The Earth was spinning faster, so there were more days in the year. The rotation has been slowing ever since. Modern atomic time clocks are adjusted by a second every year or so to make sure they match the rotation period of the Earth.

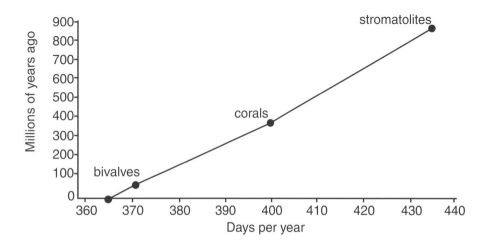

Figure 3.2 *Estimate of the decrease in number of days per year for planet Earth during the last 900 million years. As the Earth's rotation slows there are fewer days in each annual circuit around the Sun.*

This slowing is due to the gravitational pull of the Moon and to internal friction of the mobile masses in the Earth's interior. The effect is much like spinning an egg on the kitchen table. A raw egg will spin more slowly and for less time than a hard-boiled one.

The time span known as the Hadean (see Figure 2.21) encompasses the time from the formation of the planet to the unknown date when the first rocks crystallised to coherent crust. The least dense materials would have risen to the surface, forming the continental crust, which floats on the denser mantle, and the most dense materials would have sunk to the core at the Earth's centre.

The time from when these rocks formed till 2500 million years ago is called the Archaean – literally the 'beginning'. Then about 2500 Ma ago the system changed; sediments are more common and there are the first signs of life. This is the Proterozoic, or 'ancient life'. The next major change occurred 545 Ma ago, the time of the explosion of complex life-forms at the start of the Cambrian.

The total time span from 4600 to 545 Ma is called the Precambrian, and it covers 90% of Earth history. Yet it occupies just one chapter of the twelve in this book. That is a reflection of how little we know about it. Nevertheless, let us examine what data we have.

Heat loss from Earth

We know that the Earth is hotter inside, firstly because deep boreholes and mines become hotter with depth. Generally the rate of temperature increase is in the range 25–50°C per kilometre. Many deep gold, copper and diamond mines are air-conditioned, because at 1000 m depth the rock temperature is 40–50°C, and nobody can work for extended periods in that temperature. Secondly, we know from volcanic eruptions that hot magma rises to the surface and cools.

The temperature at the base of the Earth's crust is estimated to be 500–800°C, and at the surface of the outer core around 5000°C. This is mainly residual heat stored from the primordial accretion of the planet, with some radioactive heat. The main source of heat in the Earth's crust is decay of radioactive elements.

The half-life of uranium-238 (4470 million years) is about the same as the age of the Earth (4600 million years). So at the start of the Precambrian there was approximately twice the amount of radioactive uranium-238 generating heat as there is now (see Box 1.1). As a result, the planet's radioactive heat production is decreasing. The radioactive material occurs at very low concentrations throughout the planetary mass, but the cumulative effect is still measurable heating.

Because the planet is a sphere of hot rock spinning in a void of cold space, the tendency is for heat to escape outwards – but how does this occur?

Large convection cells in the Earth's mantle transfer heat from the core and the lower mantle to the upper mantle, so that convection is the main mechanism by which the deep Earth is losing heat. These convection cells are driving the movement of the less dense plates of rock on the outer part of the Earth, like the movement of scum on a pot of slowly boiling jam. In this way the movement of the Earth's crust

– plate tectonics – is a symptom of planetary heat loss. The lithosphere of the outer Earth is losing heat mainly by conduction through solid rocks.

The Earth is cooling down. Total global heat loss from the rock mass is estimated to be 4.2×10^{13} watts, whereas radiogenic heat production is 2×10^{13} watts. Even without unravelling exactly what these figures mean in terms of temperature, it is clear the Earth is losing heat, or cooling, about twice as fast as radiogenic heat is being produced.

This implies that about half the heat emanating from the Earth is produced by radiogenic decay. So the other half must be from stored heat, which is not being replenished. Eventually the Earth will cool completely inside. Then there will be no more mantle convection, no more plate tectonics and continental drift, no more major earthquakes and no more active volcanoes.

Luckily, a cooling rock mass does not affect the incoming solar heating of the surface on which our weather and life systems depend. That will continue as long as the Sun keeps shining, which on current estimates is about another 6 billion years.

The Earth's solar heating budget is balanced on a global scale; equatorial and tropical regions receive a heat excess and polar regions experience a heat deficit. The excess at the equator is transported by atmospheric and ocean circulation towards the poles.

The balance does change slightly on a long-term time scale, and this causes glacial and non-glacial periods. That these glaciations have waxed and waned confirms this balance, because the ocean and atmosphere climate system has been able to oscillate back and forth without ever permanently going to extreme overheating or to permanent glaciation of the planet. The temperature changes during these oscillations are not large: 3–5°C in tropical and temperate regions, and slightly more near the poles.

The cooling of the rock mass and diminishing amounts of uranium mean that the Earth must have been hotter in the past. The present estimate is that total heat production 2.8 Ga ago was three times the present level. Do we have evidence of this extra heat? Yes – in strange Archaean rocks called komatiites.

Archaean

Archaean rocks are the oldest rocks we know; more than 2500 million years old. They are characteristically igneous, of two types – greenstones and granites – with minor sedimentary rocks. In Australia there are three isolated cratons of these Archaean rocks: the Yilgarn and Pilbara cratons in Western Australia, and the Gawler craton, which forms much of the Eyre Peninsula in South Australia. The Moho is about 30 km deep under the Pilbara and about 40 km deep under the Yilgarn craton.

The Yilgarn craton is the most extensive of the three, extending over 1000 km north from the coast near Esperance to the inland Murchison River north of Meekatharra (see Figure 3.3). It is some 600 km wide, extending from the Darling Range inland of Perth, across the wheatbelts past Narrogin and Merredin, to the

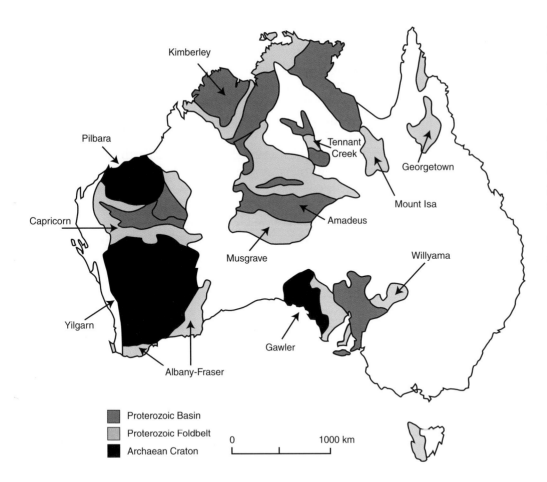

Figure 3.3 *The extents of Archaean cratons, Proterozoic basins and Proterozoic orogens (fold belts) in the present Australia. Note that the eastern edge of Australia was not formed at these times, and it is not clear whether all these Precambrian outcrops were necessarily in the present relative positions. There is evidence that the Willyama (Broken Hill) and Mount Isa regions were formed together and have been split apart.*

goldfields and nickelfields of Kalgoorlie, Kambalda and Norseman. The eastern edge is just west of Fraser Range. Most of the rock is granite and gneiss, with a thin cover of sand. Within the granite mass lie north–northwesterly faults and strips of folded greenstones. This regional grain in the rocks was well known to the early colonists, and is observed in the granites too, because of the faults.

In his book *Spinifex and Sand*, a narrative of pioneering and exploration in Western Australia including the overland camel trek from Kalgoorlie to Halls Creek and back in 1898, David Carnegie noted that in the dry regions from Southern Cross to Coolgardie early travellers had to rely on water found in small soaks around the granites:

> The interior of the Colony…is traversed by parallel belts of granite, running in a general direction of north–north-west to south–south-east. This granite crops out above the surface, at intervals of from ten to twenty or thirty miles, sometimes as low ranges and hills several miles in extent…Round the granite base a belt of grass of no great extent may be found, for the most part dry and yellow, but in places fresh and green. It is in such spots as these that one may hope to tap an underground reservoir in the rock.

Figure 3.4 *Archaean granites form low hills on the flat plain of the Yilgarn craton, Carr Boyd rocks area, about 120 km north of Kalgoorlie. Rain draining off these rocks and accumulating in soaks around the base of such outcrops was a principal source of water for early explorers in this region.*
PHOTO: BILL LAING

The Western Australian state government also spent several thousand pounds constructing drains and tanks to receive rainwater channelled off the bare granite hills.

Deep seismic profiling has shown that these granite bodies are not nearly as thick as was previously thought. Twenty years ago scientists thought most granite bodies extended tens of kilometres downward into the Earth's crust, but we are now sure the Goldfields granites overlie another fault, in this case a flat-lying regional fault (or detachment) in the rocks at a depth of 4–7 km. Below this the seismic sections show a layering in the rocks, unlike the massive nature of the seismic signature of the granites. So it seems the granites were emplaced as elongate pods into the pre-existing greenstone and gneissic belts, presumably fed upwards and along the regional faults that trend north–northwest. That is why they form the repetitive lines of outcrops observed by the early explorers.

The northern Pilbara also contains Archaean rocks, in a belt running east–west to the south of Port Hedland. Here there are sediments and basalts 2700–2800 million years old, overlying a very deformed terrane of granites and greenstones. The southern edge is clearly marked by the Chichester Range along the northern side of the Fortescue River valley, and the belt extends north to about the De Grey River. Earliest rocks of the Pilbara craton are around 3300 million years with continued accretion until the craton was essentially complete by 2850 million years ago.

The Gawler craton rocks are exposed on Cape Carnot on the tip of the Eyre Peninsula, and extend north of the Trans-Australia railway almost as far as Lake Phillipson and the Wilkinson Lakes in the southeastern Great Victoria Desert. They are mainly granites and gneiss with basaltic volcanic rocks, including komatiites. These rocks were severely deformed between 2640–2300 Ma, indicating they were emplaced before this time.

Archaean greenstones and komatiites

The greenstones contain mainly volcanic rocks, including a wide range of basaltic lavas and volcanic ashes. Among them are unusual basalts known as komatiites, which occur only in the Archaean and which are exceptionally rich in magnesium. The name komatiite comes from the Komati River in the Barberton Mountains in South Africa, where these rocks were first described. Good examples are found in Australia in the Western Australian goldfields region.

These komatiites show many of the features seen in modern basalts. They can be traced as individual flows, typically around 5 m thick, and some show columnar jointing caused by cooling contraction, or pillows where the lava has broken into bulbous masses when it flowed into water. Pillow basalts have been studied where modern lavas flow into the sea, by divers off Hawaii and by submarines in the deep ocean.

An unusual feature of the komatiites, first described and named in Australia, is the spinifex texture. The rock has long, thin crystals through it, sometimes in branching form, as if spinifex needles were embedded in the rock. It was an easy comparison to make since the komatiites abound in the arid interior of Western Australia where spinifex grass occurs. The origin of this feature can be explained by studies of cooling metals; rapid chilling causes crystals to form, and the crystals of

Figure 3.5 *Spinifex texture in komatiite. The elongate olivine and pyroxene crystals have been etched out by weathering of the outcrop.*
Photo: Mark Barley

pyroxene or olivine grow rapidly in one direction that is governed by the crystal's atomic structure, so forming needle-shaped or platey crystals. So we are sure these komatiites were cooled very quickly after erupting. Modelling also suggests that the komatiite lavas were much more fluid than modern basalts, and that they flowed quickly and turbulently from vents. The turbulence would also have allowed more rapid cooling of the lava mass.

How hot were these eruptions? There are no komatiites flowing from volcanoes today. However, experiments conducted at the Australian National University in Canberra in the 1970s showed that the temperature of such a magma at the Earth's surface would be 1650°C. Further work has suggested a range of 1400–1700°C. Modern basalts erupt at less than 1350°C, so the magmas flowing from the mantle in the Archaean could have been some 300°C hotter than the basalts of today. For this to happen the Earth must have been hotter, if not on the immediate surface then certainly at shallow depths. That is, the heat gradient into the Earth would have been higher, in which case hotter magmas could rise closer to the surface of the crust, where they could flow onto the surface through volcanoes.

We can see that these komatiite lavas were hot because there are places where the underlying rocks were melted and eroded by the hot lava flowing over them. Although it might seem strange, even solid rock that cooled and solidified at, for example, 1000°C, can be melted and assimilated by komatiite magma flowing over it at 1600°C. In the same way hot fat will melt some plastic containers on a kitchen bench.

In the Kambalda area these rocks are hosts for nickel mineralisation, and in Kalgoorlie and elsewhere for gold.

Archaean metamorphism and granite emplacement

Granites are very different from the volcanic greenstones. The granites are pale rocks with abundant quartz, and crystallised deep in the crust of continents rather than under the oceans. Some of the Archaean granites pre-date the volcanic greenstones and were the terrain onto which the komatiites were erupted. However most of the granite intruded later. In the eastern Goldfields (Kalgoorlie region), the volcanic rocks have been dated at between 2713 and 2672 Ma, and the granites at 2685–2675 Ma. Metamorphism has been dated at 3300 Ma in the eastern Pilbara, and episodes at 3700, 3300, 2800 and 2650 Ma in the Yilgarn craton. The metamorphism is typically the high temperature – low pressure type.

The granites extend over thousands of square kilometres, indicating the huge scale of emplacement of these igneous rocks. How can an area of some 600 000 square kilometres of granite be intruded into the crust? The overlying crust has long since been eroded away, exposing the granites underneath.

Part of the answer is that not all was emplaced at once. The ages of most of the granites across the Yilgarn lie in the range 3000–2600 million years. So some intrusions could be separated by tens or hundreds of millions of years. Even so, we can ask the question, What was in that 600 000 km^2 before the granite was injected?

Was there a huge hole? Unlikely. More probably the hot granite magma rose through the crust along large fractures and faults, with the overlying rocks spreading apart and also sinking downwards to make room. How did it rise? The accepted theory is buoyancy! The granite was formed by the melting of other rocks deep in the crust, perhaps close to the boundary with the mantle. Granite consists mainly of quartz and feldspar, which are relatively low-density minerals. Does this mean granite rock can rise because of its lower density, just like hot water in a cold tank or oil through water? Yes! The rate of movement will be extraordinarily slow by our standards, but it seems the only realistic explanation to date.

Archaean sedimentary rocks

Sedimentary rocks tell us a lot about the nature of the early surface of Earth. Good information has been obtained from the Pilbara area where sediments 3400–3500 Ma old are exposed. There are volcanic ashes (tuffs), and sandstones with ripple marks and bedding, all of which indicate transport and deposition by water or wind, probably in rivers or along beaches and on tidal flats.

The unusual thing in these formations is that there are almost no fossils: no shells, no bones, no signs of burrowing through the layers. These sediments were laid down before complex life-forms evolved, and they lay undisturbed. Nothing was there to make tracks across the surface or to burrow down into the sediments.

So we are sure, even at that time, that there were exposed continents with shorelines, and that there were waves and currents washing sediments onto beaches and tidal flats. It would have been strange to land on these shores and look inland where there was not a tree, not even grass to be seen. Even in wet climates the rocks would have stretched away to the horizon, like a Martian landscape.

Proterozoic

The Archaean lasted until the major change in Earth history about 2500 Ma ago. The time from 2500 to 545 Ma is known as the Proterozoic. In the Proterozoic we see evidence for the first major continents, the first glaciations, and the first signs of complex life.

Many sites around the world show a major unconformity underlying the Proterozoic sedimentary rocks. The Archaean rocks were deformed and then eroded, and younger Proterozoic sediments and volcanics were then laid down unconformably on this eroded surface. Commonly the younger sediments are flat and undisturbed, and in several places on Earth they have been dated at around 2500 Ma old. In the Pilbara this unconformity has been dated at closer to 2800 Ma, and future work may push the start of the Proterozoic back to this time elsewhere. The sandstones of the Arnhem Land escarpment, sitting on older deformed rocks, are another place where this unconformity is evident. Of course this means that the deformation, uplift and erosion of the Archaean rocks must have started millions of years before this time. The dated sedimentary rocks represent the time of sedimentation upon the older, eroded Precambrian landscape.

By around 1600 million years ago, three distinct pieces of Australia had been assembled together. We are still not sure whether each of these three pieces were close together at the time, or whether they were well separated. The Pilbara and Yilgarn cratons had amalgamated between 2200 and 1620 Ma ago to form a continental mass that now makes up most of Western Australia. Four separate orogenies have been identified, testifying to multiple episodes of compression as the two cratons were forced togther: about 2200 Ma, 2000–1960 Ma, 1830–1780 Ma and 1670–1620 Ma. The Kimberley craton was joined to rocks extending down to Tennant Creek and across Arnhem Land to Mount Isa; and more rock had been added to the Gawler craton, enlarging and extending it to the Broken Hill region.

There was also widespread basaltic volcanism in the Proterozoic: around 1800–1700 Ma in the Mount Isa region and the Arunta region near Tennant Creek, 1600–1580 Ma in the Gawler craton, and 1200–1100 Ma in the Musgrave Ranges.

The Proterozoic rocks have been affected by several phases of metamorphism, generally of the high temperature – low pressure type, commonly associated with major regional movement of subsurface hot fluids. The metamorphism has been dated at 1850 Ma in the Kimberley and Mount Isa regions, and at 1690–1670 Ma and 1570–1540 Ma in the Gawler craton, 1590 Ma at Broken Hill, and 1550 Ma in Mount Isa and Georgetown areas. The youngest is 1300–1160 Ma in the Albany region.

The large orebodies

Many of Australia's major orebodies – sources of gold, copper, lead, zinc, silver and uranium – occur in Precambrian rocks, especially in the Proterozoic rocks. The mineralisation is accompanied by drastic alteration of the surrounding rocks that is caused by the circulation of hot fluids (hydrothermal systems) in the Earth's crust. These fluids transported and then precipitated the metal-bearing minerals. Most ores are sulphides, but there are also oxides such as magnetite and hematite.

The Archaean orebodies include the gold deposits of the Kalgoorlie–Norseman area, formed around 2600 Ma ago by giant hydrothermal systems centred around granite plutons. Gold deposits at Whim Creek in the Pilbara were formed earlier, before 2900 Ma ago.

The Proterozoic orebodies host several of Australia's most important mines. The copper–lead–zinc orebodies in northwestern Queensland at Mount Isa, Ernest Henry and Osborne were formed by hydrothermal systems associated with regional metamorphism 1600–1500 Ma ago. However, there is evidence that the lead–zinc mineralisation in sedimentary deposits, such as those at Macarthur River, may have originated much earlier than this.

In central Australia, at Tennant Creek and the Tanami, the mainly gold and copper mineralisation is associated with granites and regional metamorphism dated at 1830–1820 Ma ago.

In southern Australia, at Broken Hill, there is evidence for metamorphic remobilisation of earlier gold, copper, lead and zinc at 1690 Ma. The Olympic Dam orebody follows intrusion of the granite at Roxby Downs around 1590 Ma.

In summary, there were, across Australia, a series of major regional deformation events associated with the concentration of metallic ores in the Proterozoic 1830–1500 Ma ago.

Banded iron formations (BIFs)

The Proterozoic was also the time of formation of the banded iron formations (BIFs), which are the basis of the Australian iron ore industry.

The BIFs are thinly layered sedimentary rocks composed mainly of iron oxides and silica (like quartz), with an iron content typically of 30%. The BIFs form spectacularly bedded rock faces in colours of red, dark red and dark grey, and many individual layers, even if only a centimetre thick, can be traced for a hundred kilometres or more.

The Hamersley area in Western Australia is the major place where BIFs are developed, and the stark continuity of the cliffs along gorges in the Hamersley Range is a wonderful sight.

While minor iron formations have been formed elsewhere on Earth at younger ages, thick and extensive BIFs in large quantities are characteristic of the Precambrian, mainly in the early Proterozoic. Huge amounts, making up 90% of all the iron formations, occur in the time span 2600–1900 Ma, situated in five major areas: Australia, South Africa, Canada, Brazil and Russia. The Hamersley rocks are

Figure 3.6 *Banded iron formation outcrop in the Hamersley Gorge.*
Photo: Nick Oliver

500 m thick and extend over 400–500 km, and have been dated at about 2470 Ma. It is clear that there was an episode during which immense quantities of iron were transported and deposited over vast distances, alternately with silica in very fine layers. The main theories about this episode involve accumulation in a quiet environment – either in the deep sea or in a broad shallow-water basin around the edge of a stable continent. This seems to be the only way that thin layers could be deposited undisturbed over such large areas.

Next there must have been an unusually large supply of iron and silica, and a water chemistry that allowed precipitation of these materials, and that chemistry must have oscillated rapidly to allow for alternation of iron-rich and silica-rich laminae. Finally, BIFs contain minor amounts of other sedimentary rocks, particularly thin shales, some dolomite layers, and some tuffs. This means that small amounts of other sediments occasionally reached the basins, but the major picture is of steady, finely layered deposition of iron-rich and siliceous minerals.

Igneous rocks, basalts and rhyolites occur interbedded with some BIFs, and these were formerly interpreted as subsequent intrusions into the sedimentary sequence. By and large the traditional picture has been sedimentation of the BIFs when there was little active movement of the Earth's plates or of volcanism which would disturb the quiet setting.

But in 1997, geologists working on a slightly younger BIF dated at 2450 Ma showed that the interbedded basalts and rhyolites are *not* younger, but are the same age as the BIFs. Outcrops in the Coondiner Gorge showed that the basalts were erupted onto the sea floor and then blanketed by BIFs (see Figure 3.7). The erupted

Figure 3.7 *Partly weathered basalts at Coondiner Gorge, showing rounded pillow structures formed when basalt flows into water. The top of the outcrop is a banded iron formation (BIF) and overlies the basalts, thus confirming that the deposition of the BIF happened at the same time as undersea eruption of basalts.*
Photo: Mark Barley

volume of these volcanics has been calculated to be more than 30 000 cubic kilometres. Thus the BIFs were deposited when massive volcanism was occurring. It is probable that the Earth was tectonically active, and vast amounts of iron were formed from deep-water volcanic eruptions and geysers. This iron and silica was precipitated and settled in the deep sea, and in places was swept by currents onto shallow-water shelves around the continents.

It is also possible that the increasing oxygen content of the crust and atmosphere helped precipitate iron that had previously been dissolved in water. Iron is soluble in acidic or poorly oxygenated water. When water becomes less acidic or more oxygenated, the dissolved iron precipitates as a red or orange coloured material. (The outlets and walls of irrigation systems are commonly stained red because the spraying increases the oxygen content of the water, and precipitates the red iron oxides.) Such a process may have happened on a massive scale in the Precambrian oceans as the Earth's atmosphere and oceans became more oxygenated.

The BIFs that are being mined have very high iron contents (about 65%). Such an exceptional iron content has been reached by removing other components in the original rock, especially silica. It may have been caused by prolonged weathering at the surface in the period 2200–2000 Ma, or subsurface penetration by oxygenated groundwater. The evidence is that, for either case, the waters were oxygenated by 2200 Ma ago.

Glaciations

Proterozoic rocks also show a major change in the Earth's climate: the first glaciation. In Canada there are glacial rocks – the Huronian glaciation – dated around 2200 Ma ago, although we do not have evidence of this in Australia. However, their occurrence confirms the planet had cooled sufficiently to allow widespread ice development, and this fits with the wide occurrence of water-laid sediments after 2500 Ma ago.

The first signs of glacial activity in Australia are in late Proterozoic rocks. These rocks are scattered in a broad sweep through the desert areas of central Australia extending from the Kimberley through WA, straddling the South Australia-Northern Territory border and then passing southwards through the Flinders Ranges, with another probable occurrence on King Island in Bass Strait (see Figure 3.8). Similarly aged glacial deposits have been found in southern Africa.

The glacial deposits have two characteristic rock types: thick beds of tillite, and dropstone units. The tillite is an unsorted mass of gravel, sand and mud, without any bedding or stratification, and is the sediment dumped by melting glaciers and ice-sheets. The dropstone units are beds of fine mudstones with scattered large pebbles and boulders of foreign rocks. Layers under the pebbles and boulders are deformed and bowed downwards. The only realistic theory about how these dropstone beds formed is that icebergs carried the boulders out to sea or into a large lake, then the ice gradually melted and the rocks fell into the soft sediment below.

The Kimberley tillites overlie glaciated pavements, and have boulders which

Figure 3.8 *The sweep of Proterozoic glacial rocks through central Australia.*

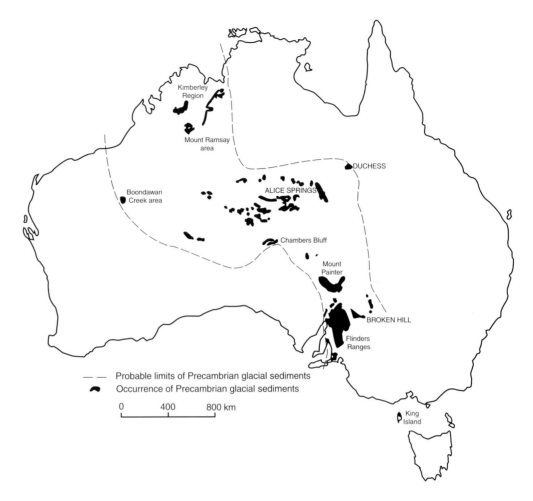

have been striated, faceted and polished by ice. Most pavements indicate ice movement from north to south. Rocks at Mount Ramsay in the Kimberley show two tillites, one about 670 Ma old and another about 750 Ma old, that are correlated with similar deposits at Mount Painter near Adelaide.

The two glacial phases occurred about 780–700 Ma ago (the Sturtian phase) and 610–575 Ma ago (the Marinoan phase). The thickest deposits are north of Adelaide in the Flinders Ranges, where the Sturt tillite is up to 5.5 km thick and the Marinoan sequence is 1.5 km thick. The boulders transported in these Sturt tillites have come from the east, so there must have been a glaciated mountain range extending from Broken Hill to Mount Painter at the time. The Marinoan sequence also includes substantial wind-blown deposits, formed by strong winds that blew off the polar regions and redistributed sand across an arid, periglacial landscape. These glacial sequences also include limestones and dolomites, other sedimentary rocks and some peculiar ironstones. The ironstones are typically laminated or massive, and contain hematite and magnetite with admixtures of sand grains and silt. Ironstones formed during waning phases of the glaciations.

Figure 3.9 *Sturt tillite overlain by dolomites, Flinders Ranges, South Australia. Note the poorly sorted nature of the tillite.*
PHOTO: VIC GOSTIN

It is thought that during glaciations the seas near continents were covered by ice shelves. The seawater became charged with organic material and almost anoxic; that is, depleted of dissolved oxygen. In such waters iron and manganese are soluble. As the ice shelves melted the waters became stirred and more oxygenated, and the iron and manganese oxides precipitated onto the seabed. Volcanic exhalations may also have added an extra supply of iron to the water.

The carbonate rocks (limestones and dolomites) were deposited as the sea level rose following deglaciation. Such carbonate rocks are typical of warm water conditions, so how did we get this intercalation of cold climate glacial deposits with warm water carbonates?

Recent work on palaeomagnetic orientations in the Marinoan rocks indicates that the South Australian deposits were formed at a latitude within 8° of the equator. This means that the whole Earth was glacial around 600 Ma ago, and there were glaciers in equatorial regions – probably the greatest icehouse this planet has known. Three magnetic polarity reversals have been measured in the Marinoan rocks. The known time frame for such reversals indicates that equatorial glaciations persisted for at least several hundreds of thousands to perhaps millions of years.

Periodic deglaciations could have restored very warm water conditions to equatorial regions even if the polar regions were still ice-bound. Accordingly, the carbonates could have been deposited before the onset of the next glacial phase, since Australia was at the time in equatorial latitudes.

Our present experience of climate variations on Earth, even accounting for the Pleistocene ice ages when ice covered much of Europe and North America, makes

Figure 3.10 *a. Uluru, showing the sandstone bedding with near vertical inclination. Erosion, mainly by rain, has etched the softer layers forming deep furrows across the rock. Kata-Tjuta is just visible in the top left-hand background.*
PHOTO: RICHARD RUDD

b

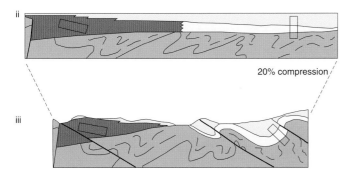

the scenario of equatorial glaciers seem bizarre. Some models of stellar evolution suggest that the Sun's luminosity was 3–6% lower 600 Ma ago than it is now, and that would contribute to a cooler Earth. However, the evidence of glacial and warm water deposits together means more research is needed to resolve the mechanisms governing climates on the early Earth.

Box 3.1 Geology of Uluru and Kata-Tjuta

The rocks at Uluru and Kata-Tjuta are sedimentary rocks deposited very close to the end of the Precambrian. These rocks date from the time of final assembly of Gondwana. The ancestral Petermann Ranges to the southwest were much higher mountains, and rivers eroding these ranges transported sediment to the north.

Kata-Tjuta was closer to the source of the sediments, and the walls of these rounded hills display steeply-dipping beds. The rocks are conglomerates, with rounded boulders of granite, gneiss and basalt, and also sandstone and rhyolite and metamorphic rocks. Many of the pieces are 0.3–0.5 m across, which indicates that strong currents in the streams transported and deposited the boulders. We think the conglomerates were deposited in alluvial fans around the edges of the ancestral ranges. Finer material would have been carried further downstream.

Uluru itself is made of sandstone tipped on edge – in particular, a sandstone containing a high proportion of feldspar. The rocks at Uluru are the downstream equivalents of those at Kata-Tjuta.

b. Sedimentation, folding and erosion to the present landscape. The two boxes represent the positions of Kata-Tjuta (brown) and Uluru (yellow).
i   During the final stages of the orogeny that uplifted the Petermann Ranges (600 Ma), alluvial fans of coarse gravel built out from the mountain edge (brown) and river sands were deposited further north (yellow). These units overlie an unconformity eroded across much older, already folded Proterozoic metamorphic and sedimentary rocks.
ii After 550 Ma ago, erosion of the ranges and sedimentation of the fan and river sediments stops, and the area was covered by shallow seas (blue).
iii Between 400 and 300 Ma ago, the area was compressed and folded during the Alice Springs Orogeny. Note the 20% shortening of the Earth's crust. The compression folded the previously flat sediments and moved them along thrust faults.
iv Continuing erosion lowered the profile of the landscape.
v   By about 65 Ma ago, some younger alluvial sediments (pale green) and then sand dunes formed across a very flat landscape.

When the sediments were laid down there was no vegetation on the landscape; perhaps only bacteria and algae existed in moist cracks and other niches. The land surface was bare, so rain resulted in rapid runoff and erosion. Since, in the soils of humid regions, weathering breaks the feldspars down to clays, we interpret the climate as warm and dry because the feldspars are not weathered to clays. The very coarse sediments are also indicative not only of higher topography but perhaps flash flooding, which can transport very coarse gravels before depositing them on the alluvial slopes as the flood waters sink into the dry ground.

Geologists know from looking at other rocks around the Uluru region that during Cambrian time the seas flooded in and deposited more sediments, so burying the sediments at Kata-Tjuta and Uluru (see Figure 3.10b). Then around 400–300 Ma ago these rocks were folded and faulted, and uplifted again. Erosion has gradually worn away the overlying rocks, re-exposing those that are now left as Kata-Tjuta and Uluru. Mount Connor to the east is another remnant of the same sequence.

There are no major rivers in the region to continue deep erosion of the rocks. The surrounding plains are formed of sediments deposited by wind and sheet flow during the past few million years. These sediments lap up against the edge of Uluru and Kata-Tjuta, leaving them as isolated remnants on the plains.

Origin of life

Experiments in the 1950s showed that electrical discharges conducted through gas mixtures involving oxygen, nitrogen, carbon dioxide or methane, and water vapour can form compounds such as formaldehyde and simple sugars. By combination with phosphates we can imagine the formation of complex molecules similar to many found in living organisms. However, even the formation of a few compounds is a long way from developing DNA with an ability to reproduce itself. This primitive stage of chemical evolution in which such molecules developed occurred well before the onset of life itself. When this happened, and how, we can only conjecture – many of the original forms may not have been preserved, and anyway the rock record is incomplete.

Archaean sedimentary rocks 3550 Ma old from North Pole in the Pilbara contain undulating, layered structures apparently formed by cyanobacterial mats growing on the original sedimentary surface. If this is the case, the structures represent the oldest organisms known. Such structures also occur in South African rocks dated at 3400 Ma.

At present it is generally accepted that the oldest cyanobacterial mats are around 2700 million years old. These layered units are called stromatolites and are thought to have formed in a similar manner to the modern algal stromatolites in Shark Bay, Western Australia (see Figure 3.11). The mats have sticky slime around the cells so that sediment grains stick to the surface. The mat then grows again, more sediment adheres, and so on, forming a layered structure that can be preserved.

These primitive cyanobacteria (or blue-green algae) had minute cells less than

Figure 3.11 Stromatolites in Hamelin Pool, Shark Bay, Western Australia.
PHOTO: DAVID JOHNSON

25 μm in diameter, without nuclei. Today they live 'on the edge', in environments with extreme temperatures and salinity, such as desert ponds or hot-water springs. Free from competition in the Precambrian, they proliferated and formed extensive slimy mats in more normal environments.

Today algae and cyanobacteria form a thin layer over intertidal areas just about everywhere, and sometimes these greenish or brownish layers can be seen coating the sediment grains immediately the tide recedes, especially in arid areas. This organic material is food for small fish, crabs, worms and shellfish that graze across the sediment surface, so it is being continually consumed. But in the Precambrian there were none of these animals to scavenge the algae, so thicker layers built up. In arid areas calcium carbonate was precipitated from the surrounding waters to cement the layers, and in some cases the algae produce the carbonate. This calcium carbonate cementation preserved the stromatolites as fossils.

Why do stromatolites occur in Shark Bay today? Hamelin Pool in the southern part of Shark Bay is partly separated from the open ocean. The hot climate, lack of freshwater influx, and limited exchange with the ocean mean that the seawater evaporates and salinity increases. In places the seawater is twice normal ocean salinity, and at these concentrations most normal marine life cannot survive – it would be like salting them down. As a result the normal crabs, fish and shellfish are absent and the cyanobacteria grow uninhibited. Standing on the edge of the Hamelin Pool tidal flats and watching the stromatolites is like going back over 2500 million years in Earth history: you can look upon a depositional surface similar to that before any of the major animal groups existed.

Increasing oxygen content

All life has the potential to leave behind some of the complex biochemical compounds which characterised the living plant or organism. Some of these compounds have been preserved in sediments, and are called biomarkers. In a sense they are fossil molecules. Sediments 2700 Ma old from the Pilbara region contain compounds derived from cyanobacteria, but also certain others (steranes) that are strong evidence for the existence of eukaryote algae.

This is a major finding: cyanobacteria are prokaryotes – primitive cells in which all the metabolic processes occur within a continuous mass of cytoplasm inside the cell wall. But eukaryotes have distinct, self-contained organelles inside the cell, each with a specific function. The nucleus holds the genetic material, the mitochondria carry out respiration and hence are the source of energy flow, and the chloroplasts handle photosynthesis.

The evolution of the eukaryote cell was a fundamental development of life. It signifies much greater energy flow generated by an oxygen-based metabolism of carbohydrates in the mitochondria, and substantial oxygen production by green plants with chloroplasts. Previous work using carbon isotopes had indicated the presence of eukaryotes in rocks from the Mount Isa region as old as 1700–1500 Ma. The Pilbara work has extended the existence of eukaryotes back to 2700 Ma ago.

It is possible that the earliest Archaean stromatolites were formed by cyanobacteria living in an atmosphere with relatively little oxygen. Many other bacteria living in anaerobic conditions, such as organic-rich sediments, produce methane. However, the stromatolites become much more abundant after 2500 Ma ago, in the Proterozoic, and it seems likely this profusion is evidence of a marked change in the composition of the atmosphere.

It is generally accepted that the original atmosphere of the Earth had less oxygen than at present. While cyanobacterial mats do produce oxygen, most of this oxygen is combined with other elements to form compounds. However, it is clear that by 2400–2200 Ma ago the atmosphere contained far more oxygen. For instance, deposition of the main BIFs after 2600 Ma is strong evidence that the atmosphere and ocean waters were more oxygenated. The marked increase in stromatolites is further evidence.

What led to this rising oxygen content? One theory is that it was the increasing profusion of oxygen-producing bacteria and algae, especially after the emergence of eukaryotes 2700 Ma ago. An alternative theory is that hydrogen escaped from the atmosphere into outer space. Methane is broken down by UV radiation to release hydrogen atoms, which escape easily into space from the upper atmosphere. Even in the Archaean there would have been a gradual loss of hydrogen. Most hydrogen on Earth is bound with oxygen into water molecules, and probably always has been. So as hydrogen is lost, the oxygen is left and therefore increases in concentration within crustal rocks and eventually in the atmosphere.

Some of the additional oxygen in the atmosphere would have been converted into ozone, which has absorbed dangerous UV radiation ever since. The higher oxygen concentrations and protection from UV radiation allowed the evolution of a wide range of organisms whose life systems were based on respiration – the intake of oxygen and the release of energy and carbon dioxide. Thus an early atmosphere without oxygen would have had less carbon dioxide (CO_2) but more methane (CH_4). Today the high oxygen content of the modern atmosphere ensures that most methane produced by animals and plants is converted to CO_2 and water, but in the Archaean the methane would remain unaltered.

The first evidence of more complex life-forms is represented at various locations around the world, but some of the best evidence is preserved in the remote Flinders Ranges in South Australia. The fossils come from a sandstone which forms great bluffs around Wilpena Pound. The fossils are the impressions of jellyfish, strange worms, and other soft bodied animals which lived in the oceans millions of years before animals with shells or internal skeletons appeared. In Europe there are fossils of primitive animals with a disc on the substrate, a stalk and a frond-like body dating back to 570 Ma ago.

No doubt these are chance remnants of a fauna which was widespread in the oceans. After all, it must take very special conditions to preserve such soft-bodied animals. But nobody knows how this change happened. Why did these complex life-forms suddenly evolve?

There had been little life for nearly 2000 million years after Earth formed, then came the stromatolites in abundance from 2500 Ma ago, followed by some soft-bodied fossils around 600 Ma ago, and finally the shelly fossils around 530 million years ago. (This appearance of complex fossils, near the base of the Cambrian is taken up in Chapter 4.)

Box 3.2 Wilpena Pound and the Ediacaran fauna

Wilpena Pound is a broad fold in sedimentary rocks, which were deposited 650–545 Ma ago. Like those at Uluru and Kata-Tjuta these rocks represent the final stages of the Precambrian, but at Wilpena they evidence the start of multi-celled organisms. This is one of the key groups of rocks on the planet. The Ediacaran fossils occur in a thin unit within quartz rich sandstones, and are around 550 Ma old.

Various combinations of tentacles and soft, chambered bodies indicates a close relation to modern jellyfish. Another is the frond-like impression of an ancient sea-pen. Yet there are also impressions of worms and crawling animals, some sort of soft-bodied animal, perhaps a trilobite-like arthropod. Many of the fossils are large – up to 0.45 m across – that is a *big* jellyfish. Some of the sea-pens were up to 1 m high.

In the late 1990s fossils of a primitive, fish-like chordate with a well-formed head and tail fin were recovered from the Flinders Ranges. These fossils underlie the main Ediacaran fossil beds, and are estimated to be around 555 Ma old, according to recent press releases.

The Ediacaran fossils have now been found in rocks of the same age elsewhere –

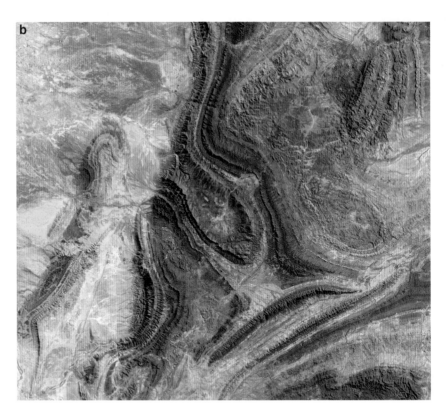

in eastern Canada, in England and Russia, and in Namibia. In Canada they occur in deep water sediments, a very different setting to the shallow, at times emergent environments indicated by the South Australian occurrences. The abundance and diverse water depths of the sediments incorporating these fossils indicates a widespread food source such as phytoplankton and possibly zooplankton.

The great significance is that these fossils represent the first clear evidence of multicellular organisms on Earth. At this stage all were soft bodied: there were no shells, nor carapaces, nor bones. Yet the shells were only 20 million years away – in the explosion of hard-shelled metazoans in the Cambrian.

Figure 3.12 *a. Bluffs at the edge of Wilpena Pound.*
PHOTO: IAIN GROVES/SERAC PHOTOGRAPHY
b. Satellite image of the Wilpena Pound area, showing the complex geology.
IMAGE COURTESY OF PERILYA LIMITED AND GEOSCIENCE AUSTRALIA, CANBERRA. CROWN COPYRIGHT ©. ALL RIGHTS RESERVED. WWW.GA.GOV.AU

Figure 3.13 *Ediacaran fossils. a.* Charnodiscus longa, *probably a free-standing frond such as a sea-pen. b.* Dickinsonia costata *thought to be a segmented worm.*
PHOTOS: VIC GOSTIN

Supercontinents: Rodinia and Gondwana

The original theory of continental drift was supported by the same fossil species being collected on separate continents. Fossil plants, such as *Glossopteris,* were found in Australia, South America and India. The presence of the same land plants across these now-scattered continents suggested they were once all a single landmass.

The German climatologist and geophysicist Alfred Wegener published in 1915 an expanded version of his 1912 book *The Origin of Continents and Oceans*. In this he was the first to suggest continental drift. He suggested that a supercontinent he called Pangaea had existed in the past, and then broke up, starting 200 million years ago. The initial scientific response was hostile. Then in 1937 Du Toit placed Antarctica as the central keystone of an early southern supercontinent he called Gondwana. He correctly argued that Antarctica stayed put while the other pieces 'fled the Pole', with Africa, India, Australia and South America drifting away to the north.

On Scott's 1912 expedition to Antarctica, fragments of primitive sponge-like organisms *Archaeocyatha* had been found in limestones, supporting these links. But more than that – these fossils were marine and they match similar fossils in marine sequences in other continents such as North America. This implies that the original Gondwana had at one time been a series of fragments with sea between them, and also that it was possible the present northern hemisphere lands were linked to Gondwana!

It is important to realise that while we say something happened in Canada or South Africa, that is just because that is where the rocks lie today. In the Archaean these fragments were not assembled as they are today. The crust of the Earth has been moving since it first formed. The atlases and geography we know today are only true for now. In the past the landmasses were totally different shapes.

The geological similarity of rocks, of the same age, in continents now widely dispersed across the globe is a key point. There are really only two possible causes. Either the same conditions were happening across the entire globe at the same time, which does not seem likely given the present range of climates and geological settings, or that the rock masses used to be together.

In Australia, the three Archaean cratons – Pilbara, Yilgarn and Gawler – appear to have been entirely separate land masses originally. How far apart they were, we do not know. However, their individual histories are different, and it is most likely that they were separate fragments of the Earth's crust. It was not until 2000–1600 Ma ago that the fragments amalgamated to form the framework for central and western Australia, and even then there was probably land mass around them. At that time the present coastline of Australia did not exist.

Rodinia

Distinctive Archaean rocks occur in very old terranes on the other continents – in particular, igneous rocks that were all deformed at the same time, 1300–1100 Ma

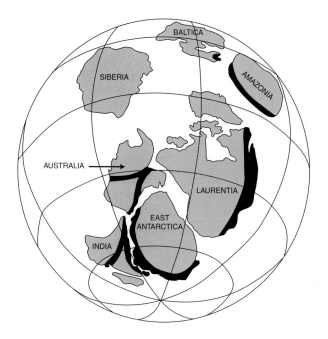

Figure 3.14 *Rodinia around 750 Ma ago, near the end of break-up of the supercontinent. The black zones represent Grenville metamorphic rocks (1300–1100 Ma) which can be traced across the old craton boundaries. Note Laurentia (present North America and Greenland) adjacent to Australia. The blank areas may have been seas, they may have been other continents which have now completely disappeared. The grey areas represent other rocks – volcanics, sedimentary rocks and granites. So there are a lot of unknowns; we are sure only that the old rocks were once joined together in this approximate shape.*

RODINIA 750 Ma

ago (the so-called Grenville series), which can be matched across reconstructed boundaries. Rocks in the Albany region of Western Australia match those on the coast of Antarctica and in parts of central India. More importantly, they are the same type and age as those along the eastern side of the present USA and Canada.

There are other data. For instance the komatiites can be found in Western Australia, in the Barberton Mountains of South Africa and in the Abitibi area of Canada. The thick banded iron formations of the Hamersley area in Australia are of similar age and look almost the same as those of Transvaal in South Africa and the Abitibi region of Canada. When the continental masses are reassembled the geology seems to match across the breaks.

From such evidence we are now sure that most of the present continents were united in a super-sized continent called Rodinia. The matching rocks show us how they fitted together like a jigsaw puzzle.

What do the folded rocks in these Grenville metamorphic belts signify? They show that in the period 1300–1100 Ma ago these continental blocks were being pushed together, and therefore that was the time of the final assembly of Rodinia.

In Rodinia the present-day Alaska probably abutted where Queensland is now, and the US west coast lay just off the present east Australian coast. The primitive fragments of Africa (Congo, Kalahari and West Africa) and of South America (Amazonia) were nearby. An alternative arrangement places southern China as well against eastern Australia, so there is more research needed to sort out the final story.

So between 1300 and 1100 Ma ago the primitive Australian mass, which would later form the core of the western half of Australia, was joined to India,

Antarctica and Laurentia (the core of North America with Greenland) to form the early supercontinent of Rodinia.

Recent radiometric dating of granites and gabbros in southern China, and other intrusive basaltic rocks in widely spaced locations in Australia, India, Madagascar, southern Africa and North America (Laurentia), shows a major period of magmatism involving two main episodes, at 830–795 Ma and 780–745 Ma. The long duration, some 85 million years, indicates a major rifting event (perhaps caused by a mantle superplume) that was instrumental in the break-up of Rodinia.

Gondwana

About 750 Ma ago Rodinia split, and the western side of Laurentia broke away from Australia. Laurentia drifted far away, opening a gap to form the first elements of what would become the Pacific Ocean. The eastern side of the US split away, forming the earliest opening of what would later become the Atlantic Ocean.

What was left was a core of Australia, India and East Antarctica. During these plate movements between 750 and 500 Ma ago, the fragments of later Africa and South America drifted around the globe towards India and Australia. By 500 Ma

Figure 3.15 *Gondwana 180 Ma. Eastern Gondwana is grey, western Gondwana white. The separation of eastern and western Gondwana created the Indian Ocean, and the break-up of Africa from South America created the Atlantic Ocean.*

ago the cores of Africa and South America had locked together with Australia, India and East Antarctica to form the new supercontinent of Gondwana. Evidence of this amalgamation can be seen in the Petermann Ranges, which consist of rocks deformed by compression around 600 Ma ago.

Gondwana was thus composed of an eastern section (Australia, India and Antarctica), which had been a core of the old Rodinia formed before 750 Ma ago, together with a western part (Africa and South America), which had joined on in the late Precambrian (750–520 Ma). The convergence of earlier cratonic blocks after 650 Ma led to mountain formation during the early stages of the development of Gondwana. Gondwana persisted for an extraordinarily long time – up until 180 Ma ago, when it began to break up in the Jurassic (see Chapter 7).

In summary, Rodinia existed from 1100 to 750 Ma ago, and Gondwana existed from 520 to 180 Ma ago. In the present phase of the Earth's atlas the continents are all dispersed widely, except for India, which has collided with southern Eurasia. Africa and Australia are both heading north, so the next phase of supercontinent development could be underway.

The Proterozoic supercontinents preserve far more of the Earth's history than the Archaean. Much of the Proterozoic rock formations occur in Australia; and we are only just starting to understood them.

Summary

The Precambrian represents the majority of Earth history, from 4600 to 545 Ma ago. The oldest sand grains that have been found are 4404 Ma old, and these must have been eroded from even older rocks.

Until 2500 Ma the Archaean shows evidence of a hotter Earth with eruptions of komatiite lavas. Dry land and oceans were present at least 3500 Ma ago. Signs of life in the form of cyanobacteria are certain from 2700 Ma on, and probably from 3500 Ma. It is probable that the Earth's atmosphere then had much methane and very little oxygen.

Increasing oxygen content was caused by loss of hydrogen to outer space and the evolution of photosynthetic algae and plants. The presence of eukaryotes indicates that widespread oxygen formation had started by 2700 million years ago.

The Proterozoic, from 2500 to 545 Ma, marks the formation of the banded iron formations (BIFs), the first glaciations in Australia (780–575 Ma), the widespread occurrence of stromatolites, and the evolution of complex life.

The higher oxygen content of the atmosphere could well have been the stimulus for the evolution of multicellular organisms – the soft-bodied animals of the Ediacaran rocks, which foreshadowed hard-shelled animals in the Cambrian.

Diverse continental masses were assembled into the supercontinent of Rodinia between 1300 and 1100 Ma ago. Rodinia broke up 830–745 Ma ago, and the next supercontinent – Gondwana – was assembled by 520 Ma ago.

SOURCES AND REFERENCES

Arndt, N.T. & Nisbet, E.G., 1982. *Komatiites*. George Allen and Unwin. 526 pp.

Barley, M.E. & Loader, S.E. (eds) 1998. *The tectonic and metallogenic evolution of the Pilbara Craton*. Special Issue *Precambrian Research*, 88. 265 pp.

Barley, M.E., Pickard, A.L. & Sylvester, P.J., 1997. 'Emplacement of a large igneous province as a possible cause of banded iron formation 2.45 billion years ago'. *Nature*, 385: 55–58.

Betts, P.G., Valenta, R.K. & Finlay, J., 2003. 'Evolution of the Mount Woods Inlier, northern Gawler Craton, Southern Australia: an integrated structural and aeromagnetic analysis'. *Tectonophysics*, 366: 83–111.

Bolt, B.A., 1982. *Inside The Earth*. W.H. Freeman and Co., 192 pp.

Brasier, M.D. & Lindsay, J.F., 1998. 'A billion years of environmental stability and the emergence of eukaryotes: new data from northern Australia'. *Geology*, 26: 555–558.

Brocks, J.J., Logan, G.A., Buick, R. & Summons, R.E., 1999. 'Archaean molecular fossils and the early rise of Eukaryotes'. *Science*, 285: 1033–1036.

Buick, R., Thornett, J.R., McNaughton, N.J., Smith, J.B., Barley, M.E. & Savage, M., 1995. 'Record of emergent continental crust ~ 3.5 billion years ago in the Pilbara craton of Australia'. *Nature*, 375: 574–577.

Carnegie, D.W., 1898. *Spinifex and Sand*. Penguin Facsimile Edition 1973. 454 pp.

Catling, D.C., Zahnle, K.J. & McKay, C.P., 2001. 'Biogenic methane, hydrogen escape, and the irreversible oxidation of early Earth'. *Science*, 293: 839–843.

Cawood, P.A. & Tyler, I.M., 2004. 'Assembling and reactivating the Proterozoic Capricorn Orogen: lithotectonic elements, orogenies, and significance'. *Precambrian Research*, 128: 201–218.

Dalziel, I.W.D., 1992. 'Antarctica; a tale of two supercontinents'. *Annual Review of Earth & Planetary Sciences*, 20: 501–526.

Dunn, P.R., Thomson, B.P. & Rankama, K., 1971. 'Late Pre-Cambrian glaciation in Australia as a stratigraphic boundary'. *Nature*, 231: 498–502.

Frakes, L.A., 1979. *Climates through Geologic Time*. Elsevier. 310 pp.

Glaessner, M.F., 1984. *The Dawn of Animal Life*. Cambridge University Press. 244 pp.

Goodwin, A.M., 1991. *Precambrian Geology: The Dynamic Evolution of the Continental Crust*. Academic Press. 666 pp.

Green, D.H., Nicholls, I.A., Viljoen, M. & Viljoen, R., 1975. 'Experimental demonstration of the existence of peridotitic liquids in earliest Archaean magmatism'. *Geology*, 3: 11–14.

Gresham, J.J. & Loftus-Hills, G.D., 1981. 'The geology of the Kambalda Nickel field, Western Australia'. *Economic Geology*, 76: 1373–1416.

Groves, D. I. *et al*., 1986. 'Thermal erosion by komatiites at Kambalda, Western Australia, and the genesis of nickel ores'. *Nature*, 319: 136–139.

Huppert, H.E. *et al*., 1984. 'Emplacement and cooling of komatiite lavas'. *Nature*, 309: 19–22.

Karlstrom, K.E. *et al*., 1999. 'Refining Rodinia: Geologic evidence for the Australia–Western U.S. connection in the Proterozoic'. *GSA Today*, 9: 1–7.

Li, Z.X., Li, X.H., Kinny, P.D., Wang, J., Zhang, S. & Zhou, H., 2003. 'Geochronology of Neoproterozoic syn-rift magmatism in the Yangtze Craton, South China and correlations with other continents: evidence for a mantle superplume that broke up Rodinia'. *Precambrian Research*, 122: 85–109.

Lottermoser, B.G. & Ashley, P.M., 2000. 'Geochemistry, petrology and origin of Neoproterozoic ironstones in the eastern part of the Adelaide geosyncline, South Australia'. *Precambrian Research*, 101: 49–67.

Myers, J.S., 1997. *Archaean Granite-Greenstone Geology of the eastern goldfields, Yilgarn Craton, Western Australia*. Special Issue, *Precambrian Research*, 83: 1–215.

Myers, J.S., Shaw, R.D. & Tyler, I.M., 1996. 'Tectonic evolution of Proterozoic Australia'. *Tectonics*, 15: 1431–1446.

Nisbet, E.G., 1987. *The Young Earth: An Introduction to Archaean Geology*. Allen & Unwin. 402 pp.

Oliver, N.H.S., Rubenach, M.J. & Valenta, R.K., 1998. 'Precambrian metamorphism, fluid flow, and metallogeny of Australia'. *AGSO Journal of Australian Geology and Geophysics*, 17: 31–53.

Pledge, N.S., 1999. *Fossils of the Flinders and Mount Lofty Ranges*. Third edition. South Australian Museum, Adelaide. 28 pp.

Powell, C.McA. & Meert, J.G. (eds) 2001. *Assembly and breakup of Rodinia*. Special Issue *Precambrian Research*, 110. 386 pp.

Reading, A.M. & Kennett, B.L.N., 2003. 'Lithospheric structure of the Pilbara Craton, Capricorn Orogen and northern Yilgarn Craton, Western Australia from teleseismic receiver functions'. *Australian Journal of Earth Sciences*, 50: 439–445.

Sohl, L.E., Christie-Blick, N. & Kent, D.V., 1999. 'Paleomagnetic polarity reversals in Marinoan (600 Ma) glacial deposits of Australia: Implications for the duration of low-latitude glaciation in Neoproterozoic time'. *GSA Bulletin*, 111: 1120–1139.

Swager, C.P. *et al*., 1997. 'Crustal structure of granite-greenstone terranes in the Eastern Goldfields, Yilgarn Craton, as revealed by seismic reflection profiling'. *Precambrian Research*, 83: 43–56.

Sweet, I.P. & Crick, I.H., 1992. *Uluru and Kata-juta – A geological history*. Australian Geological Survey Organisation.

Taylor, S.R., 1997. 'The origin of the Earth'. *AGSO Journal of Australian Geology and Geophysics*, 17: 27–31.

Thompson, R.B., 1995. *A Guide to the Geology and Landforms of Central Australia*. Northern Territory Geological Survey. 136 pp.

Vanyo, J.P. & Awramick, S.M., 1982. 'Length of day and obliquity of the ecliptic 850 Ma ago: preliminary results of a stromatolite growth model'. *Geophysical Research Letters*, 9: 1125–1128.

Veevers, J.J., 2003. 'Pan-African is Pan-Gondwanaland: Oblique convergence drives rotation during 650–500 Ma assembly'. *Geology*, 31: 501–504.

Wilde, S.A., Valley, J.W., Peck, W.H. & Graham, C.M., 2001. 'Evidence from detrital zircons for the existence of continental crust and oceans on the Earth 4.4 Gyr ago'. *Nature*, 409: 175–178.

Websites

Radiometric dating
www.asa.calvin.edu/ASA/resources/Wiens.html

The Precambrian world and Rodinia
www.scotese.com/precambr.htm

Precambrian life and Ediacaran faunas
www.geol.queensu.ca/museum/exhibits/ediac/ediac.html
www.ucmp.berkeley.edu/vendian/critters.html
www.palaeo.gly.bris.ac.uk/Palaeofiles/Cambrian/fossils/ediacara/ediacara.html

Alfred Wegener and the scientific struggle for the concept of continental drift
www.pangaea.org/wegener.htm

Geology of the Yilgarn, Western Australia
www.ga.gov.au/map/yilgarn

Illustrations

Fig. 3.2: redrawn and reproduced courtesy of the Western Australian Museum.

Figs 3.3, 3.14: redrawn and reproduced with permission of the American Geophysical Union after figs 1, 10, in Myers, J.S., Shaw, R.D. & Tyler, I.M., 1996. 'Tectonic evolution of Proterozoic Australia'. *Tectonics*, 15: 1431–1446.© 1996 American Geohysical Union.

Fig. 3.8: redrawn with permission of Macmillan Publishers Ltd after fig. 2 from Dunn, P.R., Thomson, B.P. & Rankama, K., 1971. 'Late Pre-Cambrian glaciation in Australia as a stratigraphic boundary'. *Nature*, 231: 498–502 with permission of Geoscience Australia, with additional information from Williams, I.R., 1987. 'Late Proterozoic glaciogene deposits in the Little Sandy desert, Western Australia'. *Australian Journal of Earth Sciences*, 34: 153–155. Diagram courtesy Geoscience Australia, Canberra. Crown copyright ©. All rights reserved. www.ga.gov.au

Figs 3.10 b–f: modified and reproduced with permission of Geoscience Australia, Canberra from Sweet, I.P. & Crick, I.H., 1992. *Uluru and Kata-juta – A geological history*. Australian Geological Survey Organisation, p. 13. Crown copyright © All rights reserved. www.ga.gov.au

CHAPTER 4 *Part of Gondwana*

WARM TIMES: TROPICAL CORALS AND ARID LANDS

After the enormous time span of the Precambrian, lasting 4000 million years, complex life suddenly burst forth in an astonishing array within a few million years. What types were there? And onto what sort of Earth did they evolve?

The span of this chapter covers 191 million years, comprising:

- *55 million years in the Cambrian 545–490 Ma,*
- *56 million years in the Ordovician 490–434 Ma,*
- *24 million years in the Silurian 434–410 Ma and*
- *56 million years in the Devonian 410–354 Ma.*

For this time, from 545–354 Ma ago, there was still no Australia as we know it. The present Australia was just a segment on the northern edge of Gondwana. The Australian segment did not lie east–west, but north–south, with western and central Australia embedded in the supercontinent, and being eroded. An open ocean lay across what is now eastern Australia.

Gondwana consisted of large onshore areas, huge chains of active volcanoes, with embayments of shallow seas and coral reefs, passing offshore into deep water. A possible modern analogy is the American land mass with the shallow Caribbean Sea and reefs, the volcanic arc of the West Indies, and the deep waters of the Atlantic.

The present land mass that is eastern Australia was not made at that time. There is a line that can be drawn from the present western side of Princess Charlotte Bay through southwestern Queensland and into far northeastern South Australia, with a sharp turn into New South Wales, southeast again to take in Broken Hill, and finally a dogleg west to Kangaroo Island in South Australia. This is the Tasman

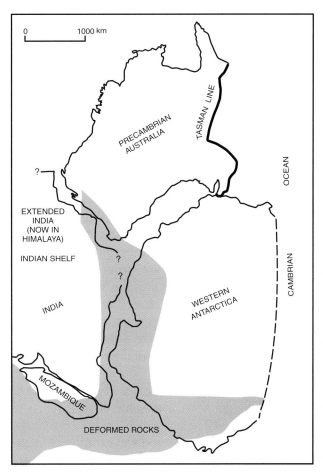

Figure 4.1 *Cambrian palaeogeography showing relationship of the Precambrian parts of Australia (west of the Tasman Line) to western Antarctica and India. All are joined in the supercontinent of Gondwana. Deformed terranes in southwestern Western Australia near Albany match similar rocks in Antarctica and Mozambique. East of Australia and Antarctica was open to the Cambrian ocean.*

Line, and its extension is also obvious in the geology of Antarctica. It marks the eastern edge of Precambrian rocks, and was the original coast of Gondwana. All to the east is younger and was not there at the start of the Cambrian; it has been added by repeated volcanism and sedimentation during the succeeding 545 million years.

Explosive radiation of life

The time from 545 to 251 Ma is known as the Palaeozoic, which means 'old life'. We think of these primitive Palaeozoic times as long ago, which they were, but they are relatively recent in terms of the Earth's age and existence. Most of Earth's history had happened by this time. Earth had been in existence for over 4000 million years, so nearly 90% of the time had passed even before the Palaeozoic started.

In one way, placing a time boundary at 545 million years and calling it the base of the Cambrian, or the end of the Precambrian, is just an arbitrary line in the Earth's history. However, geologists have placed it there because it marks a major change in the patterns of sedimentation and, within a short time, the appearance of a new range of life.

Looking back, it seems a logical progression. The Petermann Ranges in the Northern Territory were formed around 600 million years ago, as the crust was deformed and then uplifted. The rapid erosion of these mountains, which were much higher then than they are now, carried gravel and sand northwards, forming the coarse conglomerates that make Kata-Tjuta, and the sandstones that form Uluru. Evaporites accumulated in lakes.

In shallow waters further south, soft-bodied Ediacaran fossils had been buried in the sands and muds at Wilpena. That general warm climate persisted. Abundant shelly fossils start around 530 Ma ago, some 20 million years after the deposition at Ediacara.

The tectonic movements did not stop with the Petermann Ranges uplift, which may be related to the final assembly of Gondwana. Across northwestern Australia massive outpourings of basic volcanics between the Kimberley and Arnhem Land extended southeast into central Australia and western Queensland. The episode appears to have started over 600 million years ago in the area from Tennant Creek across the Barkly Tableland. Minor volcanism also occurred in the Warburton area, and there are prominent dykes of this age in the southwest of Western Australia. Most of the more extensive basalts are to the northwest and younger than 570 Ma, with the activity stopping about 555 Ma ago. One extensive dolerite dyke in the Kimberley has been dated at 513 Ma, indicating that the crustal stress and igneous activity lasted into the base of the Cambrian.

Large areas of Gondwana started to subside and the sea flooded over the continents. But we will continue the story by going back to some of the first rocks picked up in the central part of Gondwana.

In 1908 Frank Wild was in Antarctica, on Shackleton's expedition. On the coast of Antarctica he picked up some fragments of limestone containing strange

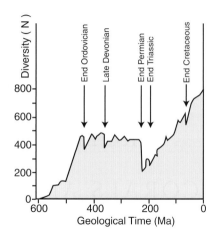

Figure 4.2 *The diversity of life appears to have increased rapidly after the start of the Cambrian 545 Ma ago, despite five major extinction events. This trend is caused partly by sampling bias – older rocks have been recycled through orogenies and eroded more than recent rocks, so that more recent fossils are more likely to be found.*

fossils. They are primitive sponge-like organisms, shaped like an inverted cone, which secreted the calcareous skeletons now fossilised and named *Archaeocyatha*. These rocks match similar Cambrian limestones of the same age in the northern Flinders Ranges of South Australia. Further, in both places there is a major unconformity where Precambrian rocks were deformed, uplifted and eroded. Then the sea flooded over this landscape, depositing marine sediments that now appear as fossiliferous limestones. These were shallow, warm, tropical seas, and *Archaeocyatha* is one of the first hard-shelled organisms we know existed on Earth.

One of the strange experiences that repeatedly occurs to geologists is the realisation that the ground being walked upon was once utterly different. The present climates of Antarctica and central Australia – one glacial and the other desert – could not be more different from each other. Yet the rocks imply that there was a close relationship between the two in the Cambrian.

What is more interesting is that we find this basal Cambrian unconformity overlain by fossiliferous marine sediments all over the world – in Britain, North America, Europe, Russia, Australia, Antarctica. Why a major tectonic upheaval in the Earth was followed by the explosion of life is still a puzzle.

There was evidence elsewhere on Earth of the appearance of new organisms. In China, sediments 525–520 Ma old have a wide range of fossils, including shells, trilobites, sponges, algae and jellyfish. In Canada the black muddy sediments of the Burgess Shale, about 515 Ma old, preserve many impressions of strange soft-bodied creatures.

Figure 4.3 *Archaeocyathids in a block of limestone (E = Ethmophyllum dentatum, Cos = Coscinocyathus irregularis, A3 = Archaeocyathus dissepimentalis). The original archaeocyathid was an inverted, hollow cone, standing on the apex and opening upwards. The circular fossils in this photo are cross-sections of that hollow cone.*

PHOTO: HOWCHIN COLLECTION, SOUTH AUSTRALIAN MUSEUM

In the huge range of fossils that suddenly arose were virtually all the main groups we know today. There is evidence of primitive vertebrates; that is, animals with a backbone. The first obvious vertebrates, fish, are also recorded in the Cambrian. There is a wide range of shells, squid-like creatures, early arthropods (segmented animals like modern crabs), corals, sea-urchins and crinoids. All were marine.

How did so much complexity develop so quickly? The answers are yet to be found. One recent theory is based on the effects of the end of the extraordinary Precambrian ice age 780–575 Ma ago. This ice age (see Chapter 3) essentially created a 'snowball' Earth, with glaciers even near the equator. The waning of this global glaciation must have produced rapid development of large areas of warm, shallow water that provided a wide range of new habitats suitable for colonisation by new organisms. The removal of ice also allowed significant light into the oceans for the first time in perhaps millions of years.

One of the most amazing developments in this early explosion of life is the eye, a light-sensitive set of cells linked to the brain, which interprets the image. Before 545 Ma ago it appears there was nothing on Earth that actually *saw* anything. The first eyes we know are those on the trilobites. Of course the first organisms able to see would have been able to more easily identify prey for food and shelter for security. The eye became an important evolutionary tool.

Fossils

The early record of the evolution of life is well represented in Australia. The main groups of fossils are marine: corals, molluscs, brachiopods, stromatoporoids, trilobites and echinoderms. There was very little life on land initially, apart from bacteria and some algae. Land plants did not evolve until the late Silurian, about 410 Ma ago, and land animals even later.

The Palaeozoic (545–251 Ma) was a time of rapid evolution. Fossil species changed over time, some becoming extinct while others evolved, so that many correspond to specific time zones.

The evolution of these fossils was interrupted by five major extinctions, the reasons for which are explored in Chapter 12. Two of those extinctions happened in the time span of this chapter: the end Ordovician (440 Ma) and late Devonian (370 Ma). The other three extinctions lie at the Permo–Triassic (251 Ma), Triassic–Jurassic (210 Ma) and Cretaceous–Tertiary (65 Ma) boundaries.

In some cases the extinctions wiped out all members of a particular animal or plant group, but in most cases some species became extinct while others took their places to recolonise that environment.

Corals

Corals are built by polyps that secrete a calcareous skeleton. The polyps grow upwards, forming a small tube, or corallite, as they grow. In some cases just one

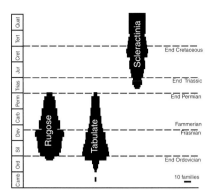

Figure 4.4 *Age ranges of tabulate, rugose and the modern scleractinian corals. Note the abrupt start, and especially the end of the rugose and tabulate corals at the Permo–Triassic boundary extinction event (251 Ma). Modern scleractinian corals are first recognised as fossils some 240 million years ago.*

polyp 10–30 mm across may form a solitary coral, but most corals comprise many polyps joined together, forming a colony of adjoining corallites. Palaeozoic corals are of two types, both now extinct: the **tabulate corals** which formed a small floor or *tabula* in the corallite under each new position of the polyp, and the **rugose corals**, which are characterised by strong radial septa, which can form ridges down the outside of the corallite, giving it a rugose appearance. Tabulate corals appeared first, in the Cambrian, and increased markedly in the Late Ordovician, coinciding with the first rugose corals. Both rugose and tabulate corals became extinct at the end of the Permian 251 Ma ago.

Modern scleractinian corals only evolved in the Triassic, some 240 Ma ago, and are of two types: shallow water reef-building corals that have symbiotic zooxanthellae, and deep water types without zooxanthellae. The symbiotic zooxanthellae are microscopic algae that live in the coral tissue and assist the coral's metabolism and ability to build the calcareous skeleton. It is the zooxanthellae that give coral tissue its bright colours. We do not know if the tabulate and rugose corals had such symbiotic algae, and whether they were brightly coloured as a result, or just a dull grey or white colour.

Molluscs

A wide range of molluscs evolved very quickly in the Cambrian, and the three main types we know today were already obvious: bivalves, gastropods and cephalopods. Bivalves are shells with two valves joined at a hinge, like oysters, pipis and many of the shells picked up on a modern beach. Gastropods are spiral shells like garden snails or the whelks in mangroves. Cephalopods are animals like the *Nautilus* that have straight or spiral shells with chambers (see Figure 6.15), unlike gastropods which have no partitioned chambers inside the shell. An octopus-like animal occupies the last-formed and largest chamber in the spiral of a living cephalopod.

Molluscs have been present since the Cambrian, although a small number of primitive bivalves existed in the very late Precambrian. There were great increases in bivalve and gastropod diversity after the Permo–Triassic and Cretaceous–Tertiary extinctions, so that these forms are far more common now than at any previous geological time. In contrast, cephalopods are much less common now. The nautiloids evolved in the Cambrian and became extremely common in the Cambrian, Silurian and Devonian. Many died out in the Devonian extinction, and most became extinct at the end of the Permian. One extinct group, the ammonites, arose in the Devonian and became extinct at the end of the Cretaceous, along with the dinosaurs.

Brachiopods

Brachiopods were a very extensive group in the geological past, though only a few species remain on Earth now. Two modern examples are *Lingula*, which burrows into shallow marine sediments (it can be found around Magnetic Island off Townsville) and *Terebratula*, which adheres to rocks off New Zealand using a strong ligament. Brachiopods evolved rapidly in the Cambrian, and were a major component of shallow marine faunas right up to 251 Ma ago. The extinctions at 251 Ma and 210 Ma destroyed most of the brachiopods, and their place in the shallow water ecological niche has been assumed by molluscs.

Brachiopods also have two shells, like a bivalve, but the type of muscles holding the shell valves together are different, the feeding mechanism is different, and the brachiopod shell is composed of calcite, rather than the aragonite shell of most bivalves.

Stromatoporoids

Stromatoporoids formed layered colonies made of calcite, which can be seen under the microscope to have millions of tiny rectangular cells. The stromatoporoid colonies can be flat or have undulating layers, which sometimes blanket corals. Other colonies have a domed or upright form, and these differences may reflect changes in the energy level of the environment.

Stromatoporoids were common from the Cambrian to the Devonian, and a few lasted until the Cretaceous. They were a dominant reef organism in the Silurian and Devonian. In these reefs the stromatoporoids encrusted and overlaid other organisms, binding them together to form the coherent reef framework. This role is assumed primarily by coralline algae on modern reefs.

Sclerosponges, which are found at water depths of 90 m on the reef fronts in Jamaica, may be the modern equivalents.

Trilobites

Trilobites are so named because of the three-lobed structure of the external carapace. There is a central ridge with lateral pieces on either side. Trilobites were arthropods, like crabs and lobsters. The carapace was flexible, so that some have been found curled up, like slaters taken into the light from a wood-pile. Underneath this carapace were rows of legs, with a head and mouth at one end and a tail at the other.

Trilobites grew by moulting, shedding their carapace, and hardening a new exoskeleton, just like modern crabs. These carapaces and pieces are common fossils.

Trilobites were common animals in the Cambrian and Ordovician, for which primitive ancestors are suspected in the late Precambrian. They suffered massive declines in the 440 and 370 Ma extinctions and were finally extinguished in the 251 Ma event at the end of the Permian.

Echinoderms

Echinoderms include starfish, brittle stars, sea-urchins, sand dollars, sea-cucumbers and crinoids. All have an exoskeleton of calcite with a characteristic five-fold (pentameral) symmetry, a feature that is most obvious in starfish but present in all members of the group. Crinoids have a cup, surrounded by tentacles, that sits on

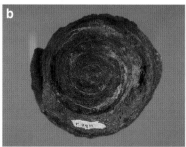
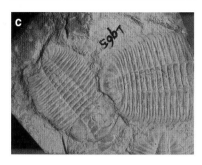

Figure 4.5 *Palaeozoic fossils:*
a. *bivalve (70 mm long);*
b. *gastropod (60 mm across);*
c. *trilobites (each 50 mm long);*
d. *a very early plant fossil,* Baragwanathia longifolia, *from Victoria (150 mm long).*
PHOTOS: DAVID JOHNSON

top of a flexible stem. Many are still present, with isolated species on reefs, continental shelves, and in the deep sea.

Echinoderms appeared in the Cambrian and were initially represented mainly by the crinoids, although a few of the others have been present since the Silurian. Sea-urchins, sand dollars and starfish all evolved rapidly after the 210 Ma extinction at the end of the Triassic.

Fish

Vertebrates dominate the oceans and land today. Primitive animals with a backbone (chordates) may date from the late Precambrian, but it is in the Cambrian and Ordovician that fish – the first obvious vertebrates – are found. The significance is manyfold: the evolution of a backbone that could carry nerves and blood vessels, the evolution of jaws, and the evolution of an internal bony skeleton that could be modified into fins and limbs.

The earliest fossil fish found in Australia, and earliest vertebrate from the Southern Hemisphere, is *Arandaspis*. It was found in sediments around 465 Ma old in central Australia. A primitive fish, *Arandaspis* probably lacked fins. The front half of its body was encased in a bony shield and the trunk was covered with oblique rows of narrow, serrated scales.

However, it is the Devonian when all the main groups of fish evolved. The cartilaginous fish like sharks and rays were present, as well as the bony fish. Sharks and rays do not preserve well as the cartilage decomposes easily, although shark teeth are well known from the Devonian onwards. Devonian fossil fish are known from widely-spaced localities in Australia, from the Taemas-Buchan area in Victoria to western Queensland and Gosse Bluff in central Australia. Devonian fish are also well known overseas, from Spitsbergen, USA, China, and Antarctica.

Australia's place in fossil fish science was firmly established with the discovery, in 1956 at Canowindra in New South Wales, of strange markings in a sandstone slab overturned by a bulldozer. The slab had impressions of more than 100 Devonian fish, about 360 million years old. Since then many more have been found, and some are on display at the Age of Fishes Museum in Canowindra. It is worth noting that these are all freshwater fish, not marine, and they were preserved in a lagoon or shallow river channel that suddenly dried up. Two types of fish are well represented in the Canowindra fish fauna: placoderms and bony, lobe-finned fish.

The common placoderms at Canowindra are *Bothriolepis, Remigolepis* and *Groenlandaspis*. Placoderms were armoured fish – primitive fish with a heavy headshield of bony plates. The mid and rear sections were flexible and probably scaly. Individual plates are up to 0.3 m across and the fish were up to 6 m long. These fish were extinct by the end of the Devonian.

The bony, lobe-finned fish at Canowindra are *Canowindra grossi, Mandageria* and *Cabonnicthys*. Specimens of these fish are 0.5–1.9 m long. The heads were reptilian in appearance, with a rounded snout, and each had strong jaws with fangs. *Canowindra grossi* had a nasal system and lungs. A primitive lungfish, *Griphognathus*, was also found at Canowindra.

Two lobe-fins survived to the present time: the modern deep-sea coelacanth, and the lungfish, which is found in seasonal, tropical freshwater waterways. Modern lungfish have a pair of lungs that enable them to breathe air and so survive droughts. Fossilised Queensland lungfish date back to over 100 Ma ago.

These lobe-finned fish have heavier and stronger fins than normal ray-finned fish. The heavier fins on the lower side are supported internally by bones and muscles. These internal bony linkages later evolved to form the pectoral and pelvic girdles of later vertebrates.

It is worth noting that these two adaptations – the ability to breathe air, and the skeleton that would later become the pelvis of land animals – had their origins over 360 million years ago.

Figure 4.6 Cabonnicthys, *a Devonian fish from Canowindra (specimen is 0.3 m across).*
PHOTO: ALEX RITCHIE

Box 4.1 How are fossils preserved?

In most environments the preservation of a fossil is an extremely rare event. There are exceptions, such as a coral reef where the reef itself is built by preserving the coral, algae and shells as fossils. However, even on the reef only a small proportion of the living organisms are finally preserved. All the soft-bodied biota, such as worms and most algae, decompose. Consider the abundance of fish living on reefs; yet they are very rarely found as fossils in reef limestones.

What is required for a fossil to be preserved? Two things: initial burial, and no removal or destruction when the rock is buried.

INITIAL BURIAL

For an animal or plant to become a fossil, the sediment surface onto which it falls must be preserved, and this means that another layer of sediment must be laid on top. Most sedimentary surfaces such as a beach are reworked regularly as the tide comes in and out or currents move material to and fro. During this reworking the object is battered and eroded by the currents, so that what is preserved may be only a very worn remnant of the original.

However, there must come a time when that particular sedimentary surface is covered for the last time, and never again uncovered. The sequence has built another layer, and everything within that layer is buried and preserved. The higher the rate of sedimentation, the greater chance of burial and preservation, and the greater chance that the fossil is a complete specimen.

Catastrophic events such as storms offshore on continental shelves or major floods on land are common causes of rapid deposition, and can bury large numbers of organisms alive. In such situations, shells such as bivalves or brachiopods are buried with both valves still joined, and on land the plants are buried in river mud. We can therefore distinguish those assemblages of fossils that are preserved essentially where they lived, from those assemblages that represent materials transported after

Figure 4.7 *Preservation of fossils. a. Difference between an original fossil and its mould. b. Petrified Permian wood showing annual growth rings (specimen is 0.3 m across).*
PHOTO: DAVID JOHNSON

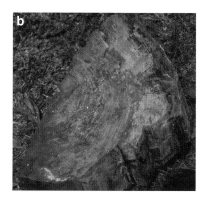

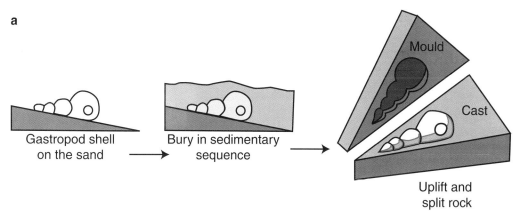

death and then accumulated elsewhere. For instance, many of the shells seen on a beach did not live there; they lived just offshore, and the separated valves were washed onshore after death.

TYPE OF PRESERVATION

Fossils are preserved either as the original form (a cast) or a reverse imprint (a mould). A mould is formed where the original fossil forms an imprint in the sand or mud where it comes to rest, and then the original fossil is dissolved or later removed. Moulds can be show the outside surface (external moulds) or the internal surface (internal moulds).

The brachiopod and gastropod shown in figure 4.5 are the original forms. The soft tissues have decomposed but the hard parts are as they were in life, millions of years ago.

In some cases the original material may be replaced by a new mineral. For example, the original calcium carbonate of a shell may have been replaced by quartz, generally preserving the fine detail of the shape. Another example is silicified and opalised wood, in which the original cell structure and annual growth bands of the wood can still be seen. Such specimens are called petrified wood, which means they have been turned to stone. Fossils can also be replaced by pyrite, siderite or hematite. The term 'silicified' simply means that silica is the petrifying agent.

PRESERVATION AFTER BURIAL

Compaction during burial will generally flatten fossil materials unless they are very rigid, or the sediment around them has become rapidly cemented. If so, the strength of the cemented rock will prevent compaction, at least for the first 2–3 kilometres of burial.

Buried sediments are not static and unaltered. A wide range of fluids, generally hot and sometimes acidic, circulate through the rocks, and these can easily dissolve many fossils. Sometimes the hole occupied by the fossil is left and can be filled later by another mineral, so the fossil's form is still there. However, in many cases the sediment collapses and all evidence of the fossil is removed.

Inevitably the fossil assemblage is only a fraction of the original living assemblage.

Warm seas with arid plains, volcanic arcs and deep troughs

Between 545 and 390 Ma ago the present land mass of Australia had two quite separate geological situations. Onshore the climate remained warm. Most of the present continent was emergent and was being eroded, the land was dry and riverine sediments were being deposited. Shallow, warm seas extended over the land mass to varying extents. However, the eastern margin fronted a deep ocean and was the site of a series of volcanic arcs that separated these shallow seas from the deep troughs offshore. The shallow seas were elongate extensional basins, so there was some tension in the crust behind the arc, at least between the compressive phases. The picture emerging is that eastern Australia may not have been a series of simple compressive arcs but part of a convergent margin with periods of extension and abundant volcanism. Eastern Australia has been called an extensional orogenic zone.

Throughout this period a series of orogenies caused compression of the crust, with consequent deformation and metamorphism, uplift and erosion. Each of these orogenies caused folding of the older rocks, which were then eroded before younger sedimentary rocks and volcanics were laid on top. A major orogeny 387–370 Ma ago, the Kanimblan, brought this phase of Australian geology to a close and marked the change from warm times to the icehouse of the Carboniferous and Permian periods (see Chapter 5).

Cambrian

There is no record of Cambrian rock in the southwestern third of Australia. The Cambrian sedimentation extended in a wide belt from the northwest to Tasmania. One edge lay along the eastern side of the Kimberley, incorporated the MacDonnell Ranges in central Australia, and the Flinders Ranges. The other side lay north of the Daly River, and extended south east to Mount Isa.

The sea began to flood this subsided central area of the continent around 553 Ma ago, forming a broad, shallow sea. The coastline appears to have been quite embayed, with deposition of shallow water limestones. On land there were sandy rivers and shallow lakes which accumulated evaporites. These rocks are exposed in the areas between Mount Isa and Camooweal, north at Thorntonia, and south to Dajarra and Urandangi. Limestones form major parts of the sequences in the MacDonnell and Flinders Ranges. The shallow sea extended through what is now the Simpson Desert and Queensland's Channel Country.

East of the general trend of the Tasman Line lay the deep ocean off Gondwana. In part this boundary was a volcanic arc which lay from near Birdsville, through far western Victoria to Tasmania. This arc, the Stavely Arc, separated the widespread shallow marine limestones and tuffs to the west from deep water sedimentation to the east.

Figure 4.8 *The palaeogeography of the Australian sector of Gondwana for representative portions of Cambrian, Ordovician and Silurian time. Australia was inundated by shallow seas in the central and eastern regions during the Cambrian and Ordovician, and the Larapinta Seaway extended from northwestern to southeastern Australia between 480 and 460 Ma ago. During the Silurian, the seas withdrew leaving a dry land mass except in the eastern part. The eastern continental margin was marked by a series of volcanic arcs with deep ocean basins further offshore. The arcs moved successively oceanwards with time.*

For most of Cambrian time, the tectonics had been mild across Australia, with slight downwarps of the crust which provided shallow basins in which the sediments were deposited. Considerable basalt eruptions testify to fracturing of the crust which permitted ascent of mantle magmas.

This Cambrian phase of sedimentation was terminated by the Delamerian Orogeny, which deformed and uplifted the sequences to form mountains from the middle to late Cambrian. The orogeny started around 520 Ma ago with tectonic uplift in some parts lasting till 480 Ma ago. Recent SHRIMP data in South Australia date deformation from 522–490 Ma followed by uplift. This tectonism also occurred throughout Gondwana, where there was widespread granite intrusion between 510–490 Ma, but only small intrusions in the Kanmantoo region of South Australia.

There was also some tectonic activity after 500 Ma ago between Carnarvon and Perth, with the first evidence of the Darling Fault, which was reactivated many times and is still evident as the scarp east of Perth. Uplifted country to the east released quartz sands into a long basin. Around 520 Ma the seas withdrew, leaving all but the eastern parts as dry land.

Ordovician

Australia still formed the northern edge of Gondwana. The general trend of the edge of the continent lay from the Normanby River in north Queensland, through south western Queensland to near Broken Hill, then Ballarat and western Tasmania.

The series of shallow embayments present in the Cambrian were extended, and central Australia became a continuous seaway. Northern Australia, from the Kimberley coast in Western Australia as far east as Einasleigh and Charters Towers in Queensland, was high, dry land. To the south it was dry land from the Pilbara to

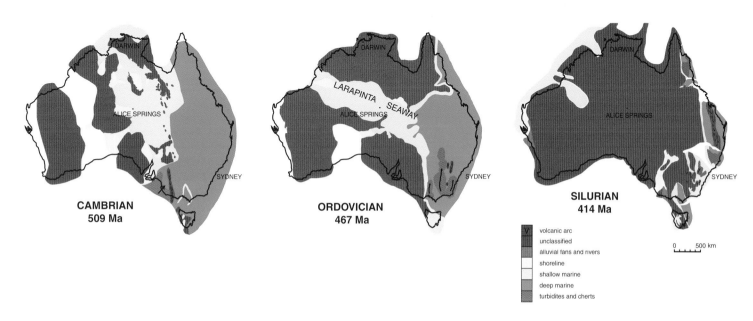

Broken Hill and south to northwestern Tasmania. However through the middle lay a tongue of the ocean. Between 480 and 460 Ma the ocean had joined from east to west forming the Larapinta Seaway through central Australia. The ocean extended from the east under the present site of the Simpson Desert, and inland from the north under the present Great Sandy Desert, and also at times around Darwin and the edge of Arnhem Land. Then starting around 460 Ma the sea started to withdraw leaving shoreline and terrestrial deposits.

The shallow water sedimentation in the shallow inland seaway produced quartzose sandstones composed of sand eroded from the Precambrian rocks, and also abundant limestones. In some areas the sedimentary basins appear to have had partly restricted circulation so that very organic rich sediments and phosphorites accumulated, similar to the present Dead Sea.

A volcanic arc – the Molong Arc – was still present, though it had moved some 600 kilometres offshore. Between 460 and 436 Ma it extended from near Dubbo south to just west of Canberra, and probably into northeastern Victoria. The volcanism has been dated as early as 460 Ma around Kiandra, but most started around 455 Ma, with a sharp cessation and erosion from about 436 Ma. The volcanic arc produced mainly andesites, such as the thick units near Cargo in New South Wales. The volcanoes occupied shallow shelves on the edge of the ocean, with carbonate sedimentation producing widespread limestones, such as those at Cudal near Orange.

In the deep water east of the arc, sedimentation was characteristically of turbidite sandstones with muddy upper layers which contain graptolites. The sandy parts of the turbidite represent rapid deposition of sediment, while the muds

Figure 4.9 *An outcrop of a beautifully graded turbidite bed of Devonian age. The turbidite is coarsest at the base and finest at the top, reflecting sedimentation as the turbidity current slows. The turbidite overlies previously deposited siltstones.*
Photo: Ray Cas

represent finer sediment from the current, and then slow accumulation of fine sediment over the hundreds or thousands of years between turbidite events. In some basins, mud accumulated to form thick shales with little sandy turbidite input.

Graptolites consist of a thin stem (nema) with cups (thecae) along each stem, each colony being 10–30 mm long and commonly branched. A single animal (zooid) lived in each of the small cups. The most likely modern equivalent is a primitive, colonial chordate *Cephalodiscus* that has been dredged off New Caledonia. Each zooid migrates daily up the stem so it can feed more easily from the water currents, and then returns to the protection of the theca.

Most graptolites were floating organisms (planktonic), though the many-branched (dendroid) forms and the modern *Cephalodiscus* live attached to the seabed (benthic). How did such delicate fossils survive on the seabed?

These graptolitic shales are generally black and finely laminated, which means the bottom waters of the ocean in which they were deposited did not have burrowing organisms. If worms, burrowing shellfish and crustacea were there, the laminae would be disturbed. Why were there no bottom-dwelling animals? The answer is that the waters had a very low oxygen content. Consequently few animals lived there and the organic material was preserved, making the shales black, and there was no disturbance of the laminae. Pyrite grains are common in unweathered rocks, and in some cases the graptolites have been replaced by pyrite.

The graptolites also evolved rapidly, and we can date the Ordovician shales using the presence and absence of different graptolite species (see figure 4.10). In Australia 30 zones have been defined, each zone spanning less than 1 million years. The graptolites are valuable fossils for dating because they were planktonic and widely dispersed, and most species are preserved in sediments that were deposited in shelf, slope and deep-sea environments. Such environments are generally quiet, so the fragile fossils were not disturbed, broken or transported after settling to the seabed.

The Ordovician sequences in central and northern Australia are relatively undisturbed. There has been faulting, but minimal folding. In southeastern Australia the sequences are strongly deformed by east–west compression, and form a north–south grain in the regional geology that can be traced from Tasmania, through Victoria into New South Wales. This deformation, followed by uplift and erosion, was identified at the end of the 1800s on the Razorback near Tallong in New South Wales, where the Silurian beds unconformably overlie the crumpled Ordovician sedimentary rocks.

Ordovician sedimentation was terminated by the 460–435 Ma Benambran Orogeny. This east–west compression formed major folds, and this is why so much of central Victoria and New South Wales is composed of steeply dipping altered rocks. Particular deformations have been dated in Victoria at 455, 440 and 425 Ma, and each corresponds with a major change in sedimentation pattern. Presumably the deformation involved mountain building or faulting which changed the river and basin configurations. The extent to which the crustal compression may be a series of rolling, discrete events rather than continuous change is still being researched and debated.

Elsewhere in the world there is evidence of glaciation during the Ordovician, so

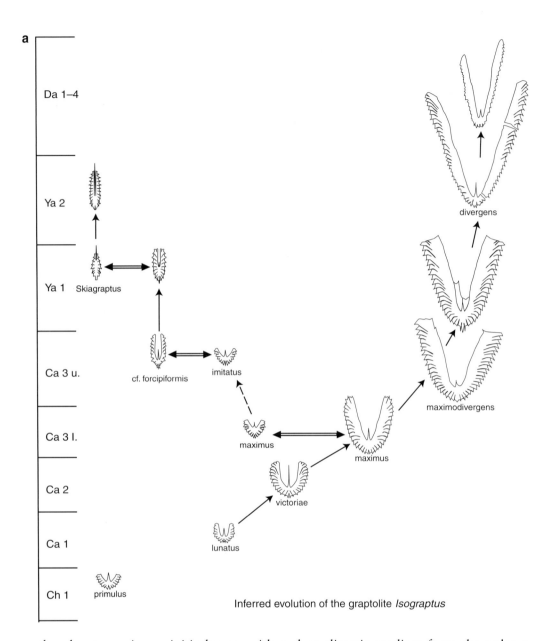
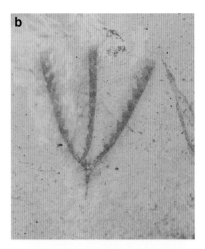

Figure 4.10 *a. Evolution of the graptolite* Isograptus *during part of the Ordovician in Australia. The lettered scale on the left-hand side represents fossil biozones. b.* Tetragraptus fruticosus *(specimen 30 mm long). c. Graptolite fragments (specimen 100 mm across) preserved in Ordovician deep-sea black shale.*

the planet was in a mini-icehouse, with a clear climatic gradient from the poles to the equator. There is every chance it was much as it is today.

The end of the Ordovician is marked by a major global extinction around 440 Ma ago.

Silurian–Devonian

The Delamerian and Benambran Orogenies squeezed together and closed the previous sedimentary basins and uplifted the land mass, and most of central and western Australia became emergent.

For the entire Silurian – 24 million years – the ocean lay just off the north-western coast, with two main depositional basins receiving sediment off the continent. Between Geraldton and almost to Exmouth Gulf, extensive deposits of fluvial sediments lay west of the Darling Fault. These red sandstones are exposed on the cliffs at Kalbarri and inland along the Murchison River gorge. They contain footprints of a giant water scorpion, which is some of the first evidence in Australia of animals taking to the land.

Further north an arm of the sea extended under the present site of the Great Sandy Desert, almost to the Western Australian – Northern Territory border. Further east this basin was filling with a thick sequence of aeolian dunes that are now the sandstones outcropping in Palm Valley, west of Alice Springs.

By the mid Devonian there was a major embayment of the sea between the Pilbara and Kimberley regions, under the present Great Sandy Desert. This basin had major reefs and intertidal carbonate environments along its northern side. These limestones are exposed inland of Fitzroy Crossing, in the walls of Geikie and Windjana gorges.

The situation was more complex in eastern Australia. Uplift had pushed much of the coast as far east as Roma, though there was still sedimentation west of this line in northern Queensland and in Victoria. The volcanic arc included a set of volcanic chains in eastern Victoria, another in the New England area, and one in northern Queensland, from west of Townsville through Chillagoe to inland of Cairns. Volcanism was primarily andesitic.

Deep-water sedimentation continued again in the Hill End, Cowra and Tumut regions, separated by a high ridge with volcanics through the Molong–Yass area. The volcanism has been dated at 420–415 Ma. Tuffs, sandstones and limestones accumulated between the volcanic sites and in small basins to the west. The limestones contain excellent coral faunas, although in areas where the rocks have been deformed, such as Chillagoe, most of the fossils have been destroyed. Each of these volcanic arcs seems to have been on the edge of a narrow steep shelf, with fans of sediment, and sometimes reefal limestone slumping and being transported by debris flows and turbidity currents into deep water to the east.

From about 415 to 383 Ma in the Devonian there was a continuous volcanic chain, the Calliope Arc, from north of Rockhampton south to Grafton. This line lies about 250–300 kilometres east of the earlier Molong Arc.

In terms of palaeogeography the old craton lay to the west, contributing sediment from central Australia. Volcanic mountains contributed sediment from the east. Thus there was a north–south basin through central and eastern Victoria, New South Wales and Queensland receiving sediment from both sides. The sea inundated this basin at various times, causing deposition of limestones, while at other times rivers advanced, depositing fluvial sandstones and mudstones. This sedimentation extended into western New South Wales across Parkes and almost to Tibooburra.

There was widespread andesitic and rhyolitic volcanism in central New South Wales, in the Snowy Mountains region and near Eden, extending northwards through New England, and also north of Clermont in Queensland. Fluvial sediments and shallow marine limestones are still characteristic, such as in the Burdekin areas in north Queensland, south of Anakie in central Queensland, and at Taemas–Buchan in Victoria.

First land plants and animals

Although some Ordovician plant remains are known, the earliest land plants found in Australia occur near Yea in Victoria and are around 412 Ma old. The plant is *Baragwanathia longifolia* (see figure 4.5d), a primitive vascular plant that grew in damp areas. By the Devonian a wide range of land plants grew along the river edges and around lakes. The most common were ferns and primitive gymnosperms, some of which formed tall trees. Such a vegetated environment would have provided habitats for land animals and encouraged their further evolution. For by now there was something on land for animals to eat.

The first land animals are recognised from this time – centipedes, primitive spiders and amphibians. The oldest insects are from Scotland and are about 400 Ma old. Insects are arthropods; that is, animals with an exoskeleton that protects the soft inner parts. They have the same body format as a trilobite, but have evolved the ability to live in air.

Late Devonian upheaval

The Tabberabberan Orogeny 390–385 Ma ago in southeastern Australia heralded the end of this last episode of marine sedimentation. Another rising arc extended from behind Mackay to near Mount Morgan and south through the western New England area. The northern section of this arc is called the Connors Arc and the southern section is the Baldwin Arc. Major outpourings of rhyolitic lavas and tuffs occurred.

In Tasmania and Victoria, and into southern New South Wales the deformation is dated around 385–380 Ma. A similar date is found in the Alice Springs area, where the rocks were folded and metamorphosed. However, northern New South Wales and Queensland in general show slightly younger ages for the deformation, 377–352 Ma, indicating that the upheaval may have progressed northwards with time.

While the familiar tectonic pattern of craton to the west and volcanic arc to the east continued, there was no marine sedimentation west of the arc. The line of volcanoes lay along the edge of the continent. Afterwards the sea was all east of the arc, and deep water sedimentation lay to the east across southeastern Queensland and northeastern New South Wales. A major river system draining the interior delivered sediment eastwards through Parkes. In Queensland the pattern is similar. The resulting sandstones typically show the large scale cross-bedding of river channel deposits. Some of the freshwater lakes dried and preserved the fish fossils at Canowindra. A similar phase of alluvial sedimentation in northern Western Australia deposited the sandstones that have been weathered into the hummocky hills of the Bungle Bungle Range.

So this long span of warm-climate marine and riverine sedimentation, which started in the Cambrian 190 million years earlier, ended. With the final upheaval of land to push the sea east of the continent, the times of shallow marine embayments that at times extended into central Australia was gone, and there was widespread erosion of the continent.

At the same time, around 370 Ma ago, there was another global extinction event, and then by 330 Ma ago the Earth's climate entered a major icehouse phase during the Carboniferous and Permian (see Chapter 5).

Granites

The repeated deformations associated with the Delamerian, Benambran and Tabberabberan orogenies compressed much of eastern Australia into a series of north–south trending folds. Granite intrusion occurred repeatedly along eastern Australia in response to these orogenies from the Cambrian to Devonian, and continued until as recently as 90 Ma ago in parallel with later tectonic events.

In southeastern Australia, through the Snowy Mountains, past Canberra and up to Wellington, the granite ages range from 435–425 Ma. On the coast, the Bega–Moruya granite batholiths are younger, at 420–400 Ma. (The Moruya granite was quarried to make the pylons for the Sydney Harbour Bridge.) These intrusions extend from the coast south of Bega, north to Wellington and Parkes, and west to the South Australian border region. There are probably more buried under the sedimentary sequences of the Murray region.

The later granites can show similar age patterns. For instance, the granites of the Herberton–Mt Surprise area in northern Queensland are 330 Ma in the west, but are as young as 280 Ma farther east. The Urannah granites between Bowen and Collinsville have ages spanning 308–284 Ma.

Most of the Cambrian to Devonian granites form elongate bodies sandwiched between faults and folded sediments and volcanics. Many of the granite suites have been shown to encompass very large age ranges across the Silurian, with minor activity continuing into the Early Devonian, while maintaining close compositional ranges.

The granite composition reflects the types of rocks from which the melt was derived. The two main types are I and S granites. I-type granites formed from the melting of older igneous rocks, and S-types from sedimentary sequences. The granites are commonly associated with metamorphism of the sedimentary and volcanic rocks, as at Cooma (see Box 4.2).

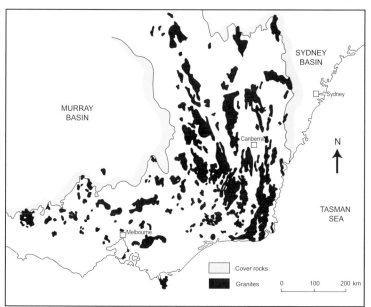

Figure 4.11 *Granites form much of the countryside in southeastern Australia.*

Box 4.2 Cooma – granite emplacement and metamorphism 435–433 Ma

The Cooma area contains rocks that have been well studied and show many features typical of a metamorphosed sequence:

1. severe deformation, commonly with up to 50% shortening as the rocks were squeezed into tight folds
2. extensive metamorphism that imposed graded mineral changes
3. injection of granitic magma, mostly derived from partial melting of the deeper sections of the deformed rocks.

The original rocks were Ordovician deepwater sediments – turbidites and shales. These were buried, deformed and metamorphosed. Detrital zircons and monazites in the metamorphosed sediments have been dated by SHRIMP of 600–500 Ma and 1.2–1.9 Ga. These dates represent the age of the igneous activity that produced the mineral grains, before they were eroded and then deposited in the sediments.

East–west compression developed five phases of deformation, which are reflected in the fold structures. There were also two late-stage tectonic movements, which then formed kinks in some of the older fold structures, making seven deformation phases in all.

There is a zoned arrangement of metamorphic rocks, centred on the Cooma Granite at Cooma township. To the east is a large fault which separates the Cooma Complex from deformed igneous rocks extending to Bredbo. To the west, metamorphic grade decreases in a series of concentric zones. Close to the contact of the Cooma Granite there are quartz-rich veins and elongate units, which indicate that the rocks had started to melt. Quartz is the last material to crystallise from a cooling melt, and is the first to become liquid as the rock is reheated. Lenses of granite meltrock occur in the deformed rocks to the west.

The mineral assemblages in the rocks at Cooma indicate low-pressure metamorphism, at a higher temperature to the east and decreasing to a lower temperature to the west. The maximum temperature in the granite zone is estimated to have been 730°C, decreasing westwards. The pressures were around 3.5–4.0 kbar, corresponding to a burial depth of approximately 12 kilometres.

Uplift and erosion brought about decreases in pressure and temperature. The interpreted temperature drop with age is shown in figure 4.12c, from 730 to 180°C in the period 435–390 Ma. The long-term average cooling rate was 12°C per million years. This cooling rate is relatively slow and probably reflects gradual regional cooling, rather than the faster cooling expected around a hot magma body. Thus it seems the metamorphism and the melting that produced the granite were both a result of regional heating, and not localised heating by intrusion of a hot magma body.

It is likely that the higher-grade rocks around Cooma itself were originally more deeply buried, and hence hotter, than those to the west. Thus the transect from east to west is actually a vertical profile up through the crust, and that section of the crust has been turned on its side.

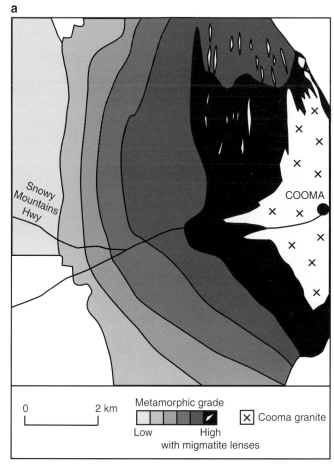
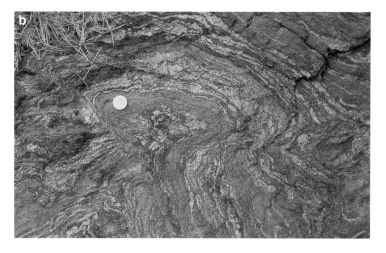
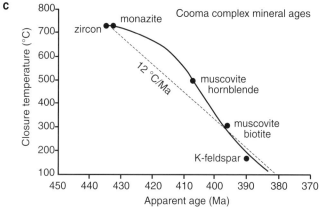

Figure 4.12 *a. The Cooma Complex, showing the faulted boundary to the east and concentric metamorphic zones to the west. b. Folded gneiss at Cooma. The paler layers are granitic material that has been melted out of the rocks as they were heated deep in the crust. Subsequent compression folded the layers into this complex structure.*
[Photo: Michael Rubenach]
c. The cooling history of the Cooma rocks, showing a long-term average decline of 12°C per million years.

Summary

Rapid evolution of hard shelled fossils established the main types of marine invertebrates by 525 million years ago. Many of the individual species are extinct and have been replaced, but the main global groupings such as arthropods, molluscs and corals were all there. Fishes were well established by the Devonian.

From 545 to 390 Ma ago there were shallow warm seas over parts of Australia, with a series of volcanic arcs and deep water sedimentation in the east. Seas flooded across central Australia, forming a complete connection – the Larapinta Seaway – from 480 to 460 Ma ago.

A series of continuing east–west compressions of the crust and granite emplacement terminated each episode of volcanism and sedimentation. These cycles of sedimentation and deformation created new zones of continental crust building successively outwards around the edge of Gondwana. These rocks form the present eastern Australia.

A final phase of erosion after the major orogeny from 387 to 370 Ma was followed by widespread erosion and then the climatic change to glacial conditions at 330 Ma.

SOURCES AND REFERENCES

Collins, W.J., 2002. 'Nature of extensional accretionary orogens'. *Tectonophysics*, 21: 6-1 to 6-12.

Collins, W.J. et al., 2000. *Granite magma transfer, pluton construction, the role of coeval magmas, and the metamorphic response: Southeastern Lachlan Fold Belt Field Guide, FP3*. Geological Society of Australia Inc. 142 pp.

Cook, P.J., 1988. *Palaeogeographic Atlas of Australia Volume 1 – Cambrian*. Bureau of Mineral Resources Australia.

Cook, P.J. & Totterdell, J.M., 1991. *Palaeogeographic Atlas of Australia. Volume 2 – Ordovician*. Bureau of Mineral Resources Australia.

Cooper, R.A., 1973. 'Taxonomy and evolution of *Isograptus* Moberg in Australasia'. *Palaeontology*, 16: 45–115.

Doyle, P., 1996. *Understanding Fossils*. John Wiley. 409 pp.

Engel, M.S. & Grimaldi, D.A., 2004. 'New light shed on the oldest insect'. *Nature*, 427: 627–630.

Harris, L.B. & Li, Z-X., 1995. 'Palaeomagnetic dating and tectonic significance of dolerite intrusions in the Albany Mobile Belt. Western Australia'. *Earth and Planetary Science Letters*, 131: 143–164.

Johnson, S.E., Vernon, R.H. & Hobbs, B.E., 1994. *Deformation and Metamorphism of the Cooma Complex, Southeastern Australia*. Specialist Group in Tectonics and Structural Geology, Geological Society of Australia Inc.

Parker, A., 2003. *In the Blink of an Eye*. Simon & Schuster. 316 pp.

Percival, I.G., 1985. *The Geological Heritage of New South Wales*. New South Wales National Parks and Wildlife Service. 136 pp.

Ritchie, A. & Gilbert-Tomlinson, J., 1977. 'First Ordovician vertebrates from the Southern hemisphere'. *Alcheringa*, 1: 351–368.

Thompson, R.B., 1991. *A Guide to the Geology and Landforms of Central Australia*. Northern Territory Geological Survey. 136 pp.

VandenBerg, A.H.M., 1999. 'Timing of orogenic events in the Lachlan orogen'. *Australian Journal of Earth Sciences*, 4: 691–701.

VandenBerg, A.H.M., & Cooper, R.A., 1992. 'The Ordovician graptolite sequence of Australasia'. *Alcheringa*, 16: 33–85.

Walley, A.M., Strusz, D.L. & Yeates, A.N., 1990. *Palaeogeographic Atlas of Australia Volume 3 – Silurian*. Bureau of Mineral Resources Australia.

Williams, I.S., 2001. 'Response of detrital zircon and monazite, and their U–Pb isotopic systems, to regional metamorphism and host-rock partial melting, Cooma Complex, southeastern Australia'. *Australian Journal of Earth Sciences*, 48: 557–580.

Willman, C.E., VandenBerg, A.H.M. & Morand, V., 2002. 'Evolution of the southeastern Lachlan Fold Belt in Victoria'. *Australian Journal of Earth Sciences*, 49: 271–289.

WEBSITES

History of life through geological time
(Department of Paleobiology, Smithsonian National Museum of Natural History, USA)
www.nmnh.si.edu/paleo

Fossils
www.ucmp.berkeley.edu/index.html

ILLUSTRATIONS

Fig. 4.1: redrawn, modified and reproduced with the permission of John Veevers after fig. 19 in Veevers, J.J. (ed.) 2001. *Billion Year Earth History of Australia Atlas*. GEMOC Press, Sydney.

Figs 4.2, 4.4: modified and reproduced with permission of John Wiley & Sons Limited after fig. 4.9 (itself modified from Sepkoski, J.J., 1982. *Geological Society of America Special Paper 190*) and fig. 13.5 in Doyle, P., 1996. *Understanding Fossils* © 1996 John Wiley & Sons Limited;

Figs 4.8a, b, c: modified and reproduced with permission from Geoscience Australia, Canberra, after fig. Cambrian 2b in Cook, P.J., 1988. *Palaeogeographic Atlas of Australia. Vol. 1 – Cambrian*. Bureau of Mineral Resources Australia; fig. Time Slice 3 in Cook, P.J. & Totterdell, J.M., 1991. *Palaeogeographic Atlas of Australia Vol. 2 – Ordovician*. Bureau of Mineral Resources Australia; fig. Time Slice 3 in Walley, A.M., Strusz, D.L. & Yeates, A.N., 1990. *Palaeogeographic Atlas of Australia. Vol. 3 – Silurian*. Bureau of Mineral Resources Australia. Crown Copyright ©. All rights reserved. www.ga.gov.au

Fig. 4.9: provided by Ray Cas, also reproduced as Fig. 10.28e in Cas, R.A.F. & Wright, J.V. 1995. *Volcanic successions: modern and ancient. A geological approach to processes, products and successions*. Allen & Unwin; Chapman & Hall, London, 528 pp.

Fig. 4.10: reproduced with permission of The Palaeontological Association, redrawn after fig. 34 in Cooper, R.A., 1973. 'Taxonomy and evolution of *Isograptus* Moberg in Australasia'. *Palaeontology*, 16: 45–115.

Figs 4.11a, c, 4.12: by permission of the Geological Society of Australia after fig. 5.0.2 in Collins, W.J., *et al.*, 2000. *Granite magma transfer, pluton construction, the role of coeval magmas, and the metamorphic response: southeastern Lachlan Fold Belt Field Guide, FP3*. Geological Society of Australia Inc. 142pp; fig. 16 in Williams, I.S., 2001. 'Response of detrital zircon and monazite, and their U-Pb isotopic systems, to regional metamorphism and host-rock partial melting, Cooma Complex, southeastern Australia'. *Australian Journal of Earth Sciences*, 48: 557–580; fig. 2 from Johnson, S.E., Vernon, R.H. & Hobbs, B.E., 1994. *Deformation and Metamorphism of the Cooma Complex, Southeastern Australia*. Specialist Group in Tectonics and Structural Geology, Geological Society of Australia Inc.

CHAPTER 5 *A glaciated continent*

ICEHOUSE: CARBONIFEROUS AND PERMIAN GLACIATION

The Carboniferous period is so named because of the extensive carbon-bearing coal deposits formed at that time in Europe and North America. The Northern Hemisphere climate was tropical, with vast coastal peat swamps and reefal limestones forming offshore. The Permian continued to be hot, with seasonal river systems and large salt lakes in North America and Europe.

The climate was very different for the land masses of Gondwana, locked near the South Pole. There the climate was cold, and in some places glacial. Permian Gondwana coals in Australia, India and South Africa formed in these very different, cold-climate situations. A chain of volcanoes along eastern Australia erupted intermittently for over 20 million years.

How long did this ice age last, and what were its legacies in the rocks?

The Australian land mass was glacial and actively volcanic – almost the complete opposite of the hot and dormant continent of today. A possible analogue is the continental mass of cold-climate Alaska, bordered by the Aleutian volcanic arc that passes into the deep waters of the northern Pacific Ocean.

Australia formed the northeastern part and coast of Gondwana. To the south lay Antarctica, to the southwest India – all parts of an ancient continental interior. An ocean edge still lay along the entire east coast, from Tasmania to east of Cape York. The warmth of the previous 191 million years had disappeared and the planet entered a phase of global cooling around 330 Ma ago – an icehouse. Central Gondwana became covered in an ice sheet and glaciers transported material outwards from the high mountain ranges, leaving moraines across the surrounding landscapes.

SHRIMP dating of Carboniferous sequences in New South Wales shows that the glaciation was underway from 325–310 Ma ago. There was also tectonic uplift which, with the glacial erosion, erased most of the Late Carboniferous geological record, and it was in the Permian, as the ice thawed, that sediments began to be deposited across this glaciated landscape.

The Carboniferous and Permian icehouse produced a planetary climatic environment probably very similar to the recent ice ages. Fluctuations in the Gondwanan glaciation probably controlled sea-level changes, which influenced patterns of sedimentation. The same association has been evident in response to Pleistocene glacial fluctuations over the last million years (see Chapter 10).

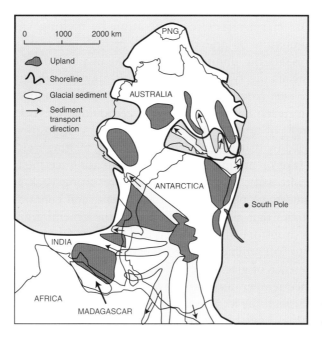

Figure 5.1 *Palaeogeography of part of the Carboniferous–Permian (302–280 Ma), showing Australia and Antarctica forming a major northeastward extension of Gondwana. The present India, Madagascar and part of north-eastern Africa are also shown. Offshore lay the Tethys Sea. The location of the South Pole at the time (for 285 Ma ago) is shown.*

The climate was glacial, with ice sheets in Antarctica and locally on highlands in Australia and elsewhere. Note the ice transport northwards from Antarctica into southern Australia.

Along the eastern margin there was a volcanic arc over 1800 km long from south of Sydney to northern Queensland. The presence of the volcanoes indicates more active tectonics along this eastern margin. The volcanism lasted from around 300 to 265 Ma ago. Layers of tuff in coals indicate that sporadic volcanism persisted until almost 250 Ma ago, although the massive outpourings of lavas and volcaniclastics stopped much earlier.

By 280 Ma ago there were deep embayments on the western coast as far south as Geraldton, and into the Great Sandy Desert, and into South Australia as far north as Oodnadatta. Seismic data collected during petroleum exploration show that these basins have deep fractures and normal faults along their margins and in the basement under their sedimentary fills. These features indicate considerable stretching of the continental crust, which allowed the central parts of the basins to drop down along the faults. In the east a long basin extended from just north of Batemans Bay in New South Wales to Collinsville in northern Queensland. This was connected to other extensive basins inland.

Indicators of glacial activity

We cannot visit the past and see the rivers and glaciers that flowed millions of years ago, but their effects have been preserved in the rocks. Signs of ice activity in modern Australia are not readily apparent, but we can list three major pieces of geological evidence that indicate glacial activity.

Firstly, striated pavements are formed where rock fragments embedded in the basal ice of a glacier scraped over the bedrock, producing scratches and grooves parallel to the direction of ice movement. Finer sand in the ice smooths the rock surface, and the combined result is a polished and striated pavement.

Secondly, tills and erratics are characteristic of deposition by melting ice, especially on land. Till is a random mixture of large rock fragments within a matrix of finer sand and mud, and represents the dumping of material in place as the glacier melts or recedes. The fragments are typically angular and may be striated, since they have been plucked or gouged from bedrock, in contrast to the rounded, water-worn gravel usually found in rivers. Erratics are very large boulders of rock, some weighing tens or even hundreds of tonnes, of radically different composition to nearby bedrock. They were carried by glaciers and left stranded by melting ice.

Thirdly, dropstones occur in lake or marine deposits where floating ice melts and drops large rock fragments into soft, fine sediment. The resulting rock is generally a mudstone or muddy sandstone, perhaps with fossil shells, and contains scattered rock fragments or dropstones that are much larger than the general grain size. Typically the laminae under a dropstone are depressed and contorted from the impact of the heavy rock falling into soft sediment.

Two factors determine the presence of glacial ice today. One is the nearness to the poles, where the Earth's surface temperatures are lowest. Polar glaciers have temperatures well below zero all through the ice mass, so there is no water present, and the ends of the glacier break off as huge blocks from precipitous ice cliffs.

Figure 5.2 *Glacial dropstone in Permian sedimentary rocks, south of Wasp Head, north of Batemans Bay, New South Wales. The dropstone is much larger than the grain size of the surrounding units and could not have been carried in by local currents on the Permian seabed; it was dropped from a floating iceberg. The sediment layers underneath were compressed, and there appears to have been some scour of the sediments around the sides while it sat on the seabed. Then more sediment covered the block, and later compaction during burial of the sedimentary sequences has bowed the layers over the top.*
PHOTO: KEITH CROOK

The second factor is higher altitudes, where the air is colder. There are many mountains in temperate regions with forests or jungles at the base and glaciers at the peaks, such as the Cascades mountains in western USA, Kilimanjaro in Kenya, and mountains in Irian Jaya. These temperate glaciers have a temperature of just zero through the thickness of the ice, and contain both ice and water. The terminus is a rounded slope, typically with a meltwater stream issuing from the base. These streams transport glacial debris and form fluvial and marine sedimentary sequences downslope of the glaciated mountain ranges.

Evidence of glaciation

Southern Australia

The earliest reports of glaciation in Australia were made by the visiting geologist Alfred Selwyn as part of geological notes included in the Adelaide parliamentary papers in 1859. He compared a glaciated rock in South Australia to features he had frequently seen in northern Wales. The Fleurieu Peninsula is the most significant area, where the outcrops at Hallett Cove and in the Inman Valley inland of Victor Harbour were described in detail by Walter Howchin in the 1920s.

The most important exposures are those in the valley of the Inman River. The present river is excavating the Permian glacial tillite, part of an old valley some 8 km wide carved by a glacier. Erratics of granite several metres across that were eroded from the tillite along the banks now lie in the present river bed. The granite boulder in the bank above the pavement near Glacier Rock Tearooms is very similar to granites outcropping at The Bluff near Victor Harbour on the coast. The rock

Figure 5.3 *Glacier Rock, a striated pavement in the Inman Valley, South Australia. The ice moved from the lower right to the upper left of the picture. The lines that trend left to right are fractures imposed during metamorphism and deformation of this Cambrian quartzite.*
PHOTO: DAVID JOHNSON

pavement scoured by the ice is composed of Cambrian quartzite some 500 Ma old, and is now a polished, striated and grooved pavement.

At Hallett Cove, on the northwestern side of the Peninsula, there are fine exposures of glaciated pavement overlain by tillite, and the same tillite has been recovered from numerous bores along the Yorke Peninsula. Offshore to the west there are glacially polished and striated surfaces overlain by tillites along the northern coast of Kangaroo Island. Again the ice movement was to the northwest.

This area is quite extraordinary, and a geological monument for it represents a land surface formed before 280 Ma ago, which you can see today. It was buried and is now being uncovered. There are many other striated surfaces in Australia overlain by Permian sediments.

But where did the ice come from? The direction of the ice flow was north to northwest, as evidenced by the dispersion of glacially transported material in that direction and by the orientation of striations gouged into bedrock by the moving ice. Now this is puzzling, since the ice was clearly moving towards central Australia, out of the Southern Ocean! Yet the record of the rocks is quite clear and can be read by a geologist as clearly as if it were upon the printed page. So where were the mountains and ice caps?

Certainly they must have lain south of the present coastline along southern Australia. The ice thickness must have been more than 600 m, judging from the vertical height from the scoured valley to the tops of the local hills, where evidence of the ice is preserved. However, there must have been a much larger ice field farther south to provide the impetus for the glaciers to move and to sculpture the hills and valleys. The glaciers have left behind scattered erratics up to 8 m across. Many of these are made of granite, indicating erosion of an uplifted mountain range.

In his accounts of the area around the Inman River valley in South Australia, Walter Howchin wrote in 1918 (p. 411):

> The snowfield which formed the chief gathering ground for the ice-flood lay to the south of the present continent, and may have had a more direct relationship with Antarctica than the landmasses of the present day, probably by way of the submerged banks to the south of Tasmania.

Later writers such as Laseron contemplated the vastness of the Southern Ocean, noting (p. 87) that 'somewhere beneath this expanse of water lies the lost continent of Gondwanaland, a land never seen by human eyes, but whose existence can readily be demonstrated.' We know now that the truth is that the lost ice fields of Gondwana were in present-day Antarctica, though even there the ice has not been continuous (see Chapter 7).

Australia was once linked to Antarctica as part of Gondwana (see Figure 5.1), and the split between them occurred about 90 Ma ago. Since then Australia has drifted northwards towards the equator, towards an eventual collision and merging with Asia (see Chapter 7). The Antarctic mountains were the source of the ice and much of the Permian glacial debris now strewn across southern Australia. The Southern Ocean contains some remnants of this break-up – small fragments of each continent broken off and left in the wake of Australia's northward movement – but most of the ocean floor is basalt erupted from deeper in the Earth's mantle, injected into the ever-widening gap.

In Victoria there is excellent evidence for glacial activity in the Bacchus Marsh area, and also in the Wangaratta, Heathcote and Coleraine districts, where Permian sediments around 290 million years in age overlie striated pavements cut into metamorphic rocks of mainly Ordovician and Devonian ages. The glacial striae at Bacchus Marsh trend east to northeast, and small basement outcrops show the typical pattern of being smooth and striated on the southwestern sides and broken and uneven on the northeastern side. Such a pattern is to be expected if ice moves northwards, planing off the southern, upflow side of rock prominences and plucking fragments as it overrides the northern, downflow side.

A very fine striated pavement known as Dunn's Rock after its discoverer (in 1892) lies on the northern side of Lake Eppalock, in the Heathcote area, though it is exposed only when the lake's level is low. Striations here and near Duck Creek show a northerly trend. These striae and other evidence point to the same situation as in South Australia: that the ice came from the south and moved northward across the Australian land mass.

There are also tillites exposed around Wynyard on the northern coast of Tasmania. Striated and grooved floors underlie the tillite. Boulder pavements on the Wynyard strand confirm ice movement to the north and northeast.

There is plenty of evidence that these glaciers were not just on land but that the deposits formed around the edge of a continent, with ice discharging into the ocean.

In all three areas – South Australia, Victoria and Tasmania – the sediments overlying the pavements are partly the gravels and sands deposited in river outwash from the glaciers, and partly marine sediments with fossils and dropstones. These dropstones are up to 2 m across, although most are only 0.10–0.50 m, and they sit within beds of mudstone that commonly contain abundant shelly fossils.

The clear implication is that southern Australia at this time was close to the edge of a major ice-bound land mass to the south, and the ocean filled embayments in the present Australian land mass.

Western Australia

While the thickest and most extensive deposits of glacial tillite occur along the southern margin, it is clear that icy conditions extended over most of the Australian continent. Permian rocks with dropstones and river sediments containing striated pebbles are common in Western Australia, with especially thick units in the Geraldton and Lyons River areas.

Rocks with glacial grooves where ice has carved channels in soft sediment are known in the Grant Range south of Derby. The sediments contain pebbles of the iron formations now being mined in the Pilbara, which indicates ice transport some 40 kilometres northwards. There is an isolated example of a striated pavement on the northeastern edge of the Pilbara, although the direction of ice transport is more easterly. It seems that the Pilbara (and perhaps part of the Kimberley) areas were then higher mountain ranges with glaciers directed off the edges into a basin now occupied by the Great Sandy Desert.

In southwestern Australia it seems that much of the Yilgarn region was planated and had valleys cut by the glaciers. These may be partly the origin of depressions now filled by coal at Collie, and salt lakes near Kalgoorlie.

Eastern Australia

The earliest known evidence for glaciation is Carboniferous, and occurs as a striated surface on top of a unit known as the Paterson Toscanite, erupted around 330 Ma ago near Maitland (inland of Newcastle). Toscanite is a very hard, pale lava, like rhyolite, typically formed by the melting of continental crust. Overlying this striated toscanite are sediments deposited in glacier-fed rivers and also in lakes as varves. Varves are finely laminated sediments comprising couplets – a layer of sand or silt deposited during the spring thaw and summer when the rivers were flowing quickly, followed by a layer of clay that settled from the still lake water during the winter when the rivers stopped flowing and the lake may have been frozen over.

This is about the same time that the earlier parts of the glaciated pavements in Tasmania formed. About and just after this time, several deposits of very coarse gravels filled old valleys running eastwards. The valleys are sometimes incised into other Carboniferous sediments but are also cut into much older rocks. Clearly this represents the end of a major phase of uplift and erosion in eastern Australia; a mountainous region in central western New South Wales was the source for the rivers and gravel. These old river deposits can be identified in the Tamworth area

Figure 5.4 *Varved lake deposits. The annual layers are in couplets: the darker, coarser layers represent sand deposited during the summer thaw when the streams are flowing strongly, with larger gravel fragments dropped from floating ice on the lake. Thin paler layers are mud that settled during the winter when the streams and perhaps the lake were frozen over. Compaction has forced the finer layers to drape over the coarse gravel fragments.*
PHOTO: CHRIS HERBERT

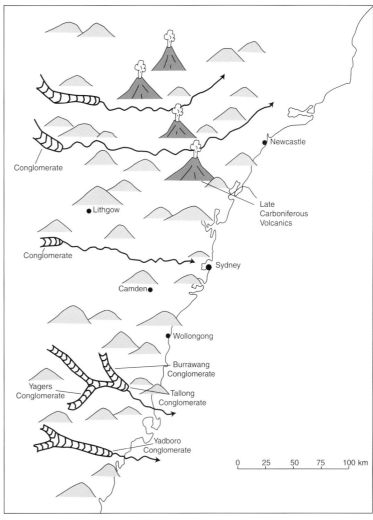

Figure 5.5 *Carboniferous gravelly river systems in New South Wales. Most of the landscape was being eroded at this time, and little sediment is preserved. However, these deeply incised channels containing conglomerate extended down from the hinterland into what would later be the Sydney Basin.*

ICEHOUSE 105

and extend down towards Gosford and Newcastle and also much farther south, crossing the coast just to the north and south of Jervis Bay (see Figure 5.5).

Two other types of rocks are present. Firstly, there are more varved lake deposits in which many of the gravel fragments show striated surfaces where they have been scratched by moving ice. It is clear that the gravelly rivers represent the outwash material from alpine glaciers to the west. Secondly, there are plenty of tuffs through the sequence, and in places thick lavas, so this glaciated landscape also had active volcanoes. A comparable modern example would be Alaska.

The glaciation was at a maximum around 280 Ma ago. At this time the southern and western parts of present-day Australia were part of the icy uplands of Gondwana. Ice covered and eroded the old Precambrian Yilgarn and Pilbara blocks, and extended through central Australia, across the Flinders Ranges and into central Victoria and Tasmania. Permanent ice had receded by 275 Ma ago, although there were isolated valley glaciers and plenty of evidence for seasonal ice conditions. For example, sedimentary rocks near Kiama in southern New South Wales show palaeosoil profiles 258 Ma old, with features indicating permafrost conditions. The layers show the same disturbance seen in modern frozen soils in Canada and Alaska.

Fossil evidence of the cold climate

The marine fossils from the Cambrian to Devonian sequences, described in Chapter 4, were warm-water organisms. The fossil assemblages are dominated by corals, stromatoporoids, calcareous algae, molluscs and brachiopods, and reefs were common. In contrast, the Carboniferous and Permian marine fossils from Gondwana are dominated by different brachiopods, bryozoans, molluscs and echinoderms. Only one solitary coral is known, and reefs are absent. These represent a cold-water fauna. Further support for the cold-water origin, apart from the evidence of glaciers, is contained in these marine shells: analyses of the oxygen isotopes in the calcium carbonate making the shells indicates that the shells formed in cold waters (see Box 7.2, page 154).

The plant fossils are a distinctive assemblage known as the Gondwana flora, which represents the first cold-climate vegetation that evolved on Earth (see Box 5.1).

The volcanic arc

It is sometimes hard to appreciate just how different the Australian continent has been in the geological past, compared to its present form. Let us consider, for example, a few spots along the eastern edge of Australia.

In North Queensland there is evidence of volcanism extending for some 80 million years, between 350 and 270 Ma ago. The volcanics occur as scattered outcrops extending from the Prince of Wales and Horn Islands in Torres Strait, to the Pascoe River north of Cooktown, along the western edge of the Atherton Tablelands, inland of Townsville, and then either side of the Bowen Basin in central Queensland.

Figure 5.6 *Permian sedimentary rocks exposed on a sea-cliff, Maria Island, Tasmania. The beds contain abundant fossils, mainly brachiopods and molluscs, and there is a dropstone in the lower left corner.*
PHOTO: DAVID JOHNSON

Figure 5.7 *Bombo latite (a type of basalt) showing columnar structure formed during cooling, and overlying red sandstone at Kiama, New South Wales.*
PHOTO: DAVID JOHNSON

One major arc, called the Eungella Arc, extended from Collinsville through Eungella and farther south. The country between Herberton and Chillagoe and farther northwest along the Featherbed Range contains an almost complete volcanic province. There are circular calderas some 10 km across and another 30 km across, which must have produced enormous eruptions.

In New South Wales, rhyolitic volcanism 286–266 Ma ago around Boggabri and Currabubula shed material eastwards towards the New England region. In the Hunter region, volcanism continued through the deposition in the peatlands, forming tuff bands within the coal seams as recently as 251 Ma ago. At Kiama, on the southern coast, volcanism formed basaltic lavas about 264 Ma ago. Now there is a famous blowhole where the waves rush into a sea cave at the base of the cliff and, if the tide is right, trap a pocket of air that blows seawater up through a hole in the roof of the cave. To the north is Bombo Head, where a quarry provided most of the road metal and railway track foundations for this region. It too consists of solidified lava. These lavas flowed northward from an emergent island volcano east of the present coastline and some tens of kilometres southeast of Kiama.

As well as the widespread extrusive volcanism already mentioned, intrusive magmatism also occurred. Towards the end of the Carboniferous the granite batholith around Bathurst was emplaced. Permian granites now form most of the high country of the New England and Queensland border regions.

In northern Queensland there are Permo-Carboniferous granites extending from inland of Mackay past Townsville, onto the Atherton Tablelands and farther north up Cape York. Many of these granites intrude rhyolitic volcanics, so they represent the underground magmas, which were the source of the extruded volcanics.

Development of the coal basins

The period from 325 to 295 Ma ago was a time of glaciation and erosion as the crust was uplifted. The crustal extension and subsidence that began around 295 Ma in northwestern, central and along eastern Australia formed basins in which thick sedimentary sequences could be accumulated. This section of Gondwanan crust was then under great tension.

While the ocean remained permanently around the eastern and some of the northwestern coast, between 266 and 256 Ma ago the sea flooded into these onshore basins. This flooding may have been partly due to melting of the ice sheets, though most of that had happened 10–20 million years earlier. So it seems there was increased crustal stretching and subsidence of the basins, allowing the seas to flood in. Continued sediment supply by the surrounding rivers gradually filled in the basins, pushing the sea back and re-establishing dry land. Onshore rivers delivered sediments after the ice withdrew, and lush peatlands accumulated the plant materials that would later become the black coals.

Peatlands

At this time peatlands occurred across much of Australia, from Cape York down to Tasmania and west to Perth, and in central Australia. There might well have been others that have since eroded, because there is no evidence for marine sediments within the continent, so the ocean existed only around the edge. Cool, humid

Figure 5.8 *Peatlands around the coastal plain bordering James Bay in northern Canada. These cold climate peatlands are frozen in winter, and are forming in climates comparable to those peats that are the Permian coals of present-day Australia and Gondwana.*

PHOTO: DAVID JOHNSON

climates were clearly ideal for plant growth and peat accumulation, and most of the peats were buried sufficiently deeply to be converted into black coal.

We commonly think of tropical environments with lush vegetation as the obvious environments to accumulate peat. Certainly there are large peat swamps in Indonesia, Burma, and the Amazon basin. However, the biggest peat accumulations on Earth are in the subpolar regions, especially the large land masses in high latitudes in the northern hemisphere – Scandinavia, Russia, Alaska, Canada. In these regions the annual growth of plant material is much lower than in the tropics, but the whole environment is refrigerated through a long winter every year. In such conditions there is minimal breakdown of the organic materials. This is in great contrast to the situation in the tropics, where plant material is very quickly broken down by fungi and bacteria and recycled into the standing forest. Only in flooded swamps where the organic material is under water is the rate of decomposition slow enough to allow peat accumulation in the tropics.

What amount and age of peat makes an economic coal seam?

Fresh peat is very spongy and porous; water-filled spaces commonly contribute up to 90% of the mass. Observations on ancient rocks indicate that peat is compacted by a factor of 10 in the formation of black coal. If we consider that the original peat has to be compacted 10 times just to remove the original porosity, then this figure seems acceptable. It means a workable 2 m seam of black coal would have required an original peat thickness of 20 m.

The accumulation of peat is slow; in cold climates it is around 1 mm per year, which is 1 m of peat per thousand years. Assuming a compaction ratio of 1: 10, 1 m of coal would take around 10 000 years to accumulate. The 9 m thick Middle Goonyella Seam at the Goonyella mine in central Queensland would have taken about 90 000 years to form. The thickest Permian black coal seam we know in Australia, the Big Seam at Blair Athol, is up to 31 m thick. Such a coal would have required some 310 m of peat and taken maybe 310 000 years to form.

Even a 2 m coal seam requires 20 000 years of steady accumulation. The groundwater levels must stay high to prevent the peat drying out, and to assist its preservation. There must be no influxes of sediment, which would make the coal dirty and uneconomic. For a good coal this peat accumulation would have to extend over a considerable area. So it is clear that thick, clean coals require a special set of circumstances, and that is why they have not formed everywhere.

The Carboniferous–Permian time period represents one of the great times of global accumulation of organic carbon as coal. There were the Northern Hemisphere Carboniferous coals and then the Gondwanan Permian coals. In Australia, the peat laid down forms the coal mined in New South Wales in the Illawarra, Hunter and Lithgow areas, in the Bowen Basin of Queensland, and at Collie in Western Australia. The buried organic matter also produced the natural gas in the Cooper Basin in northeast South Australia and southwest Queensland.

At Hartley, near Lithgow in New South Wales, there is a deposit of carbon-rich mudstones that contained a very high proportion of algal material. This is an oil shale, known when it was mined in the early-mid 1900s as kerosene shale.

Box 5.1 *Glossopteris* and the vegetation of the cold-climate peatlands in Gondwana

In brown coal the original plant materials and pollen are preserved so we can determine the vegetation of the peat swamp. However, the change from brown to black coal involves compaction and chemical alteration, which destroys the botanical features used to identify plants. Examination of black coal under a microscope can show whether the coal was made of fresh wood or altered wood, or whether pollen or algae were important contributors, but it cannot distinguish the plant species.

The other problem is that the fossils are only fragments of the former plants. We have only leaves, or twigs, or stumps – we cannot see the whole plant.

Three methods are used to reconstruct the vegetation of the Permian peatlands. Firstly, there are plant fossils preserved in the sediments immediately above or below the coal seam. It is possible these are species that grew on stable ground and are different to the species that grew in the waterlogged peatlands. However, many plant fossils are preserved in thin mudstone layers intimately associated with the coal, and we can be fairly sure that they are the same as grew in the peat. Secondly, the peat is sometimes petrified – that is, replaced by a mineral such as siderite or silica – and this preserves the plant microstructure. Thirdly, pollen and spores can be recovered from sediments in the coal measures, and the characteristic form of pollen and spores can be tied to particular plant types.

The first observation we can make is that all the Permian petrified wood shows very strong annual growth bands. This is the case for most trees that grow in climates with marked seasonal change to a cold winter. The wood cells formed in the spring and early summer are large and become smaller during the late summer, with a tight band of small cells formed in the autumn. No growth occurs during the cold winter.

The second observation is that leaf fossils are commonly concentrated in layers, with sediments above and below this layer having very few fossils. This is interpreted as representing autumnal leaf fall, which is further evidence of markedly seasonal climate.

The third observation is there is a major change in the flora between the Carboniferous and the Permian plant fossils.

The Carboniferous flora was dominated by lycopods such as *Lepidodendron*. These primitive plants were generally spore-bearing, although some were trees up to 40 m high. The characteristic fossil impression is like a tyre-track – a series of diamond scars left on the bark where the leaves have detached. There were also abundant ferns and mosses in the peatlands.

The Permian vegetation is characterised by the presence of the seed fern *Glossopteris*, a primitive gymnosperm (see Figure 5.9). Gymnosperms include the

Figure 5.9 Glossopteris *leaves*.
PHOTO: DAVID JOHNSON

conifers, and all have seeds that are unprotected; that is, not contained within the ovary wall as is the case with angiosperms, the flowering plants.

Glossopteris is a very characteristic plant with its tongue-shaped leaf and strong vein pattern. Other Permian vegetation included ferns, conifers, cycads, horsetails and mosses. Detailed studies of fossil plants and sediments in coal measures have shown that different plants occurred in different environments. There was a zoned vegetation, the same as we see on any modern landscape. Some plants live in moister situations and some in drier ground.

It is also worth noting that the earliest land plants are Silurian – the *Baragwanathia* from Victoria, about 400 Ma old. The Carboniferous flora dates from around 350 Ma ago, very soon after the evolution of land plants began. The widespread development of seed ferns in the Permian, some 300 Ma ago, was a major step in plant evolution because it marked the use of seeds as a mechanism of plant reproduction. There were signs of seeds in the Devonian, but the main method of plant reproduction was by spores, and this required moist conditions to enable the spore to survive and germinate. The evolution of seeds provided a way for plants to survive dry or freezing periods. It was not for another 160 million years, when primitive angiosperms evolved, that full use was made of this capability. The angiosperms then radiated into the full range of environments across the planet.

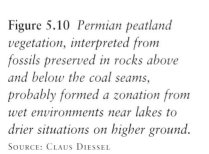

Figure 5.10 *Permian peatland vegetation, interpreted from fossils preserved in rocks above and below the coal seams, probably formed a zonation from wet environments near lakes to drier situations on higher ground.*
SOURCE: CLAUS DIESSEL

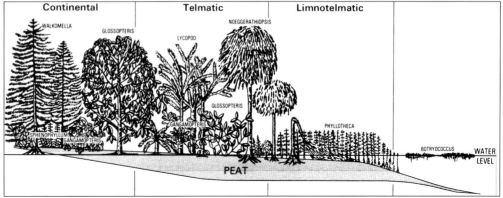

The peatlands developed in a range of environments, since plants had evolved to the stage where they could occupy several positions in the landscape, all of them wet. The peatland environment can be determined from the sediments on either side of the coal seam. The three commonest environments are alluvial fans, river floodplains and coastal marshes.

Alluvial fans

Alluvial fans form around the edges of mountain ranges, and are composed of coarse gravels and sands shed from a rapidly eroding landscape. Fast-flowing rivers strip the sediments and deliver it through gorges onto fan-shaped mounds. The

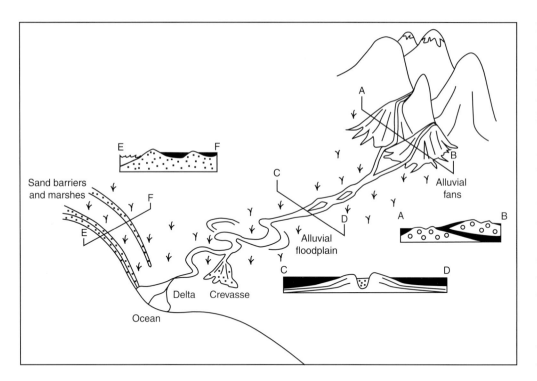

Figure 5.11 *Three peatland environments: alluvial fan, river floodplain, coastal barrier and marsh. Given suitable climate and groundwater conditions, a peatland can develop in each of these three depositional settings. In the alluvial fan setting, the sediments are coarse gravels which form wedges of conglomerate between split coal seams. Alluvial environments have channel sands and levee silts forming a channel belt, with the peats accumulating on floodplains. The resulting coals tend to be extensive on either side of the channel belt. Near the coast, peat can accumulate in marshes protected from the sea by sand barriers, and can also form on the deltaic plains.*

rivers tend to flow very quickly in the gorges and then spread out over the fans in a series of braided channels. The flow slows as it spreads out and the coarse bedload is dropped, while some finer sediment is carried farther downslope. It is typical for these fans to grow outwards for a while, and then for the flow to switch to the side, forming another fan. Such alluvial fans are common along the edge of mountain belts such as the New Zealand alps and the Sierra Nevada in the western USA.

The resulting deposits are wedges of conglomerate and coarse sandstone, which are thicker near the mountain source. Consider a peatland forming in the valley, with a mountain range shedding sediments as fans along its edge. As a fan grows it intrudes onto the peatland, overlying the peat. When the flow switches, and as subsidence continues, the peat will regrow over the fan surface. The result is a coal seam that splits at its edge, incorporating the coarse gravels of an alluvial fan. The Newcastle Coal Measures in the Hunter Valley are a good example of this type of sedimentation.

River floodplains

Rivers flowing across broad plains with low slopes normally occupy a channel belt. The channel at any one time may be 500–1000 m wide, and it moves within a belt several kilometres wide. To the side of this channel belt are normally broad floodplains tens of kilometres wide, which extend towards the next major river system, perhaps 100 km away. The channels are normally bounded by levees composed of fine sands and silts. These levees are elevated ground, and confine the river. Peat can accumulate on the floodplains if conditions are wet enough to support lush vegetation and preserve organic material.

Major shifts of the river send the channel sands and river levees over the old peatland. Floods that break the through the levees discharge muddy water onto the floodplain, and some sand is also carried along small channels. Lobes of such sediment deposited during floods are called crevasse splays.

Many coals were deposited in floodplain environments. A good example is the coal at Goonyella in central Queensland. The coal seam is overlain by a series of thick channels, with levee and floodplain sediments. This was formed when a major channel changed course and deposited sediment over a previously-existing peatland.

Coastal plains

Coastal plains are typically very flat areas, such as deltas or the broad areas landward of tidal flats. Extensive peat forms in freshwater, not in seawater, so the peatlands occur only above the tidal zone, normally maintained by strong freshwater flow from regular rainfall.

The sediments forming coastwards of such peatlands are deposited in tidal flats, or beach dunes, or the estuarine ends of river systems. As subsidence occurs, the shoreline moves in over the land and the peatland will be overlain by these coastal sediments. Thus the coal seam will have marine or brackish water sediments on top of it. Such a situation is very different to the two cases above, where freshwater fluvial sediments overlie the seam.

The marine nature of these sediments can be recognised because they can contain marine fossils and the sediments are burrowed. Interlaminated sandstone and mudstone is common. The sand was deposited by strong tidal currents and the mud settled at slack water – the turn of the tide, when currents are negligible.

The coal at German Creek in central Queensland, at Greta in the Hunter Valley, and in the upper part of the coal measures at Collinsville are good examples of coals formed on coastal plains.

One effect of this marine influence is that seawater may overlie the peat as the sequence subsides. This seawater contains sulphate, which can be reduced to sulphide by the organic matter in the peat. It is common for this sulphide to be precipitated as pyrite at the top of the seam. These brassy tops are a real problem because the sulphur is an unacceptable contaminant of the coal. Fortunately for Australia most of our black coal was formed in freshwater fluvial environments and has a low sulphur content.

Box 5.2 Burning mountain: Mount Wingen

Mount Wingen lies between Scone and Murrurundi, in the upper Hunter Valley in New South Wales. It was first seen by Europeans in 1828. The discovery was described thus:

> A settler, while on a shooting excursion on the opposite range of mountains, about 12 miles distant, observing the ascent of smoke in that direction, enquired of the Aborigines, who were in company with him, whether or not the bush had been set on fire by some of their people. They replied in the negative, and signified that it had been burning for a great length of time.

The Reverend Charles Wilton wrote about the 'Burning Mountain of Australia' in an article published in the *Sydney Gazette* on 14 March 1829. He drew attention to the many opinions and theories advanced, that this might be a *real* volcano, a mere seam of coal on fire, or a mass of ignited sulphur.

Wilton had visited the mountain and reported that there was no volcanic mouth, and that the mountain was made of sandstone not lavas. Then the fire was raging over about half an acre, through which there were several openings ('chasms' in his words) that were beautified with efflorescent sulphur. Furthermore, coal was well known from the surrounding country, and would be a suitable source of combustion.

It is clear from Wilton's descriptions and summary that yet again this was something that marked Australia as very different to the Europe they had come from.

> I have compared the phenomena presented by this mountain with written descriptions of volcanic action and subterranean fire, in other portions of the Globe, but can discover no exact similarity between them. The Burning Mountain of Australia may, I think, be pronounced as unique – one other example of nature's sports – of her total disregard, in this country, for those laws which the Philosophers of the old world have long since assigned her.

The mountain is still burning. Geological mapping has shown that the mountain is composed of Early Permian marine and coal-bearing sediments, and although we cannot know what started the fire in prehistoric times, coal is the fuel source for the burning mountain.

The burnt-out zones underground have caused subsidence of the overlying materials, so that fractures and pits occur on the surface. In places the surface is extremely hot, with acidic fumes emanating. Some of the sediments immediately above the burnt zone have been thermally metamorphosed to crystals, which indicates temperatures up to 1700°C.

Permo-Carboniferous palaeogeography

For most of Carboniferous time, 354–298 Ma ago, there was very little sedimentation preserved on the Australian landmass. It was a period of uplift and intense glacial erosion. Thin fluvial deposits do show some plant fossils and even some thin dirty coals were formed.

However it was not till the continent started to deglaciate around 290 million years ago that substantial sedimentation occurred. Several large sedimentary basins had been established by around 285 million years ago. While the basins had subsided sufficiently for extensive, shallow marine embayments to develop along the

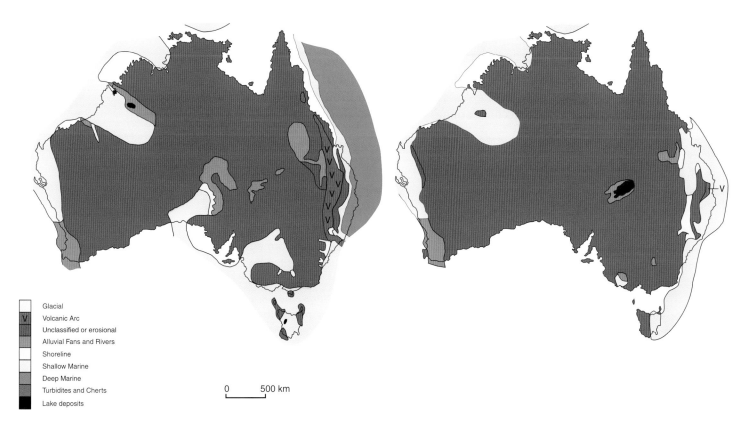

Figure 5.12 *The present extent of Permian sedimentary sequences in the Australian part of Gondwana. It is highly likely that thin units of Permian sediments covered more of the continent and connected the isolated outcrops in northeastern and southeastern Australia, and that these were subsequently eroded. However, none of these units would have been deep basins, so the brown areas correspond more or less to what were land areas at the time.*

a. Around 276–272 million years ago the volcanic arc lay in deep water to the east of Australia, stretching from Townsville to Newcastle. To the west and south were shallow marine embayments with the shoreline (yellow) and fluvial (tan) sequences at their edges. In many places these terrestrial sequences were removed by later erosion, so that the marine units now occur adjacent to bedrock. The white areas represent zones where the situation is unclear, although most were probably dry land.

b. By 260–265 million years ago the volcanic arc had all but ceased activity. There was minor activity near Gympie and in the New England region of New South Wales, and one major locale in the Yarraman area west of Brisbane. A locally cold climate in western Victoria and west of Sydney produced glacial dropstones, which are seen in marine sediments of the Branxton area in New South Wales. Shallow seas had withdrawn from inland southern Australia, but persisted along the west coast.

western side, northeastern and central Australia were still dry land. The Sydney area was a shallow embayment. The volcanic arc from Townsville to Newscastle was producing major lavas and ash falls, which filled the eastern side of the developing Sydney and Bowen Basins (see Figure 5.12).

Around 275–270 Ma ago the sea level rose and marine conditions developed in most of the coastal basins. Then the seas withdrew and the fluvial coal measures developed over the basins. This latest phase from 270 million years onwards represents the major stage of coal accumulation.

Summary

During the Carboniferous and Permian periods, Australia was still part of Gondwana. Glaciation occurred from 325 to 310 Ma ago, with erosion by ice extending into the Early Permian some 290–300 Ma ago. Crustal extension and subsidence from around 295 Ma in northwestern and eastern Australia formed basins in which thick, coal-bearing sedimentary sequences could be deposited as the climate thawed.

The Gondwanan flora, characterised by the seed fern *Glossopteris*, marks a major step in plant evolution with the ability to reproduce by seeds rather than spores.

SOURCES AND REFERENCES

Bourman, R.P. & Alley, N.F., 1999. 'Permian glaciated bedrock surfaces and associated sediments on Kangaroo Island, South Australia: implications for local Gondwanan ice-mass dynamics'. *Australian Journal of Earth Sciences*, 46: 523–531.

Brakel, A.T. & Totterdell, J.M., 1993. 'The Sakmarian–Kungurian palaeogeography of Australia', pp. 385–396, in R.H. Findlay, R. Unrug, M.J. Banks & J.J. Veevers (eds) *Gondwana Eight*. Balkema, Rotterdam.

Brakel, A.T. & Totterdell, J.M., 1995. *Palaeogeographic Atlas of Australia. Volume 6 – Permian*. Australian Geological Survey Organisation, Canberra.

Cas, R.A.F. & Wright, J.V., 1987. *Volcanic Successions: Modern and Ancient*. Chapman & Hall, London. 528 pp.

Crowell, J.C. & Frakes, L.A., 1971. 'Late Paleozoic Glaciation: Part IV, Australia'. *Geological Society of America Bulletin*, 82: 2515–2540.

Diessel, C.F.K., 1992. *Coal-Bearing Depositional Systems*. Springer-Verlag. 721 pp.

Gould, R.E. & Delevoryas, T., 1977. 'The biology of *Glossopteris*: evidence from petrified seed-bearing and pollen-bearing organs'. *Alcheringa*, 1: 387–399.

Herbert, C. & Helby, R. (eds), 1980. *A Guide to the Geology of the Sydney Basin*. Department of Mineral Resources, Geological Survey of New South Wales Bulletin No. 26. (*See especially Chapter 14 on the evidence for glaciation.*)

Howchin, W., 1918. *The Geology of South Australia*. Education Department. 543 pp.

Laseron, C.F., 1954. *The Face of Australia*. Angus & Robertson. Second Edition. 244 pp.

Martini, I.P. & Johnson, D.P., 1987. 'Cold-climate, fluvial to paralic coal-forming environments in the Permian Collinsville Coal Measures, Bowen Basin, Australia'. *International Journal of Coal Geology*, 7: 365–388.

Percival, I.G., 1985. *The Geological Heritage of New South Wales*. New South Wales National Parks and Wildlife Service. 136 pp.

Retallack, G., 1980. 'Late Carboniferous to Middle Triassic megafossils floras from the Sydney basin', pp. 385–430, in C. Herbert, & R. Helby, (eds) *A Guide to the Geology of the Sydney Basin*. Department of Mineral Resources, Geological Survey of New South Wales Bulletin No. 26.

Roberts, J. *et al.*, 1995. 'SHRIMP zircon age control of Gondwanan sequences in Late Carboniferous and Early Permian Australia' in R.E. Dunay & E.A. Hailwood (eds) *Non-Biostratigraphical Methods of Dating and Correlation*. Geological Society Special Publication No. 88: 145–174.

Veevers, J.J. & Powell, C., McA., 1987.' Late Paleozoic glacial episodes in Gondwanaland reflected in transgressive-regressive depositional sequences in Euramerica'. *Geological Society of America Bulletin*, 98: 475–487.

Ward, C.R., Harrington, H.J., Mallett, C.W. & Beeston, J.W. (eds) 1995. *Geology of Australian Coal Basins*. Geological Society of Australia Inc., Coal Geology Group Special Publication No. 1. 590 pp.

WEBSITES

Permian Period
www.seaborg.nmu.edu/earth/Permian.html

Glossopteris
www.ucmp.berkeley.edu/seedplants/pteridosperms/glossopterids.html

ILLUSTRATIONS

Fig. 5.4: redrawn by permission of the New South Wales Department of Mineral Resources after fig 2.6 in Herbert, C. & Helby, R. (eds) 1980. *A Guide to the Geology of the Sydney Basin. Department of Mineral Resources*, Geological Survey of New South Wales Bulletin No. 26.

Fig. 5.12: redrawn and modified with permission from A.A. Balkema Publishers, the Netherlands, after figs 2 and 4 in Brakel, A.T. & Totterdell, J.M., 1993. 'The Sakmarian–Kungurian palaeogeography of Australia' in R.H. Findlay, R. Unrug, M.J. Banks & J.J. Veevers (eds) *Gondwana Eight*. Balkema, Rotterdam.

Warm plains and then seas

CHAPTER 6

MESOZOIC WARMING: THE GREAT INLAND PLAINS AND SEAS

The vast cold-climate peatlands that formed the Australian black coal deposits started to dry up, and a similar situation existed elsewhere in Gondwana. We can tackle this warming phase of Australian geological history in two parts.

The older and longer part lasted from 251 to 140 Ma ago. Ice had begun to diminish about 255 Ma ago and there was a great warming, at first forming an arid landscape dominated by inland rivers and lakes.

The second started 140 Ma ago and lasted 40 million years. A rapid sea-level rise formed great inland seas, with a volcanic arc down eastern Australia. The seas withdrew from 117 Ma ago, leaving an essentially dry continent by 100 Ma ago. Then began the break-up that led to the final disintegration of Gondwana.

A warming across the plains

Our interpretation of the climates is hampered because we are not sure of the sizes of the old mountain belts. As we know from today, mountains tend to capture atmospheric moisture, forming wetter environments for plants and animals and commonly creating drier areas on the leeward side. By and large it seems that at these times there were no major mountains in inland Australia, and the main divides were to the east. However, the sedimentary rocks and fossils do provide many clues to the course of events.

Across the present land mass of Australia there was erosion, and sediments

Between 255 and 251 million years ago the world emerged from the icehouse. Why did the ice and snow disappear?

What can we say about the 140 million years during which the dinosaurs roamed Australia?

The final volcanic arc along eastern Australia extended from northern Queensland to the Otway region of Victoria. What terminated its eruptions?

Figure 6.1 *The Three Sisters at Katoomba are erosional remnants of the sandstone escarpments around the Blue Mountains west of Sydney. These sandstones were deposited in early Triassic river systems.* PHOTO: DAVID JOHNSON

118   THE GEOLOGY OF AUSTRALIA

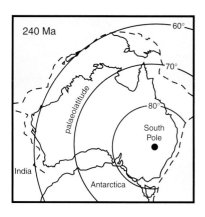

Figure 6.2 *The South Pole was located at the present site of Bourke 240 Ma ago.*

kept accumulating in river channels and on floodplains. However, the muds changed from grey (containing finely divided plant material) to brown and red (without plant material). Peat no longer accumulated. With a hotter climate the plant litter dried up and decayed. These red-brown mudstones, called 'redbeds', often contain abundant mud cracks, indicating that the mud was wet and then dried out under the sun. It is the same as the difference today between the dark grey muds in wet mangrove swamps and on the dry red plains of the inland.

Over a period of 111 million years this change continued; sometimes it was wetter, but never so icy. The extraordinary thing is that this warming was occurring at a time, 240 Ma ago, when the South Pole lay near Bourke in northern New South Wales. There were no polar ice-caps on the planet. Why did the ice and snow disappear? We simply do not know.

This part of Gondwana was almost all dry land; the climate was warm and arid, becoming more humid and still warm after about 230 Ma ago, when peats again started to form.

Box 6.1 The great extinction of life 251 Ma ago

The global warming that preceded the Triassic was closely followed by the greatest extinction of life on Earth. Over 90% of species in the oceans and about 70% of vertebrate families on land were wiped out, although land plants seemed to continue without much damage. Compared to this, the better-known extinction of the dinosaurs and other life some 65 Ma ago was a paltry affair.

Fortunately after the extinction there was a massive radiation of life, when thousands of new species evolved to take up the ecological niches left empty. With the cold climates gone it was the dinosaurs and mammals that developed on land, and the cephalopods in the sea.

How long did the extinction take to happen? We cannot date the rock layers of this age very precisely, though it is clear the extinctions occurred within estimated time spans of 8000 to less than 500 000 years. So it was not an overnight catastrophe, but certainly occurred within a small geological time frame. Just imagine seven out of ten species of land animals dying out and nine out of ten families of marine animals – that would be an unparalleled ecological disaster today.

HOW DID THE EXTINCTION HAPPEN?

There are several theories. Firstly, this was the time of the eruption of the Siberian basalts, which covered some 1.5 million square kilometres in less than 1 million years, one of the greatest outpourings of magma in the Earth's history. There was also a long period of volcanic eruption in South China. Clearly some major movement was afoot within the Earth. It also coincided with major enlargements of the land surfaces elsewhere and perhaps constriction of oceans, which may have then become putrid and uninhabitable for most life.

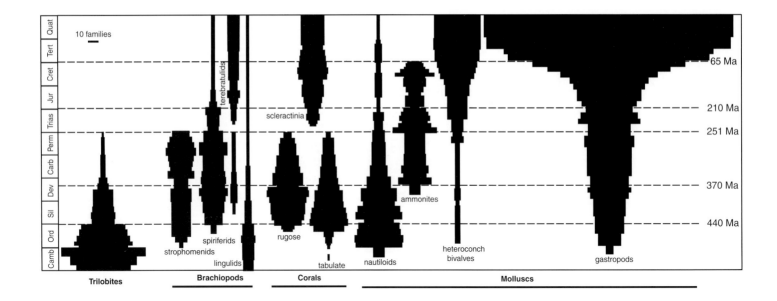

Figure 6.3 *Changes in the numbers of families of major marine organisms over the last 545 million years. The five major extinctions are marked: at 440, 370, 251, 210 and 65 Ma. Note the extinction of major groups at the 251 Ma event, and also how the diversity of other groups such as the bivalve and gastropod molluscs increased rapidly after the 251, 210 and 65 Ma events, filling the available environmental niches.*

There is very recent evidence for a major impact that could have enveloped the Earth with dust clouds, blotting out the Sun and killing much of the life. Microspherule layers have been discovered, and geochemists have also identified fullerenes – spherical molecules of 60–200 carbon atoms, known as 'buckyballs'. More significantly, these fullerenes contain trapped helium and argon whose isotope compositions do not match those of the gases normally found on Earth – but do match the extraterrestrial isotope signatures found in meteorites. There is also evidence from China that the Earth was enveloped in a sulphurous cloud, perhaps caused by an impact that released sulphur-rich material from a fracture reaching to the mantle.

The evidence is thus pointing strongly towards a meteorite impact as the cause of the 251 Ma Permo–Triassic extinction.

Finally, we cannot be sure that the planetary orbit and the rotation of the Solar System around the galaxy (every 250 million years) did not substantially alter the heat radiation received by the Earth. The answer awaits some determined scientific research.

The great inland plains

Arid fluvial plains of the Triassic (251–235 Ma)

As the Australian land mass warmed, rivers flowed across a new landscape that would become the home of the dinosaurs and the ancestors of all the major animals in Australia today. Much of Gondwana appears to have been dry land that was

Figure 6.4 *a. A reconstruction of vegetation on a Triassic coastal landscape, with an elongate river channel extending into a bay. The sides of the block diagram show the resulting sedimentary sequence. The distant hills are thought to have had coniferous forest, while the foreground vegetation includes* Dicroidium. *Note the labyrinthodont swimming in the channel.*
b. A reconstruction of the Triassic plant Dicroidium. *The black leaf can be up to 1 m long. The stippled items are reproductive structures at natural size. The top inset is a pollen (*Allosporites*), and the lower inset a seed.*
ILLUSTRATION: GREG RETALLACK

being eroded. The sediment was transported into a few large basins – the North West Shelf, some isolated deposits in central Australia and Tasmania, and the extensive Bowen–Sydney Basin along the margin of Gondwana.

The vegetation was more diverse than in the cold climates of the Permian, perhaps because of a warmer climate and the opportunity to colonise environments left devoid of plants after the extinction. The large, herb-like *Dicroidium*, apparently formed widespread heaths. On surrounding hills were trees, including conifers and ginkgos; horsetails, ferns and lycopods formed heaths; and there were reed-beds around water-logged areas. Many ferns had very large fronds, typical today of plants in hot tropical regions. While these warm and wet conditions prevailed in the east, it seems the inland plains were much more arid, with little preservation of organic material.

Most of this section of Gondwana was dry land, with a few small volcanoes in southern Queensland, but the main features were huge riverine plains. The old river channels are well exposed in the freeway cuttings north and south of Sydney and around the harbour cliffs. The sandstones have units commonly up to 3 m thick, with layering and cross-beds. This layering forms as a sand bar migrates downstream, when sand is carried along the top of the bar and avalanches down the face of the bar. In places there are irregular mudstone units, representing finer sediment that settled out after floods.

Similar sandstones form large mesas in central Queensland; for example, near Glenden. Further north, on the road to Mount Isa west of Hughenden, there is a

Figure 6.5 *Jurassic sandstone at Cobbold Gorge, Queensland, formed in an ancient river channel. The cross-beds indicate that the palaeocurrent ran from right to left. The cross-beds are 0.3 m thick, indicating the migrating megaripples/bars were about this high. Figure 6.6 shows how such cross-beds form.*
PHOTO: DAVID JOHNSON

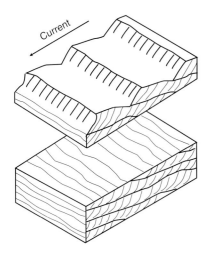

Figure 6.6 *Migrating ripples form cross-beds as sediment is transported up the back of the ripple and avalanches down the lee face. The resulting cross-beds (in the lower block) dip down-current.*

small parking area close to the point of the Great Divide that separates easterly from westerly flowing rivers. On the southern side of the road are small gullies cutting down through white quartzose sandstone very like those on the Illawarra Plateau south of Sydney. Further north are the sandstone bluffs of Mt Mulligan, northwest of Mareeba.

All these sandstones are remnants of a series of river sands that were shed by immense braided river systems along the western side of mountains rising to the east. Modern examples are the braided river systems of the Brahmaputra River that drain the southern side of the Himalaya. These modern river deposits can be 100–200 km across. Similar river systems delivered sands into rift valleys forming along the line of the coast of Western Australia. There is also some evidence these sands were being delivered, in the Sydney area at least, into shallow marine embayments, and that some of the sands may be shoreline deposits.

Tectonic compression of the basins 233–227 Ma ago (Bowen Orogeny) uplifted the eastern margins and thrust them westwards. This renewed sediment source is reflected in the sandstones, which contain more rock fragments than the quartzose sandstones previously deposited.

Amphibians, reptiles and mammals

For most of their history the riverine plains of Gondwana were too dry to accumulate vegetation. Peat did not accumulate until 228 Ma ago, when the climate became more humid. Then a series of small basins accumulated sediments and coal at Fingal in Tasmania, Ipswich, Callide and Tarong in Queensland, and Leigh Creek in South Australia.

Volcanism was absent, apart from igneous activity in the New England area of New South Wales. White quartzose sandstones further south and west characterise the sediments. The relief across most of the rest of the continent was relatively low. Rivers draining west and southwest delivered large amounts of the sediment into an immense basin covering much of central Queensland and extending into northern New South Wales and South Australia.

Despite the fact that most Triassic sediments are quartz sandstone and red mudstones devoid of many fossils, it is absolutely clear that this is because the landscape was commonly hot and dry. There was life, but it has not been preserved in these sediments. The mudstones show evidence of rootlets, soils and plant debris, although identifiable plant fossils are rare. By comparison, imagine the sedimentary record of anywhere on the western plains of eastern Australia, or the wheat belt and mallee areas of southern and western Australia. There is plenty of vegetation: not lush rainforest, but plenty of trees, shrubs and grasses. The animals are plentiful – kangaroos, bandicoots, abundant birdlife, snakes and lizards, frogs and insects – especially near water. It has been this way for thousands, even millions of years, and yet if you dig up a spadeful of soil there is no sign of all that lived there before. There are not layers of bird skeletons, insect fragments, branches of trees and parts of frogs and snakes. All have been eaten, dissolved, oxidised and returned to the living systems.

So it was in the past; only occasionally do the animals fall into a lake or deep channel where they are covered and preserved for us to see. For instance, at Knocklofty in Hobart, amphibian and reptile skeletons have been found in the rocks. These lived in lakes surrounded by scouring rushes, seed ferns, maidenhair trees and club-mosses, with larger conifers nearby.

Many of Sydney's brick houses were built from these soft mudstones, excavated from the brickpits at St Peters and Ashfield and moulded and fired in the brick ovens. The mudstones were the old lake deposits and they did preserve some of those Triassic animals and plants. One was an almost complete skeleton of a labyrinthodont amphibian, *Paracyclotosaurus davidi*, which measured over 2.25 m from nose to tail tip. The skin was covered in scales, the teeth were not large, and the appearance was of a giant salamander. Abundant fish fossils were also found, indicating that the lakes were teeming with life. Such remains have been found in many other localities in New South Wales, Queensland, Western Australia and Tasmania, though generally only as fragments.

In 1996 the discovery of another salamander-like amphibian occurred when a slab of sandstone broke open while being positioned for a garden wall near Gosford, north of Sydney. Quite apart from these amphibians and fish, numerous reptile remains have been found near Rewan in central Queensland and near Hobart, and reptile tracks have been found in the Sydney area.

The lesson is that just because the sediments are not full of fossils does not

Figure 6.7 *The Triassic giant salamander* Paracyclotosaurus davidi, *which inhabited freshwater lakes in the Sydney region around 235 Ma ago. The animal was 2.25 m long.*
ILLUSTRATION: PETER SCHOUTEN, FROM *Prehistoric Animals of Australia*.

mean there was no life. Such large amphibians must have been part of a complex ecosystem of plants, insects and other animals, just like ecosystems today.

Dinosaurs had evolved by about 230 Ma ago. The principal development, beyond previous vertebrates such as fish and early reptiles, was a specialised ankle joint that enabled the animal to walk upright. This new posture, which would be the basis for later primates and humans, was only possible because the dinosaurs had large tails that balanced the chest and head as the animals moved.

The period of warming from 255 to 65 Ma ago was the world-wide age of the dinosaurs, and Australian dinosaur remains are mainly in Cretaceous deposits, especially from central Queensland and the south coast of Victoria.

Humid tropical forests of the Jurassic

Around 180 Ma ago the climate became much wetter, although still warm and humid to subhumid. Most of the plants characterising the Triassic coals had disappeared and were replaced with completely new plants. The flora was varied, with large conifer trees, and smaller cycads, lycopods, tree-ferns, and the ground cover of mosses and smaller ferns.

Similar animals roamed the forests, and while they have not yet been found in Australia, we know from elsewhere in the world that this Jurassic life included the first mammals. We commonly think of the dinosaurs and amphibians as ancient life, far pre-dating the modern mammals, and it is true the dinosaurs have disappeared, but the mammals were with them at the start. Elsewhere in the world mammals emerged 190–200 Ma ago, though they stayed small till after the extinction of the dinosaurs 65 Ma ago. Recent discoveries in China, where an ash shower 125 Ma ago preserved an entire ecosystem, have produced a spectacular range of fossils, including feathered dinosaurs and the earliest known marsupial. It is clear the early mammals shared the same world as the dinosaurs. So far we have not found the fossil evidence in Australia, but we need to keep looking!

Extensive peatlands with anastomosing river channels formed in southeastern Queensland and northern New South Wales. These coal-bearing sequences are mined in the Rosewood–Walloon area west of Brisbane, and extend south to Grafton. Only isolated occurrences of the sandstones deposited in these river systems are preserved further north, for instance at Cobbold Gorge near Forsayth in Queensland. The coals seem to have formed mainly from large conifer trees similar to the *Araucaria* (kauri pine) whose pollen is abundant in the coal, with ferns also present.

Isolated lake sediments have been preserved; the Talbragar beds near Gulgong contain fish and pine leaves, indicating that this was a shallow lake in a forest.

A major igneous episode can be seen as dolerite sills in Tasmania, which are exposed on Mount Wellington near Hobart. Similar rocks occur in the Karoo in South Africa, and stupendous examples, for those lucky enough to see them, in the Antarctic mountains. Together these Tasmanian, South African and Antarctic intrusions and flows are known as the Ferrar Dolerites, and they total some 2.5 million cubic kilometres. They are very similar geochemically and were formed

184–177 Ma ago. These igneous injections indicate that the Earth's crust was starting to move and magma was ascending from the mantle to fill the spaces.

Volcanoes were erupting periodically along eastern Australia, and layers of ash (tuff) in the coal mark times when the whole peatland was showered with volcanic ash. Under the microscope the grains in the sandstones are clearly fragments of volcanic rocks and ash. A major chain of volcanoes must have lain somewhere east of the present coastline, spraying ash across the landscape and providing volcanic debris for the rivers to transport westwards. During these episodes much of the vegetation was probably killed, but then the plants grew again to re-establish the peatland.

Igneous activity is also evident in the Sydney area. Intrusive basaltic rocks were mined at Prospect for road aggregate, and there are similar rocks in the Mittagong area. Road aggregate was also quarried from several vertical pipes filled with layered volcanic breccia, ash and sediments. Such pipes, called diatremes, are not volcanic necks containing the remnant magma, but the subsurface parts of volcanoes that vented ash and gas into a landscape with a high groundwater table. Volcanic debris settled back into the pipe, often with plant materials that grew nearby between the eruptive phases. The sediments have yielded pollen that is of early Jurassic age.

Ninety-five known and a further 60 possible diatremes have been mapped in the Sydney Basin, extending from Bondi on the coast northwest to the upper Hunter Valley. Examples are the former quarries in the Sydney suburbs of Hornsby, Erskine Park, Marsden Park and Angorawa.

In 1994 a living plant relic from this age was found. A project officer with the National Parks and Wildlife Service abseiling into a remnant rainforest pocket in a

Figure 6.8 *Wollemi Pine and fossil leaves. The fossil leaves are from the Jurassic Talbragar beds near Gulgong and have been known for over 100 years. The modern plant was discovered only in 1994.*

Photos: Jaime Plaza, Wildlight

canyon in the Wollemi National Park found a 40-metre-high tree with fern-like leaves on the ends of the branches. Fewer than 50 of these trees are known. It is now officially called the Wollemi Pine. Never before found, its leaves match exactly the Talbragar plant fossils recovered near Gulgong from sediments deposited 150 million years ago. This is a plant the dinosaur communities knew well.

And what animals lived on this Jurassic landscape? Several sets of tracks showing the paths of four-footed reptiles have been found. One set is exposed in the roof of old diggings on a hill behind the Mount Morgan gold mine near Rockhampton.

Freshwater lakes held long-necked reptiles, the plesiosaurs, which were among the earliest in the world, as well as fish and small shellfish. Presumably there were others, such as frogs, salamanders, insects and worms, that completed the food chain. On dry land, roaming close to dense vegetation, were dinosaurs such as the plant-eating sauropods. One of these, *Rhoetosaurus brownei*, was found in 1924 near Roma in southern Queensland. A giant beast, it was about 17 m long and 3 m high at the hips, probably weighing around 20 tonnes. (In comparison, a large African bull elephant weighs between 4 and 6 tonnes.) How much did this sauropod eat, and how did it keep cool? We think it ate prodigious amounts of the lush vegetation, probably stripping soft leaves from trees, and that it rested during the warm day in the shade or immersed in water before foraging at night.

Figure 6.9 *The Jurassic dinosaur* Rhoetosaurus brownei, *which roamed central southern Queensland around 170–180 Ma ago. The animal was about 17 m long and 3 m high at the hip, and probably weighed around 20 tonnes. It browsed on the heavy vegetation of conifers and seed ferns.*

ILLUSTRATION: PETER SCHOUTEN, FROM *Prehistoric Animals of Australia*

Box 6.2 The Sydney Basin

The Sydney Basin is composed of sedimentary rocks extending from Durras, just north of Bateman's Bay, to Port Stephens and inland to Lithgow. The Basin is part of a much larger depositional area, that extends through Gunnedah to central Queensland as far as Collinsville, incorporating the Bowen Basin, and west to the Cooper Basin in South Australia (see Figure 6.10a). The Sydney Basin also extends offshore under the continental shelf.

The first recorded discovery of coal was by a group of escaping convicts, on the banks of a creek about 36 hours sailing north of Port Jackson on 30 March 1791. Coal was also discovered at Coalcliff south of Sydney in June 1797 by survivors of a shipwreck. Early explorers then found more deposits to the north and west of Sydney.

Figure 6.10

a. Sedimentation in eastern Australia was widespread during the Permian and Triassic.

b. Geological cross-section from Mount Canobolas near Orange, across the Blue Mountains to the New South Wales coast at Palm Beach. The Sydney Basin overlies an unconformity cut into folded Ordovician to Devonian rocks. These older rocks had been buried and folded for millions of years before the Bathurst granite was intruded during the Carboniferous. Then uplift and deep erosion formed the unconformity. At least 5 km of rock had to be stripped off to expose the granite. Subsidence during the Permian and Triassic allowed deposition of sediments in the Sydney Basin. Again there was subsidence, in the order of 2 km, to form the black coals of the Basin and then a 2 km uplift in the Tertiary. Eruption of basalts, rhyolites and trachytes formed Mount Canobolas and probably covered much of the Blue Mountains, though only a few peaks such as Mount Wilson and Mount Tomah remain. The geological history revealed by this cross-section confirms the enormous scale of repeated movements of the Earth's crust.

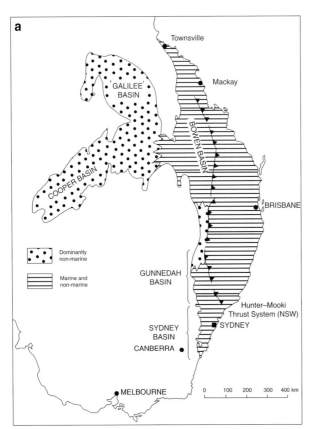

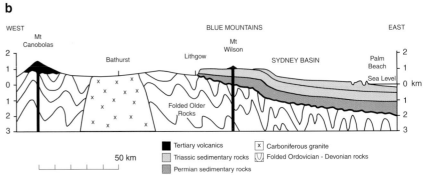

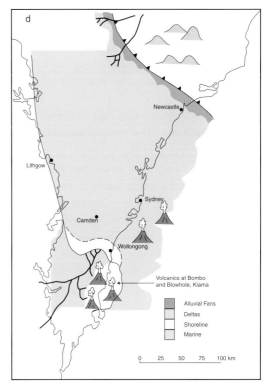
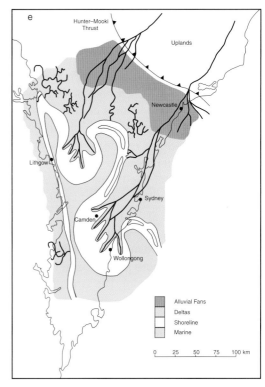

c. Unconformity at the base of the Sydney Basin, exposed in Ettrema Gorge. Flat Permian sandstones overlie steeply dipping and eroded Devonian rocks.
d. Permian palaeogeography around 264 Ma ago, when much of the Sydney Basin was a marine embayment. Volcanics were being erupted at Gerringong, Kiama and Bombo. The uplifted New England region was starting to shed sediment into the present Hunter Valley region.
e. Permian palaeogeography around 255 Ma ago during deposition of the Newcastle Coal Measures, when deltas had prograded southwards as far as Wollongong. Rising uplands northeast of the basin shed coarse gravels and sands which are preserved as conglomerates interbedded with coal in the Newcastle area.

PHOTO (C): BOB YOUNG

The coal measures dip southward from the Hunter Valley on the northern edge, yet northwards from the southern Illawarra edge and eastwards from Lithgow. It was therefore proposed that they should also lie at depth under Sydney, the centre of the basin. A bore at Helensburgh in 1884 reached economic coal, and in 1890–91 the first bore at Cremorne reached coal at 840 m, though the seam had been turned to cinders by an igneous intrusion. Finally a colliery was established at Birchgrove, where the seam was 864 m underground. Sinking of the shaft took five years, from 1897 to 1902. The mining extended under Sydney Harbour from Mort Bay towards Goat Island and Balls Head, and supplied coal to the old Balmain power station. However, long-term economic viability was prevented by the large proportion of coal that had to be left to support the roof, and by the difficult working conditions, including floor instability, high methane gas levels, and hot rocks that were over 30°C at the face. After several closures and problems, mining ceased in 1931.

The saucer-shaped Sydney Basin is some 350 km across and is filled with Permian and Triassic sediments. A bore in 1937 at Balmain confirmed that there were Sydney Basin sedimentary rocks extending to a depth of at least 1480 m. The total basin thickness was probably much greater at one time, as Jurassic sediments that were also deposited on top have been eroded. The lowest units in the Basin are Permian, except for the northern edge where some Carboniferous volcanics, glacial deposits and dirty coals are preserved.

The southern and western margins were relatively stable during deposition, and the gently dipping rocks overlie an unconformity eroded into folded and metamorphosed Palaeozoic rocks. This unconformity can be seen on the beach cliffs at Durras, in Ettrema Gorge, on the Kanangra Walls of the Blue Mountains and at Capertee near Lithgow.

The northeastern margin was tectonically and volcanically active during and after sedimentation. This tectonism is expressed in three ways: firstly, more subsidence accumulated a much greater sediment thickness in northeastern part of the Basin; secondly, the sediments contain tuffs and coarser conglomerates shed from the rising volcanic mountains north of the basin margin; and finally, continuing tectonism compressed the northeastern margin and thrust older Palaeozoic rocks southwest over the sediments.

The line of this thrust fault, called the Hunter–Mooki Thrust, runs from just north of Branxton to Quirindi, and in the subsurface as far north as Boggabilla. To the east of this line lie older rocks of the New England region, with undeformed sedimentary sequences to the west.

Permian sedimentation within the basin was mainly marine, and consists of thick units of nearshore sands (as at Nowra) and offshore silts, with some coal around the margin (as at Clyde). The Greta coals formed at this time to the north, but were soon drowned in marine sediments as the sea level rose. Consequently the seawater introduced sulphur, which formed the pyrite-rich layers at the top of the Greta coal seam. Later the Basin became filled with fluvial sediments, mainly shed from the northeastern edge, so that rivers and deltas extended as far south as Wollongong during the formation of the coals at Newcastle.

By the Triassic, coal formation had ceased and there was renewed volcanism in the Gerringong area. Rivers carried sands across most of the basin, except where there are some fossiliferous marine units under Sydney and near Terrigal. The huge sandstone cliffs of Sydney Heads were deposited by these rivers. Finally, muddy shales covered the basin as rivers became more sluggish and dropped their mud loads across broad floodplains. These shales sit on top of the Razorback and were mined in the brickpits of Ashfield and elsewhere.

Volcanism continued into the Jurassic, as evidenced by the numerous basaltic diatremes and dykes. Subsidence also must have continued, because the peats deposited among the sediments had to be buried for them to be heated and transformed into coal. Otherwise the Sydney Basin would have brown coals, as is the case in the Latrobe Valley of Victoria.

So it is estimated the Sydney Basin coals that are now mined at the surface must have been deeply buried, and then uplifted to their present position. Studies show the black coal formation was the result of high heat fluxes from the Earth's interior, leading to higher temperatures in the upper crust during the Cretaceous break-up of the continent and the opening of the Tasman Sea. Tectonic uplift from the end of the Cretaceous 65 Ma ago caused an estimated 2000 m of rock to be eroded off the top of the Basin, exposing black coals at the ground surface.

Development of inland seas

Well after 160 Ma ago, much of the present Australian area was being eroded. The eastern part of the continent was limited by an arc of volcanoes shedding sediment into vast river plains, which turned northwards towards an unknown sea somewhere past the present site of Papua New Guinea.

Underneath all this relatively quiet landscape, changes were developing. Though not apparent on the surface – apart from a few isolated volcanic eruptions, near Bunbury in southern Western Australia and on the Exmouth Plateau – there were major intrusions of basaltic magma in rock sequences all around southern and western Australia.

In northwestern Australia unusual igneous magmas were injected as pipes into the crust, and these pipes carry the diamonds now mined at Argyle. From about 154 Ma ago came the first signs of rifting and the creation of new sea-floor off the Kimberley region during the separation of this part of Gondwana. It was the precursor to a greater breaking of the crust that followed over the next 100 million years.

Marine sediments had filled a narrow trough off northwestern Australia. This trough would have been a narrow tongue of the ocean like the modern Red Sea. By 120 Ma ago the rift had propagated southwards and a seaway extended as far south as Perth. The Bunbury basalt has been dated at 135–128 Ma and overlies an erosional surface, thus marking the break-up unconformity and volcanism in southwestern Australia. It has been traced in the subsurface southwards, and also offshore to the northeast. Seismic evidence under the continental shelf between Bunbury and Perth shows that the basalt flowed down an old valley incising the continental margin.

This first break, in which India separated, left most of the Gondwanan land mass intact, so that Antarctica, Australia, New Zealand, New Guinea, Africa and South America still formed an immense continent. Gondwana itself did not finally disintegrate for another 70–80 million years.

The early European explorers of inland Australia, such as Charles Sturt, sought 'the great inland sea, towards which all the big rivers were flowing'. From their knowledge of wet European countries they had good reason to expect vast fertile lands lay beyond the horizons. But they were just 120 million years too late.

The sea levels rose, and 140 Ma ago the sea began to spread inland from the developing Indian Ocean to the west, and from the north (see Figure 6.11). The ocean flooded over the Gondwanan continent, until nearly half of what is now Australia was covered. By 117 Ma ago, there were elevated areas of dry land in southern Western Australia, the old Yilgarn craton, the Kimberley region, and along a volcanic arc to the east. Most of the rest, except Tasmania, was under water.

The weight of this water helped the crust to subside gently, allowing further invasion by the sea. The shallow seas covered the ancient landscapes in sand and mud, emphasising the vast flatness of the continental interior. The thickest accumulations cover most of western Queensland, forming the Great Artesian Basin with its vital groundwater supplies.

At this time the Earth was still in a very warm climatic phase, though there is evidence of local glacial activity. In northern South Australia and northwestern New South Wales, at the southern edge of this great inland sea, the sedimentary rocks that formed between 144–132 Ma ago contain tillite and finer sediments with dropstones, providing evidence of glaciation. At the time this region lay at about 66°S latitude, close to the present Antarctic Circle. This cold phase may represent a short-lived global climatic fluctuation, or may be the result of a local ice cap, the other evidence for which has long since disappeared.

The sea withdrew rapidly in the late Cretaceous and had cleared the continent by 99 Ma ago. At this time a new, narrow seaway opened from the west along the present southern margin, and at times extended under the present Nullarbor region. This was the next sign of the future break-up of Gondwana and the separation of Australia from Antarctica.

Volcanic debris pours into the inland seas

A new chain of volcanoes, called the East Australian Arc, developed to the east of Australia around 120–105 Ma ago. Massive volcanic eruptions, primarily of andesite, extended from northern Queensland to Tasmania. There is some evidence of similar igneous activity persisting till around 95 Ma ago. Rhyolites have been cored on the Lord Howe Rise, which could have been on the other side of that arc, and igneous rocks at Mt Dromedary in southern New South Wales are of this age.

The best remaining evidence for these volcanics to the north is in the

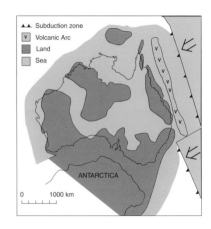

Figure 6.11 *The maximum extent of Cretaceous seas 117 Ma ago. The seas had started flooding over the continent 141 Ma ago, reached a maximum 117 Ma ago, and had withdrawn by 99 Ma ago. The western side of Australia had formed, but Australia was still part of Gondwana, although there were signs of rifting south of Victoria and South Australia. East of Australia lay a subduction zone and volcanic arc.*

Whitsunday Group islands, and to the south in the rocks along the southern coast of Victoria. Inland the sediments filled a basin that covered western and central Queensland and probably extended into South Australia and New South Wales. It has been estimated that more than 100 000 cubic kilometres of materials were erupted. Where are the volcanic mountains now? Much has been eroded, and the main part has been lost – rifted away when the Tasman and Coral Seas opened and now sunk beneath the waves of the Pacific Ocean further east. The uplift of eastern and southern Australia, which began around 95 Ma ago, accentuated the erosion. The higher mountains enabled a wide range of older rocks, as well as the volcanics, to be rapidly eroded. More than 400 000 cubic kilometres of the sediment fills basins south of Victoria.

Rainforests on land, ammonites in the sea

Around 110 Ma ago the climates were temperate, with temperatures ranging from frosts to 12°C in southern Australia and across present Antarctica, and high annual rainfalls (750–1150 mm). Conditions were probably warmer to the north, and became warmer with time across the continent. The landscape supported a range of coniferous forests containing araucarian pines, maidenhair trees and podocarps. The understorey included mosses, ferns, cycads, and plants like the zamias.

Modern araucarians include the kauri, hoop and Bunya pines, and podocarps include the celery-top and Huon pines. The podocarps are pine-like trees that do not have cones. All these trees are typical today of rainforests from Tasmania to New Guinea, and the evidence is these forests covered very large areas of the future Australian continent around 100 Ma ago. While we commonly think of these as warm and tropical during the day, anyone who has climbed the higher peaks can attest to the chilling cold that envelops them at night and on cloudy days.

In Australia there was a change in the flora about 110 Ma ago, when the first signs of angiosperms appear in coal measures near Maryborough in Queensland. Vegetation across the continent indicates a drier climate developing inland, perhaps due to the withdrawal of the seas.

Modern angiosperms include all the flowering plants, such as the colourful trees, the garden flowers and wildflowers. A very primitive plant from China, aptly named *Archaefructus*, dates back earlier than 140 Ma and is the oldest known angiosperm. Rocks from Antarctica show a rapid rise in angiosperm species after 120 Ma ago. But in Australia the 110 Ma old Maryborough coals are the first evidence of such plants.

Deposits in Queensland have yielded the *Muttaburrasaurus* dinosaur. While our imagination is commonly led by artist's drawing of dinosaurs grazing on open grasslands, many were probably shouldering their way through rainforests.

Dinosaur fossils are common, and major deposits occur in central Queensland and on the Otway coast of Victoria. Trackways left by Cretaceous animals have been found near Broome and Winton. At the Lark Quarry park near Winton in Queensland is a set of footprints preserved in rock formed from the soft mud deposited around a freshwater lake. The tracks show that a large carnivorous dinosaur had attacked a herd of smaller herbivorous animals. While pursuing one of them, the rest had scattered in panic around the sides of the attacker. After the mud dried, sediment covered the tracks, more deposition occurred, and the sediment was buried and turned to rock. Uplift and erosion has exposed these tracks again.

In southeastern Australia some higher mountains were forming, with more basaltic volcanism until 99 Ma ago. Extensive river plains and lakes covered the areas that are now Gippsland and Bass Strait, extending around the Otway coast. Lake deposits preserve lungfish, and also leaves, some with evidence of leaf miners, indicating that there must have been butterflies at this time. On damp ground were impressed the tracks of numerous dinosaurs: small bipedal dinosaurs, four-footed forms, and the predatory *Allosaurus*.

An extraordinary lake deposit at Koonwarra in Gippsland, which formed 118–115 Ma ago, preserves tantalising remains: the feather of a modern bird. The remains are unmistakable – delicate impressions of a true feather over 100 million years old. Yet we have no other evidence of birds until around 30 Ma ago. Over 70 million years passed, yet no birds were preserved.

This lake deposit includes a wide range of animals and plants that lived in the lake, or were washed in by streams, or fell into it while flying over the water. There are fish and lungfish, shells, plant leaves and a wide range of larval and adult insects, including a mayfly nymph. Amphibians and dinosaurs are also known from the same area. Unfortunately we do not have similar deposits in other parts of Australia, which would have told us whether these animals were restricted to the southeast or were more widespread.

There are estuarine deposits at Lightning Ridge and near Julia Creek in Queensland, slightly younger at around 110 million years, which include many dinosaur remains, turtles, crocodiles, and mammals, including a platypus.

What lived in the seas? Certainly missing from these seas were corals and many of the shells and calcareous algae that would have indicated warm, tropical waters. The main fossils are fish, amphibians, reptiles (including crocodilians and turtles) and ammonites.

Large marine reptiles, the icthyosaurs and plesiosaurs such as *Kronosaurus*, must have been fearsome animals in these seas. *Kronosaurus queenslandicus* was up to 17 metres long and had a skull 4 metres long with huge pointed teeth, and paddle-like arms and legs rather than feet and hands. There is no evidence that the marine mammals, such as seals and whales, existed at this time.

Figure 6.12 *Dinosaur tracks about 93 million years old, from Lark Quarry park, near Winton in Queensland. The tracks show a larger theropod dinosaur and a herd of smaller dinosaurs that scattered. The large footprints are each around 0.5 m long.*
PHOTO: GARY CRANITCH QUEENSLAND MUSEUM

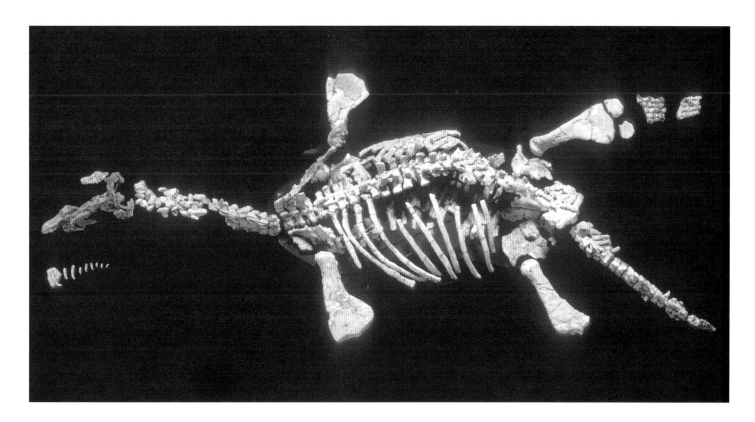

Figure 6.13 *'Eric', an opalised fossil pliosaur recovered from Cretaceous sedimentary rocks about 110 Ma old near Coober Pedy. Pliosaurs were marine reptiles and probably had a life similar to that of today's seals – preferring shallow waters and estuaries, and feeding mainly on fish. The bones of a small fish were found in Eric's fossilised stomach. The skeleton is about 1.5 m long.*
PHOTO: CARLO BENTO, NATURE FOCUS, AUSTRALIAN MUSEUM

Figure 6.14 *Cretaceous marine sedimentary rocks on the southern side of Bathurst Island, Northern Territory.*
PHOTO: BOB HENDERSON

Ammonites are spiral molluscs, although unlike snails they have internal chambers, and are most closely related to the modern pearly *Nautilus*. The octopus-like animal lived in the final, largest chamber, with tentacles trailing outside to catch food (presumably fish). Some ammonites were small, 10 to 20 mm across, but others were as large as 1 m. Although these large fossils are heavy, in real life they would have been hollow with gas-filled chambers, and capable of floating or swimming, adjusting their buoyancy as does the modern nautilus. When ammonites died their coiled shells fell to the seabed, where they could be covered with mud and preserved.

On the seabed lived a range of shells, algae and presumably worms and crustaceans. Among the shells was a large shell, like an enormous mussel some 0.50–1.0 m wide, called *Inoceramus*. This may have lain on its side, since we have no evidence that it was upright like a clam.

Under the microscope we can also find small shells of animals belonging to the order Foraminifera. These forams make up much of the finer sand on present-day coral reefs, but are also present in open-ocean seabed sediments everywhere, even in the Antarctic. It is useful for geologists that the types vary with water depth, water temperature, and also whether the water is fully oxygenated or is stagnant. Detailed studies show that these inland seas were cool and commonly just below normal ocean salinity, and that at some times the seas became more landlocked and stagnant, accumulating organic matter from algal blooms and detritus washed in by rivers. The resulting deposits are the thin and extensive oil shale deposits near Julia Creek in northern Queensland.

In a relatively short time the seas withdrew, and by about 99 Ma ago the mainland was exposed. The sea occupied an embayment that is now the Gulf of Carpentaria, and there may have been some narrow opening between Australia and Antarctica.

More recent erosion has stripped off much of the sedimentary cover, and along with it most of the evidence of these ancient times, especially over the far northern and western regions. However this erosion does reveal the rocks that underlay the inland seas.

Figure 6.15 *An ammonite (a) and a modern* Nautilus *(b), both cephalopods. Specimens 120 mm across.*

PHOTOS: DAVID JOHNSON

The tectonics of eastern Australia

On our normal historical scale little seems to change in the landscape. But on a geological time scale the changes are continuous and startling. Uplift, volcanism and subsidence continue unabated. The Earth is in constant motion, and the movements of the crust are evidence of its internal churning. The Sydney and Bowen Basins experienced extension and subsidence, then strong compression and volcanism, then another round of extension. The eastern side of Australia continued to be tectonically active from the Permian to the Tertiary.

Beginning around 300 Ma ago, the crust under present eastern Australia experienced extension. Subsidence formed a long depression from southern New South Wales to northern Queensland – the Sydney and Bowen Basins. Subsidence was greatest along the eastern, tectonically active side, where there was a line of volcanoes. A greater thickness of sediment accumulated there because the subsidence kept providing more room to accommodate the debris eroded from the surrounding higher landmass. The western boundary – from near Lithgow, and from Comet to Collinsville in Queensland – was more stable, and only thin sequences were deposited. The initial sedimentation was terrestrial, but as subsidence gathered pace the sea flooded in and marine sediments accumulated. Then subsidence slowed and the terrestrial sediments prograded, filling the basins and depositing the main Permian coal-bearing sequences and then the fluvial sediments of the warmer plains in the Triassic and Jurassic.

By 233 Ma ago, in the mid-Triassic, the extension of the crust had ceased. The eastern margin had begun to compress. Gradually eastern parts were being thrust westwards over the Basins, especially in the south. This extra weight loaded the eastern side of the Basins, and pushed terrestrial sedimentation westward. By 120 Ma ago, in the mid-Cretaceous, a new, major compressional event was under way. The volcanism that extended from Queensland to Tasmania was an expression of this crustal movement.

In the late Cretaceous the system changed again, and extension started in the south as the Tasman Sea began to open. The Sydney Basin began to lift from around 100 Ma ago as the crust domed, preparatory to rifting. By 84 Ma ago a new sea-floor of basaltic ocean crust had started to form east of Tasmania and southern New South Wales. This break in the continental margin started to enlarge. The opening of the Tasman Sea was under way.

Summary

Following a great extinction of life 251 Ma ago the Earth emerged from the icehouse and became much warmer. For the period 251–140 Ma, the present area of the Australian continent was covered by vast, arid riverine plains. Later, humid conditions allowed peatlands to form along the eastern margin, where there was lush vegetation surrounding wet areas.

Dinosaurs, reptiles and perhaps even primitive mammals roamed the land, and there were large amphibians, reptiles and fish in the freshwater lakes. Around 154 Ma ago the continent started to break up as India split away, forming the West Australian coastline.

About 140 Ma ago the seas began to rise, inundating most of the continent. The climate was still warm and humid, with abundant rainforest vegetation in which the dinosaurs lived. At sea there were giant marine reptiles, ammonites and fish, and crocodiles in the estuarine swamps. Modern birds with feathers were present. The extent of the sea was greatest around 117 Ma, and by 99 million years ago it had withdrawn from the continent.

To the east between 120 and 105 Ma ago, the East Australian Arc stretched from North Queensland to Tasmania. Perhaps in excess of 100 000 cubic kilometres of volcanic debris was erupted and washed westwards into the inland seas. The tectonic setting was extensional during the formation of the main basins, then volcanic and compressional during the Late Cretaceous, and finally extensional during continental uplift and opening of the Tasman and Coral Seas.

SOURCES AND REFERENCES

Alley, N.F. & Frakes, L.A., 2003. 'First known Cretaceous glaciation: Livingstone Tillite member of the Cadna-owie Formation, South Australia'. *Australian Journal of Earth Sciences*, 50: 139–144.

Archer, M. & Clayton, G. (eds) 1984. *Vertebrate Zoogeography & Evolution in Australasia*. Hesperian Press. 1203 pp.

Becker, L. *et al.*, 2001. 'Impact event at the Permian-Triassic boundary: evidence from extraterrestrial noble gases in Fullerenes'. *Science*, 291: 1530–1533.

Bryan, S.E., Constantine, A.E., Stephens, C.J., Ewart, A., Schon, R.W. & Parianos, J., 1997. 'Early Cretaceous volcano-sedimentary successions along the eastern Australian continental margin: Implications for the break-up of eastern Gondwana'. *Earth and Planetary Science Letters*, 153: 85–102.

Dettmann, M.E. *et al.*,1992. 'Australian Cretaceous terrestrial faunas and floras: biostratigraphic and biogeographic implications'. *Cretaceous Research*, 13: 207–262.

Drinnan, A.N. & Chambers, T.C., 1986. 'Flora of the Lower Cretaceous Koonwarra Fossil Bed (Korumburra Group), South Gippsland, Victoria'. *Memoir of the Association of Australasian Palaeontologists*, 3: 1–77.

Elliott, L.G., 1993. 'Post-Carboniferous tectonic evolution of eastern Australia'. *APPEA Journal*, 215–237.

Erwin, D.H., 1994. 'The Permo-Triassic extinction'. *Nature*, 367: 231–236.

Exon, N.F. & Colwell, J.B., 1994. 'Geological history of the outer North West Shelf of Australia: a synthesis'. AGSO *Journal of Australian Geology and Geophysics*, 15: 177– 190.

Faiz, M.M. & Hutton, A.C., 1993. 'Two kilometres of post-Permian sediment – did it exist?' *Twenty-seventh Newcastle Symposium*, 2–4 April 1993, 221–7.

Fielding, C.R., 1993. 'The Middle Jurassic Walloon Coal measures in the type area, the Rosewood-Walloon coalfield, SE Queensland'. *Australian Coal Geology, Journal of the Coal Geology Group of the Geological Society of Australia Inc.*, 9: 4–15.

Frakes, L.A. & others, 1987. 'Australian Cretaceous shorelines, stage by stage'. *Palaeogeography, Palaeoclimatology, Palaeoecology*, 59: 31–48.

Gould, R.E., 1976. 'The succession of Australian pre-Tertiary megafossil floras'. *Botanical Review*, 41: 453–483.

Henderson, R.A. *et al.*, 2000. 'Biogeographical observations on the Cretaceous biota of Australia'. *Memoir of the Association of Australasian Palaeontologists* 23: 355–404.

Herbert, C. & Helby, R., 1980. *A Guide to the Geology of the Sydney Basin*. Geological Survey of New South Wales Bulletin No. 26. 603 pp.

Hergt, J.M. & Braun, C.M., 2001. 'On the origin of the Tasmanian dolerites'. *Australian Journal of Earth Sciences*, 48: 543–549.

Long, J.A., 1991. *Dinosaurs of Australia*. Reed 87 pp.

Quirk, S. & Archer, M., 1983. *Prehistoric Animals of Australia*. The Australian Museum. 80 pp.

Retallack, G.J., 1977. 'Reconstructing Triassic vegetation of eastern Australia: a new approach for the biostratigraphy of Gondwanaland'. *Alcheringa*, 1: 247–278.

Twidale, C.R., 1994.' Gondwanan (Late Jurassic and Cretaceous) palaeosurfaces of the Australian craton'. *Palaeogeography, Palaeoclimatology, Palaeoecology*, 112: 157–168.

Veevers, J.J. (ed.), 1984. *Phanerozoic Earth History of Australia*. Clarendon Press. 418 pp.

Ward, C.R., Harrington, H.J., Mallett, C.W. & Beeston, J.W. (eds), 1995. *Geology of Australian Coal Basins*. Geological Society of Australia Inc. 590 pp.

Websites

Fossils of the Sydney Basin
www.austmus.gov.au/is/sand/fossils.htm

The Permo-Triassic extinction event
www.hannover.park.org/Canada/Museum/extinction/permass.html

Cretaceous sea and animals
www.humboldt.edu/~natmus/Exhibits/Life_time/Cretaceous.web

Illustrations

Figs 6.2, 6.11: redrawn and reproduced by permission of John Veevers after figs 10, 33 in Veevers, J.J. (ed.) 2000. *Billion Year Earth History of Australia*. GEMOC Press, Sydney; fig. 6.2 with palaeolatitude data for 240Ma courtesy of D.A. Clark and P.W. Schmidt of the CSIRO Division of Exploration and Mining, North Ryde, NSW.

Fig. 6.3: combining with permission of John Wiley & Sons Limited information from figs 8.10, 8.11, 9.12, 10.9, 12.2 and 13.5 (originally compiled by A.L. Hodder from Benton, 1993, *The Fossil Record* 2, Chapman & Hall) from Doyle, P., 1996. *Understanding Fossils* © 1996 John Wiley & Sons.

6.10b, d, e: redrawn by permission of the New South Wales Department of Mineral Resources, fig 6.10b is a schematic geological cross section based on Pogson, D.J., 1972. *Geological Map of NSW*, scale 1:1,000,000. Geological Survey of NSW, Sydney; figs 6.10d, e after figs 2.12 and 2.17 respectively in C. Herbert, & R. Helby (eds) 1980. *A Guide to the Geology of the Sydney Basin*. Department of Mineral Resources, Geological Survey of New South Wales Bulletin No. 26.

CHAPTER 7   *Australia emerges*

BIRTH OF MODERN AUSTRALIA: FLOWERING PLANTS, MAMMALS AND DESERTS

So when did Australia as we know it start to form: the unique shape, the animals and birds, the eucalypt forests and deserts? Why are there fossil crocodiles and turtles around Lake Eyre? When did modern birds appear? When did Antarctica become covered in the present ice sheets that cool the southern oceans?

No longer part of the Gondwanan supercontinent, the margin of Australia became a shallow shelf, formed by sediment shed from the inland mountains and by volcanoes along the rifts where Gondwana had broken into its modern fragments.

This chapter traces the emergence of the Australian landscape, plants and animals as we know them today. There are two parts to this story, spanning together some 100 million years. The first took 85 million years, during which Australia separated from Antarctica, and the landscapes of both continents were dominated by rainforests, and in Australia later by eucalypt forests. The second started around 15 Ma ago, when the Antarctic ice sheet formed and Australia began to dry out, creating the vast arid interior with its deserts and salt lakes.

All through the geological history up till this point there was no Australian land mass as we know it today. The piece of rock we know as Australia was simply part of much larger continents, first Rodinia and then Gondwana. These supercontinents had experienced submergence beneath vast seas, explosive tremors of long chains of volcanoes, and the uplift and growth of high mountain belts. At times there were ice and snow, at other times tropical seas and corals. The same conditions occurred across those supercontinents, irrespective of whether the individual part would later be India, Australia or Antarctica.

Then the mantle within the Earth moved, so that large upwellings developed under Gondwana. The reasons for such a change are not completely understood. One very plausible theory is that the extensive crust of the supercontinent acted as a large blanket, trapping heat moving upwards from the Earth's centre, and that eventually this heat build-up led to uplift and fracturing of the crust.

Smaller continents can move more easily and heat can be diverted to crustal processes at their edges. This is not the case for the immensely larger supercontinents. In a way, the amalgamation of crustal plates into a supercontinent may create the very situation – the trapping of the underlying heat – that eventually leads to its break-up.

The northwestern coast of Australia had already formed 154 Ma ago, and by 120 Ma ago the entire western coast was an ocean border. The eastern coast also began to fracture, starting in the south and propagating northwards, and by 84 Ma ago the Tasman Sea was opening east of Tasmania. By 56 Ma ago the Coral Sea had split open and the Queensland Plateau had separated from the northeastern coast of Queensland. This eastern break-up continued until 48 Ma ago. So the Tasman Sea formed over some 36 million years, and then spreading stopped. New Zealand has been in its present position relative to Australia since then. Why did the Tasman open for just this period and then stop? Why did New Zealand not keep drifting further off into the Pacific? We do not know.

The southern boundary break-up dates from 99 Ma ago, and the final separation from Antarctica was under way by 84 Ma ago. Recent work has suggested that it may have been even earlier. Tasmania started to separate, but did not break away, so that Bass Strait is underlain by a thick sedimentary sequence overlying stretched crust. The Australian land mass began to break away completely from Antarctica, forming the curve around southern Australia that fits so neatly

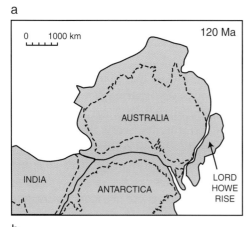
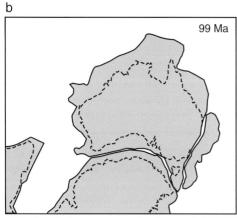
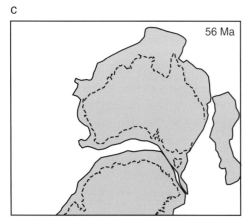
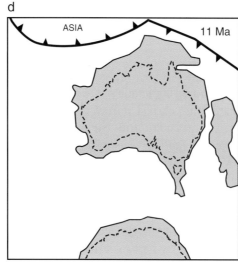
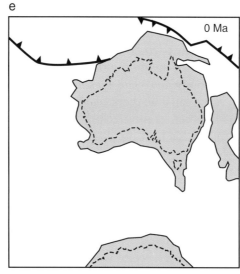

Figure 7.1 *The separation of Australia from Gondwana.*

a. The process started in the northwest with the break-up starting 154 Ma ago, and then propagating southwards until the sea had opened as far south as the present site of Perth by 120 Ma. The land originally off the Kimberley region has been lost, and the area formerly off the west coast (between India and the dotted line in b) is now compressed in the Himalaya.

b. By 99 Ma India had broken away from the edge of Antarctica and started its rapid movement north, eventually colliding with Asia to form the Himalaya. The first signs of separation along the southern Australian margin were apparent.

c. Along the eastern coast of Australia, the Tasman Sea started to break open about 84 Ma ago and by 56 Ma the break had propagated northwards so that both the Coral and Tasman Seas were opening. The Lord Howe Rise, once part of the Australian section of Gondwana, was now well to the east in the Pacific Ocean. The southern margin had opened slightly. By 48 Ma the spreading along eastern Australia stopped and so the extents of the Tasman and Coral Seas have remained the same for the last 48 million years.

d. After about 40 Ma ago the spreading rate south of Australia increased, and by 11 Ma the separation between Australia and Antarctica was over 2000 km. The cold circumpolar ocean current was established leading to increasing glaciation of the Antarctic and drying out of Australia. This increased northward drift meant that Australia started to impinge on Asian regions, and initiated the subduction zones south of Indonesia and north of Papua New Guinea.

e. Today Australia is moving more or less northeastwards as part of a single larger plate. Active subduction is evidenced by the common earthquakes and active volcanism in Indonesia and Papua New Guinea. The distance to Antarctica has opened up to almost 3000 km.

against the circular shape of the Antarctic land mass. This southern plate boundary lies midway between Australia and Antarctica, and is still actively spreading. Recent measurements of the present movement using GPS show that the plate is travelling about 35° east of north at about 67 mm/year.

Originally the movement of Australia away from Antarctica was more easterly, and then 43 Ma ago the direction changed towards the north. India had split from Antarctica earlier than Australia and drifted quickly north. Then at 43 Ma it became locked once more against the Australian part of the plate, so both continental masses began to move together, as they are today. A change in the spreading direction of the Pacific plate had occurred, and the direction changed from easterly to more northerly.

If we could have time lapse movies on a million-year time scale we would see the ancestral continent vibrate with earthquakes, then fractures appear where the land mass was buckling upwards. We would see the west open at the northern end, then India fleeing northwards, right off the map. Antarctica would be stationary. Then land to the east would break away, leaving the easterly bulge between New South Wales and Queensland, and we would see the breakaway part drift eastwards, sinking beneath the waves to become the Lord Howe Rise, an area of shallower seabed on which that island is perched. Then Australia would move northwards from the curved break between the Great Australian Bight and Antarctica.

After rifting the mainland gradually subsided as the crust cooled, so that the original line of break is now under the ocean near the edge of the continental shelf. Sediments eroded from the land continued to be deposited along the coasts, and when sea levels were lower would have moved onto the shelf.

The previous Cretaceous andesitic volcanic arc off eastern Australia had stopped erupting by 105 Ma and was split away, to disappear beneath the sea. However, extensive basaltic volcanism started about 70 Ma ago along eastern Australia, persisting until very recently (see Chapter 8).

What is left? A chunk of land with the characteristic shape we know today, which continues to move northwards towards Asia. A perspective view of the continent showing the steep continental margins dating from the break-up, and the contorted crust to the north, is shown in Figure 1.7 (page 12).

This was our separation from the parent! In geological terms, Australia was born.

Great age of much of the Australian landscape

The western part of the continental interior persisted as a vast flat landscape with erosion draining to the western coast, and also south and east towards the southern margin. The valleys cut into the bedrock have gradually filled up with old estuarine and shallow marine sediments as the margin subsided, and are now chains of salt lakes. The valley fill sediments have been dated as older than 35 Ma, so the erosion of these buried valleys must have occurred before then. Similar valleys drained southwards towards the present Nullarbor Plain.

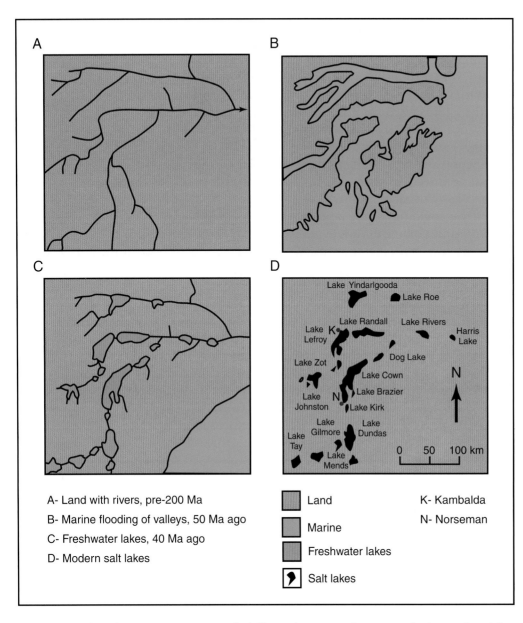

Figure 7.2 *Old drainage in the Kalgoorlie–Norseman area of Western Australia. The valleys were temporarily flooded by the sea around 50 Ma ago, then became a series of freshwater lakes (40 Ma ago) and are now just lines of salt lakes connected by shallow depressions.*

A similar drainage is preserved differently in northern South Australia. The erosion of Cretaceous sediments 30–40 million years ago left a drainage channel filled with sand and gravel that was later cemented along the edges by silcrete. The silcrete formed in this region some time in the period 15–30 Ma ago. The surrounding sediments were much softer than the silcrete and were eroded away, leaving the silcrete as a series of hard caps on mesas, following the course of the old channel. Such a situation is termed an inverted topography, because the old channel is now perched on the hilltops. Known as the Mirackina Palaeochannel, the course runs southeast for 200 km, passing just west of Arckaringa and northeast of Coober Pedy.

Figure 7.3 *a. Location of the Mirackina Palaeochannel in the Arckaringa region of northern South Australia. The palaeochannel is not well dated, although it is around 20 million years old, with later silicification. The palaeochannel now exists as a line of mesas 30–40 m above the surrounding plains. All the surrounding alluvial sediments have been eroded, leaving only the cemented channel gravels and sands.*

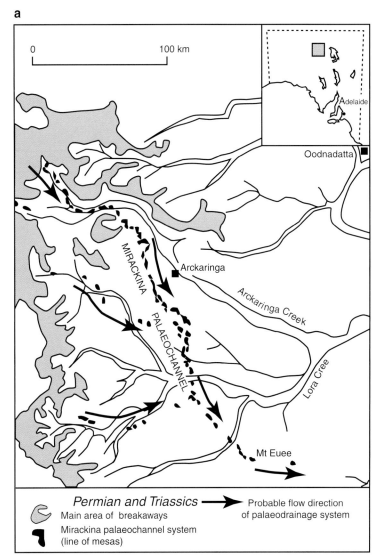

b. A mesa capped by the Mirackina Palaeochannel silcrete.
PHOTO: GREG MCNALLY

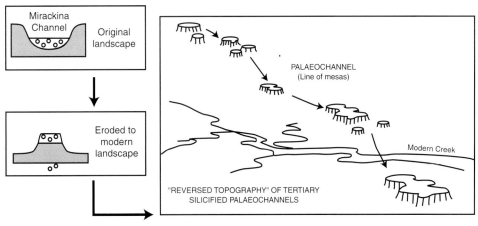

BIRTH OF MODERN AUSTRALIA   143

Much of the continent shows evidence of long erosion and very deep weathering. In Western Australia weathering of the rocks extends down for 100 m, which implies a great time for such deep chemical alteration of the host rocks. Many of the land surfaces visible across central and northern Australia are very old. They formed by erosion some 150–140 million years ago. and while they have certainly been modified by the subsequent erosion, the general form is still obvious.

In the Arnhem and Kakadu regions, Cretaceous sedimentary rocks overlie a planation surface with a few deeper old valleys, composed of much older Precambrian rocks. So these older rocks must have been uplifted and eroded down to this essentially flat surface, over 100 million years ago, before deposition of the Cretaceous units. On the Kakadu lowlands these units lie around 50 m above sea level (ASL), whereas those on the top of the Arnhem Land plateau are 100–400 m ASL. Subsequent erosion has removed much of the Cretaceous material, re-exposing the old planation surface. Thus the present land surface is really at least 100 million years old.

In the Mount Isa region, hills composed of folded Precambrian rocks have a bevelled surface cutting across them, which is clearly visible from the air. This surface can be traced laterally and is overlain by Cretaceous sediments. Clearly the surface must have formed before those sediments were deposited upon it. Similar planation surfaces can be seen in the Mt Margaret Plateau and the Peake and Davenport Ranges of northern South Australia. In these places you can sit or walk on a land surface which was part of Gondwana greater than 150 million years old – you are where the dinosaurs roamed. The brown, weathered rocks were made from the soil formed by the Jurassic rain and sunlight.

In places this old land surface has a laterite capping that extends under the Cretaceous sediments. This tells us that the weathering and laterite development happened before the Cretaceous, and it is generally accepted that the oldest laterites date back to the Jurassic some 125 Ma ago. However, the fact that much younger sediments and basalts are also lateritised means that this process has also occurred much more recently.

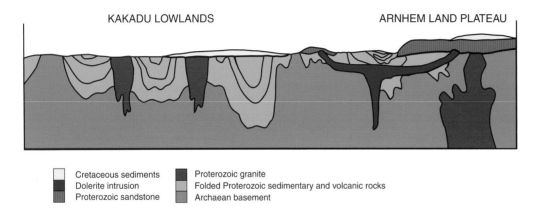

Figure 7.4 *Cross-section across northern Australia showing Cretaceous sedimentary rocks uncomformably overlying Proterozoic rocks, which in turn overlie an almost flat unconformity eroded into older Proterozoic and Archaean rocks. At least 500 million years of Earth history are missing at the upper unconformity. The Cretaceous sedimentary rocks originally covered the landscape and the two patches shown are remnants after erosion. This land surface, re-exposed after removal of the Cretaceous rocks, formed over 100 million years ago.*
THIS MATERIAL IS USED WITH PERMISSION OF UNIVERSITY OF CHICAGO PRESS

Laterites were once thought to form only under conditions of deep tropical weathering. However, Jurassic and Cretaceous climates, although wet, were not necessarily tropically hot. So laterites are now thought to form in both temperate and tropical climates.

Basalts flowed down the river channels in these old landscapes, over 30 Ma ago in some places – so the erosion of the valleys must have occurred before then. At Bungonia, basalts probably 30–40 Ma old have filled parts of the caves, so the caves must have formed before then. Recent work in southeastern Australia has also indicated that some of the land features there may be similarly old. For instance, the 'high plains' in the Mt Bogong and Dargo areas of Victoria may be remnants of these very old surfaces.

Laterites formed by weathering of basalts near Inverell, in the New England region of New South Wales, formed soon after eruption of the basalts around 20–25 Ma ago. Detailed analysis of samples from the ferricrete layers shows they preserve palaeomagnetic orientations with both normal and reversed polarity. This means that the laterites took thousands and perhaps millions of years to form.

Even where the land surfaces are younger, the soils may be quite old. In the Torrens Creek area of inland northern Queensland the yellow and brown earthy soils have been matched with basalts, indicating that these soils formed in three phases dating back to 2 Ma ago.

Australia under rainforest

Where did all the sediment go? Erosion over the past 100 million years since the withdrawal of the Cretaceous seas has removed plenty of rock. Most has been delivered to the continental shelf (see Chapter 9).

The eastern part of the continent was modified by the rise of the Great Divide down the eastern margin. Small rivers are directed eastwards, but much larger rivers run inland. Three major basins received this sediment: to the north, the depression between Australia and Papua New Guinea, which is now flooded by the sea forming the Gulf of Carpentaria; in central Australia, the basin that is now the salt lake of Lake Eyre, surrounded by windblown sands of the Simpson and Great Victoria Deserts; and further south, sediment was shed from the highest part of the country, the mountains of southeastern Australia, westwards to accumulate in the Murray Basin.

Drilling in the Murray Basin shows that it extended inland almost as far as Griffith and that it contains some 600 m of sediment dating back 50 Ma. Extensive shorelines around 6 Ma old are preserved at the inland edge, 65 m above the modern sea level.

However, for most of the period from 90 to 15 Ma ago the inland landscape was not as we see now. It was heavily forested and had abundant streams and lakes. Flowering plants increased rapidly after 120 Ma ago and were a major component of floras during the last 100 million years of Australia's development.

Initially, before the final break-up of Gondwana, the floras were much the same as the Jurassic gymnosperm forests – dominated by conifers and seed-ferns, with abundant bryophytes (mosses and their allies). Most of the species can be closely matched to existing plants, and so good estimates can be made of the climate. While individual species are now extinct, many were very close to surviving genera, such as the podocarps and the araucarians (kauri pines). The ginkgos were locally abundant in some parts of southeastern Australia, but the modern *Ginkgo biloba* is the sole remaining representative of what was a widespread group.

The earliest convincing angiosperm pollen is from units 132–127 Ma old in England and Israel, and pollen and leaf fossils from Australia, Argentina and Antarctica have been dated to 127–99 Ma ago. Rapid evolution and expansion into a wide range of environmental niches occurred at the same time as the final phase of the Gondwanan break-up and a series of significant sea-level changes after 90–80 Ma ago. It is likely that there was interchange between the isolated land masses in the period soon after break-up, when more equable global climates prevailed and there were still short distances between land masses. Thus some genera on the now widely dispersed continents are closely related.

Certainly many of the main angiosperm families are well represented in the rainforests of Tasmania and northeastern Australia, such as the Anacardiaceae, Cunoniaceae, Lauraceae, Proteaceae, Rutaceae and Sapindaceae. As continental drift continued, the plants evolved differently forming the diverse endemic floras of, for example, the Cape region in South Africa and southwestern Western Australia.

Some of the species may be the same as today. Recently, a newly described tree from rainforests in northern Queensland, *Eidothea zoexylocarya*, has been identified as the same plant that produced the large nuts known as *Xylocaryon lockii* described by Ferdinand von Mueller in 1883 from the Tertiary deposits of Victoria.

There is plenty of evidence that these were warmer times. Boulders left by glaciers at McMurdo Sound in Antarctica contain warm-water shells and shark teeth, yet they are dated at 55–40 Ma old. So the Antarctic region was surrounded by warm water until, and perhaps later than 40 Ma ago.

Studies of the pollen preserved in muddy sediments and brown coals show that southern parts of this young Australian landscape 80–65 Ma ago carried a diverse range of plants, forming forests with both tall trees and understoreys, scrublands, and wetland swamps. Many of the plants were very similar to those in the cool rainforests of Tasmania and New Zealand, or in the high-altitude tropical rainforests of northeastern Australia. In places where the rainfall was lower there were more open forests. Temperatures on the mainland were around 13–18°C. On the other side of a narrow sea, the Antarctic, the temperatures were 11–13°C, and large forests that included *Nothofagus*, the Antarctic Beech, were flourishing.

Between 65 and 55 Ma ago the vegetation in southeastern Australia was cool temperate rainforest. Mean annual temperatures were 14–20°C and rainfall 1200–2400 mm. Around 45 Ma ago the climate cooled further with large amounts of rainforest araucarian and podocarp pollens apparent in the sediments.

The Murray and other rivers such as the Darling, Lachlan and Murrumbidgee flowed into a vast basin that stretched from the Mount Lofty and Flinders Ranges in South Australia to the Victorian alps. The ancestral valleys were in place over

50 Ma ago. While this was generally dry land, there were periods when the sea level rose and this basin became a huge marine embayment. Gradually it filled with all the sediment being poured westwards from the eastern highlands, pushing the coastline out to its present position at The Coorong.

Further inland, around the southern shores of Lake Eyre, there was an extended period of erosion until around 45 to 50 Ma ago, when alluvial sediments began to accumulate, burying masses of leaves and pollen as well as logs. Pollen from the rainforest trees is present though not as abundantly as in the deposits in coastal areas in southeastern Australia. Instead there are also proteacean pollens, from plants such as banksias, grevilleas and woody pears (*Xylomelum*), and grasses or reeds. *Eucalyptus* pollen is common, and this is probably the first sign of the gum trees that we regard as a national hallmark. Some of these pollens and leaves are from species now extinct, but many can be identified very closely with plants found today.

The inland was humid, and probably warmer than the coastal regions. The general impression is of rainforest and swampy vegetation within the main valley systems and eucalypt vegetation covering the hills and ridges. The landscape may have been very similar to parts of the tropical inland of northern Australia today. That it existed around the Eyre basin tells us that this was not a salt lake 50 Ma ago, but a vast area of freshwater lakes and billabongs – probably like Kakadu is now. The animal fossils tell us it was populated by fish, freshwater dolphins, crocodiles, flamingos and other birds, turtles, cuscuses and possums, tree-kangaroos and ground kangaroos.

Between 35 and 30 Ma ago the shoreline was near the edge of the present continental shelf, except in southeastern South Australia and the break through present-day Bass Strait, which was almost complete.

Sea levels rose to a maximum around 17 Ma ago The shoreline was close to the present coast all around Australia, though the present Nullarbor Plain was a huge embayment and the sea extended inland across the Murray region as far inland as Mildura. The marine sediments deposited at this time now underlie the Nullarbor Plain, and line the lower Murray River, which is why the river cliffs near Murray Bridge contain shells and echinoid fossils.

Brown coals

Onshore in the south, thick peats accumulated (later forming the brown coals of Victoria), and there was some basaltic volcanism involving lava flows and tuffs. Humid times around 20 to 30 Ma ago allowed the accumulation of thick peats in the Latrobe Valley area, and smaller accumulations elsewhere along the southern margin of the continent. In Western Australia there is an extensive basin of such sediments and lignites, from just east of Mount Barker and Albany past Bremer Bay to inland of Hopetoun.

It is difficult to imagine this landscape today. We are so used to seeing pictures of red sands and sparse vegetation, and hearing of the lack of rain, across central

Figure 7.5 *Brown coal deposits at Morwell in Victoria.*
Photo: David Johnson

Figure 7.6 a. *Modern wet temperate rainforest dominated by* Nothofagus cunninghamii, *in northwestern Tasmania. Forests like this covered much of southern Australia, and perhaps central and northern Australia, during and after the break of Australia from Antarctica. Antarctica itself was also forested in this manner.*

b. *Early Oligocene fossils of a leaf (*Nothofagus lobata*) and cupule (*Nothofagus bulbosa*), probably of the same biological species (scale for cupule is 5 mm). These plants are very closely related to the modern-day* Nothofagus cunninghamii *and* Nothofagus gunnii *of southern rainforests.*
Photos: Robert Hill

Australia. However, the fossil evidence tells us that Australia was once covered in forests, with thick rainforest on many parts and great inland waterways teeming with life. This existed for over 100 million years, initially while Australia was still joined to Antarctica as part of Gondwana and for some 75 million years afterwards as Australia drifted north. Note also that this rainforest existed as we were separating, that is, while Australia was within the Antarctic Circle. At that time there was no ice and snow down there. Antarctica too was covered in rainforest.

Box 7.1 Pollen data from brown coal and other Tertiary deposits

The brown coals of Victoria are interbedded with river sediments and some basalts, which indicate nearby volcanoes. These coal seams are up to 130 m thick. The plant material fell down and became submerged in the waterlogged ground, so it did not decompose. More peat accumulated as the basin subsided, and shallow burial squeezed out some of the water and compressed the material into brown coal.

Studies of the larger plant remains as well as pollens show that many of the plants that formed the peat are still living today. There was evidently a transition from conifer swamps containing ancient representatives of modern trees such as kauri (Agathis), Araucaria, and Lagarostrobus, passing up to mixed angiosperm and banksia trees with she-oaks (Casuarina) on the drier ground. Many of the same plants that formed the coal grow today in the eastern highlands of Australia and in New Guinea, though no modern forest containing the same species has yet been found.

This moist climate was not confined to the southern margin, near the narrow sea separating Australia from Antarctica. The oil shale deposits of northern Queensland, at Rundle and Condor near Proserpine, formed at this time, indicating that the humid conditions were widespread across the continent. In lakes where the influx of fine muddy sediments and other plant detritus was accompanied by prolific algal growth, a rich mixture of organic mud accumulated– the oilshales. These oilshales can be heated to produce oil. Some deposits are now buried offshore near Cape Hillsborough, and others are inland at Duaringa near Gladstone and at Nagoorin.

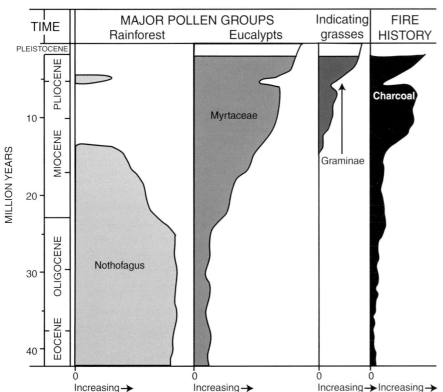

Figure 7.7 *Pollen profiles from Tertiary deposits in southern Australia. Many of the genera can be matched with those in modern cool temperate rainforests of Tasmania shown in figure 7.6.*

Earliest Australian fauna

Australia has a very special set of animals and birds, and we would like to know where they came from. The problem is that the sedimentary deposits do not form continuously year by year, preserving all that happens. There are long breaks, and in many cases whole sequences have been eroded and their record of life destroyed forever.

One of the oldest animals we know is the platypus. Teeth had been found in sediments 15 and 25 Ma old, in South Australia and also at Riversleigh in Queensland. Then in 1984 a spectacular opalised platypus jaw was found in sediments 110 Ma old at Lightning Ridge. This pushed the age of the modern Australian fauna back much further than had been ever suspected. There was a later discovery of platypus teeth 61–63 Ma old in Patagonia. This country was also linked to Antarctica as part of Gondwana, close to Australia. There are many similarities between the fossils from this time in South America and Australia, and when we can explore Antarctica more thoroughly we can expect more to be found. These findings confirm the evidence of the vegetation studies, those rainforests with platypus in the streams, stretched from Australia across Antarctica and into South America.

Fossil birds have been found in sediment deposited at least 20 Ma ago, but the record is frustratingly incomplete. A feather from the Koonwarra deposits in Victoria is around 115 Ma old, and there is a solitary penguin fossil aged 55 Ma, then more plentiful remains of a diverse range of birds after 25 Ma ago. The feather cannot have been from a sole bird species that existed then, and there must have been birds to link into the succession 50 Ma later. None of these have been preserved in any of the deposits so far found.

Elsewhere in the world, the earliest bird-like fossils date from 75 and 90 Ma ago. There has been much debate on the origins of birds, with recent evidence indicating a link to a small group of mobile bird-like dinosaurs that survived the great extinction of the dinosaurs, a theory first proposed by Thomas Huxley in 1868. This of course implies that dinosaurs were warm-blooded, not cold-blooded like reptiles and amphibians.

What did the birds eat? Presumably a range of insects, pollen, nectar and small vertebrates, just as modern birds do. Yet few of these are preserved. But there is some evidence.

The oldest frog fossil from Australia comes from Murgon in southeastern Queensland and is 54.5 Ma old. Yet this is much younger than the earliest frog fossil in the world, which comes from Argentina and is 160 Ma old. The oldest Australian python skeleton is 20–30 Ma old, from Riversleigh, so presumably both small and large reptiles were part of the ecosystems.

All we know is that as the Australian continent finally separated from Antarctica around 84 Ma ago, and following the worldwide disappearance of the dinosaurs 65 Ma ago a new and distinctly Australian fauna dominated by mammals was in evidence. We need to find the fossils that will tell us if the whole fauna dates back, like the Platypus, to over 100 million years old or whether the mammals are much younger. Whatever the case, it is sobering to watch a Platypus in a natural pool and realise that its ancestors had been reproducing and living successfully with little change for over 100 million years before any people appeared in this country.

The last 15 million years: cooling and growth of the ice-caps

From a global climate viewpoint the most important development was a sufficient separation of Australia from Antarctica (as shown in figure 7.1) to allow a complete circumpolar current to form around Antarctica. Initially this current was warm, and the Antarctic and southern Australia still supported temperate rainforest vegetation. However, after about 30 Ma ago the water and air temperatures cooled rapidly, leading by 15 Ma ago to glacial conditions in Antarctica.

This cooling process inevitably led to colder, glacial conditions in the Antarctic mountains and by 30 Ma ago there was only alpine vegetation on the continent. Cores taken in the deep sea off Antarctica show changes in the pollen types which indicate major cooling of the Antarctic around 34–37 Ma ago. Drilling in the Lemonthyme area of Tasmania has found tillites that formed from local mountain glaciers 30–35 Ma ago. At this time Tasmania was close to Antarctica. However, a full ice sheet that discharged icebergs into the Southern Ocean probably did not develop until after 15 Ma ago.

Recent mapping of mountain peaks protruding through the ice in parts of Antarctica has shown that the typical polar landscape with very sharp peaks and ridge lines has been forming only since 7.5 Ma ago. This is further evidence for the continuing refrigeration of the Antarctic since that time. It is strange to realise that the polar ice sheets are relatively young, and for most of the last 250 million years Antarctica was warm, with prolific forests, the home of dinosaurs and the ancestors of much of the animals and plants we have today.

Some time after 15 Ma ago the climate in Australia started to dry out. The *Nothofagus* largely disappeared, and there was an increase of plants typical of drier forests and woodlands. There was a dry season and fires. Rainfall was down to 1500–1000 mm a year – too little to support most rainforests.

There were comparable, major changes in other parts of the former Gondwana. For instance, in central and northern Africa, the cradle of hominid evolution, a burst of evolution in the apes began around 15 Ma ago, and plenty of human fossils exist from 3–4 Ma onwards. The earliest hominid skull, from Chad, is dated at 7 Ma. Yet when this hominid was alive the area was a lush forested lake margin, not the desert of today.

This drier regime has continued in Australia till the present, except for a brief wet interlude between 5 and 2 Ma ago. This wet phase initially supported rainforests and then wet sclerophyll forests in southeastern Australia. It is significant because the extra rainfall may have been responsible for the formation of the caves on the Nullarbor Plain and elsewhere.

About 10 Ma ago the sea level dropped significantly, so that Australia was dry except for the completed seaway of Bass Strait and a marine embayment in the Murray region. Tectonic uplift to the west finally forced the withdrawal of the sea, though a large lake persisted in the Murray region until around 700 000 years ago.

Since 5 Ma ago Australia has been within 1–2° of its present latitude. The shape and shorelines have been much the same as today, except for the oscillations induced by the Pleistocene ice ages. We know much less of the climate in far northern Australia in this period. Once the great Cretaceous seas drained away the landscape eroded, and there was little accumulation of sediments containing pollen or plants remains that might give us clues.

The global climate and polar ice caps have not remained constant, but have fluctuated regularly. The climatic record for the past million years is now well known. There were periods of extreme glaciation when the ice sheets migrated south to cover much of Europe and North America. At the same time, ocean waters cooled and sea levels dropped to 100 m below the present levels.

Recent investigations of the seabed seaward of the Ross Ice Shelf in Antarctica show that the ice extended several hundred kilometres further northward until it started to recede some 11 000 years ago. It withdrew at a rate of around 100 metres per year until reaching its present position 6000 years ago.

Ice accumulated on the highlands of southeastern Australia and in Tasmania as recently as 20 000 years ago, leaving sculptured landscapes and deposits of till. These glacial periods alternated with interglacials, when the polar ice sheets withdrew to about their present positions, the ocean waters warmed and sea-levels rose. There were sea level highs at around 340 000, 220 000, 120 000 and 6500 years ago. However, for most of the time the sea levels were lower than at present, corresponding to cooler periods when ocean water was locked in the ice caps. The typical pattern is a slow irregular cooling to a glacial sea level low, followed by a rapid warming and sea-level rise. We are now in an interglacial period.

The Australian megafauna and its extinction

The oldest kangaroos found are about 15 Ma old, from Tommo's quarry near Lake Tarkarooloo in South Australia. They were only the size of rabbits. The sizes increased steadily as the fauna evolved reaching a maximum size about 1 Ma ago, when they were a quarter to a third larger again than the present species.

Many of the present forms of Australian animals are smaller species, the fossil forms often being much larger. Sediments deposited after 2 Ma ago contain what is known as 'megafauna', reflecting the giant size of many of these animals. Apart from the kangaroos, there were wombats about twice the size of living examples, giant ground birds like emus 2 to 3 m high, the goanna *Megalania* that was at least 7 m long – twice the size of the Komodo dragon in the Celebes – and the giant echidna *Zaglossus*, whose modern relative still lives in mountains of New Guinea.

There were also very large beasts with no living relatives, such as the diprotodons. These were marsupial plant-eaters some 2 m high at the shoulder, the

Figure 7.8 *A model of a diprotodon. This animal was around 3 m long and 2 m high at the shoulder, and was a herbivorous marsupial. Diprotodons were first recognised in 1838 from jaw and tooth fragments.*
Photo: Nature Focus, Australian Museum

size of a modern rhinoceros. They looked like a cross between a wombat and a bear. Another strange animal was *Palorchestes*, which had massive forearms with sharp claws up to 120 mm long, and possibly with a long snout. There was the carnivorous marsupial lion, a leopard-sized creature with the ability to climb trees, and possessing strong claws and teeth.

A recent re-evaluation of dates for these animal fossils indicates a continent-wide extinction about 46 000 years ago. Twenty-three of the 24 genera of Australian land animals weighing over 45 kg died out. Previous work had indicated that some of these persisted till more recently, perhaps as late as 15 000 years ago. What was the cause of this Australian extinction?

Some think that the hunting pressure brought by Aboriginal people contributed to the demise of the animals; the latest estimate of human arrival is 56 000 years ago. However, it is possible that humans arrived in Australia as long as 100 000 years ago. Others think it was the continuing climate change with drying of the continent, especially increased aridity at the last glacial maximum between 23 000 and 19 000 years ago. Then when the sea level rose 20 000 years ago, vast areas of what was the extended Australia (covering some 3 million square kilometres) was drowned in seawater very quickly.

This megafaunal extinction was a global phenomenon, though it occurred in Australia well before similar extinction events elsewhere. In North America, where 28 genera of mammals including elephants, camels, horses, sloths and sabre-toothed tigers disappeared, the last wave of extinction happened 11 000 years ago. Yet the final ice sheets withdrew between 5000 and 10 000 years ago. Similarly, the giant lemurs in Madagascar and the moas in New Zealand disappeared less than 15 000 years ago.

Box 7.2 Evidence for climate change

How do we know when the ice sheets advanced, oceans cooled and sea levels dropped?

OXYGEN ISOTOPES

There are two isotopes of oxygen that are stable; that is, they do not undergo radioactive decay. The most common is oxygen-16 (O-16), the heavier oxygen-18 (O-18) being rare. In 1947 Harold Urey at the University of Chicago had determined that these two oxygen isotopes occur in different ratios, depending on the temperature. During the evaporation of seawater, the lighter O-16 isotope is more easily released into vapour in the air. This water vapour forms rain and, in glacial periods, snow and ice. During glaciations this lighter isotope is progressively locked up, frozen as ice, and the seawater becomes enriched in the heavier O-18 isotope. Any organism growing a shell also tends to include more of the heavier O-18 isotope if the surrounding water is colder. So two processes are enriching the heavier O-18 isotope in the shells of organisms that live in cold waters: the colder ocean water contains more O-18, and organisms incorporate more O-18 when growing in colder water.

The end result is that shells growing in cold waters have higher O-18/O-16 ratios than those in warm waters. We can measure the oxygen isotope ratios and estimate the ocean water temperatures. This has been done repeatedly and there is now a well-established sequence of oxygen isotope stages back to 1 Ma ago, which matches other evidence of glaciations and sea-level changes.

Cores of the sediments deposited on the deep ocean floors, recovered by the Ocean Drilling Program, extend back in time many

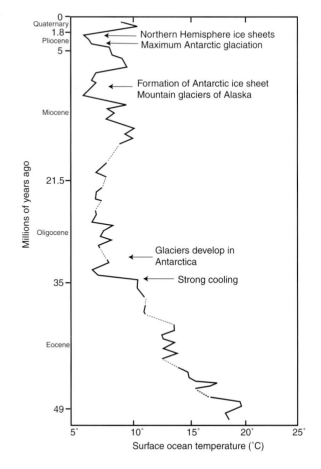

Figure 7.9 *Long-term seawater surface temperatures interpreted from oxygen isotope measurements of planktonic foraminifera from deep-sea sediment cores recovered near Antarctica.*

Around 50 Ma ago the water temperatures off Antarctica were close to 20°C, similar to that at modern beaches in northern New South Wales. There was steady cooling over the next 15 million years, primarily due to the increasing strength of the circumpolar current once Australia had split well away from Antarctica, with a sudden increase in cooling after 35 Ma ago as glaciers developed.

Climates fluctuated over the next 30 million years, with the major ice sheet evident in Antarctica after 15 Ma ago and the glacial maximum in the Pliocene. At this time the ice sheets developed in the Northern Hemisphere. A warmer period started around 1.8 Ma ago, interrupted by the recent glacial episodes or ice ages. This record clearly illustrates that the present global climate is at the icy end of the possible spectrum.

THIS MATERIAL IS USED WITH PERMISSION OF JOHN WILEY & SONS, INC.

millions of years. Minute animals and plants with calcareous skeletons are continually raining down on the ocean floor, so that when oceans are warmer the skeletons reflect this, and when the oceans turn colder the animals change and different skeletons accumulate in the next layer on the seabed. We know from modern samples which are the warm-water forms and which are the cold-water forms. Furthermore, we can also analyse the oxygen isotope ratios in the skeletons and thus calculate the seawater temperatures.

GREENHOUSE GAS BUBBLES IN ICE

Long holes have been drilled through the Antarctic ice sheet, and the cores of ice stored in cold-rooms so they will not melt. The ice core Vostok Hole 4G reached 2456 m depth, an estimated age of 220 000 years. Minute gas bubbles in the ice contain methane and carbon dioxide, and warmer, interglacial periods have higher levels of both gases than glacial periods. That is why the presently rising amounts of carbon dioxide and other greenhouse gases are causing concern, because if they continue they indicate a gradual warming of the Earth will occur and sea levels will rise over the next 400–500 years. Analysis of the hydrogen isotope deuterium in the ice indicates the temperature increase in central Antarctica since the glacial maximum has been about 8°C.

POLLEN EVIDENCE

Cores through lake sediments and peatlands can be sampled for pollen. Especially in the Northern Hemisphere, samples taken down through peat bogs displayed changes in the accumulated vegetation corresponding to warmer and colder climate plant communities over the past 250 000 years. When the climate is warmer, the pollen rain reflects the warmer climate vegetation covering the landscape, and any cooling is reflected in a change in the pollens blown into the lake or over the peatland. The analysis of the leaf materials in some cores is also used to estimate rainfall variations on a finer time scale, as scientists search for records of the El Niño Southern Oscillation (ENSO) effects.

Studies of cores from the crater lakes on the Atherton Tablelands in North Queensland have shown similar results over a shorter, more recent time frame. In particular they place the recent expansion of the rainforest at around 7500 years ago, before which time most of the area had been covered by drier eucalypt forests.

Australia's arid interior

From around 15 Ma ago (mid to late Miocene), just as the Antarctic ice sheet took hold of that continent, the climate became warmer and drier in Australia. The change was first apparent in sediments below the Murray plains in western New South Wales and then in younger sediments on the highlands. While rainforest still cloaked much of the highland areas in the southeast, farther inland towards central

Australia there were open eucalypt woodlands and grasslands around huge lakes. Gradually the interior became more arid, and the forests disappeared and were replaced by deserts and salt lakes.

Why did this occur? The growth of the Antarctic ice cap and the deserts in Australia is too much of a coincidence to be just chance. As the Earth cooled down, more and more water became frozen into the great ice sheets that cover Greenland, the ice across the North Pole and the Antarctic ice sheet. The average ocean temperatures fell, so there was much less evaporation and therefore less water in the atmosphere to form rain. Furthermore, during this time Australia was drifting northwards into hotter latitudes, so that rainfall was more easily evaporated. On an old eroded land mass without high mountains to catch rainfall, the interior became progressively drier.

The drying had become very obvious by around 8 Ma ago, and deserts developed within the last 5 million years. Remnants of the earlier vegetation survive in wet valleys such as Palm Valley near Alice Springs and Carnarvon Gorge in Queensland, where springs keep the vegetation alive all year round. But these small pockets of vegetation could not sustain the animal populations.

This drawn-out drying had three main effects on the fauna. Some of the animals became extinct. Some adapted to the new and harsh conditions – not only the heat of day, but the extreme dry cold of the desert night and the unpredictability of water and food – and became the highly specialised desert fauna. Some simply followed the vegetation, so that the high-altitude rainforests of northeastern Australia and especially of Papua New Guinea now support many of the descendants of those occupying the Lake Eyre region and central Australia over 10 million years ago.

The isolation of the Australian continent ensured the evolution of many plants and animals that are distinctly Australian. Although they were originally part of the same supercontinent and although both have a similar range of geology and climates, the flora and fauna of Australia and of southern Africa are vastly different and quite individual. Why they have evolved as they did – the large cats, elephants and antelopes in Africa, but marsupials in Australia – is not known, but it is certain that the differences are a result of the differing geological paths the two continents took over 90 Ma ago.

The deserts and sandplains

About one-third of Australia is now arid land comprising sandy and stony deserts. The largest area of sandy deserts, composed of the Great Victoria, Great Sandy, Gibson and Tanami Deserts, extends from South Australia and the Northern Territory across Western Australia to the northwestern coast between Port Hedland and Broome. The sand dunes are oriented mainly east–west and are now stabilised by spinifex and other vegetation. In places the dune crests are still commonly mobile, although the Great Victoria Desert is well stabilised and probably no longer mobile. The other large area of sand, and the more arid, contains the Simpson, Tirari and

Figure 7.10 *a. The situation 18 000 years ago, showing both sandy and stony deserts, and the dune orientations. The aridity started around 15 Ma ago with the refrigeration of Antarctica which began storing vast quantities of water in the ice-sheet. More recently there were periods of increased aridity during the sea-level lows associated with the Pleistocene glaciations. The solid line shows the shoreline at the Last Glacial Maximum (LGM) when the sea level was about 115 m lower than present and the Australian shoreline was on the present continental shelf. Note that at this time, the Australian mainland, Tasmania and Papua New Guinea were a single continuous landmass. Snow and ice covered parts of the highlands of Tasmania, northern Victoria and Papua New Guinea.*

The arrows on the map indicate the trends of longitudinal sand dunes, which parallel the wind direction, in the Australian deserts. These dunes were mobile during the drier phases but are now largely stabilised by vegetation.

b. Longitudinal dunes in the Simpson Desert.
PHOTO: RICHARD RUDD

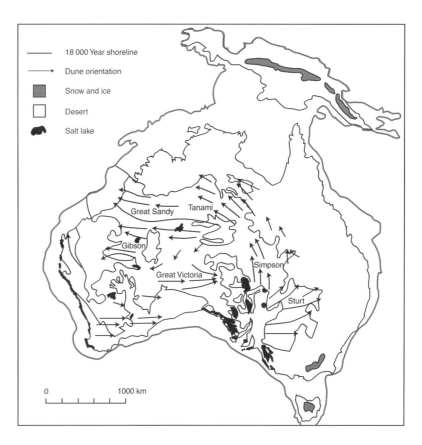

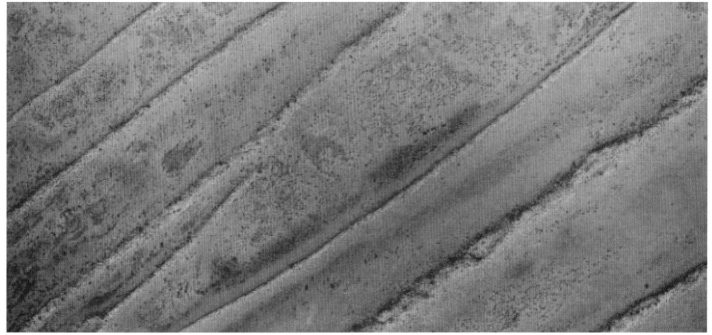

Strzelecki Deserts east of Lake Eyre. In these the dunes trend more northerly. Overall, the dune orientations in Australia describe a large, anticlockwise wind pattern.

In Western Australia there are both coastal and inland sandplains. The coastal sandplain extending from Bunbury to Geraldton and (as cemented sandstones) to north of Carnarvon, is related to coastal deposition. These are sands brought down by rivers and then spread out along the coast by waves and currents. Much of the sand has sea-shells in it, indicating a marine origin.

Further inland are sandplains that may contain sand blown from inland Australia. However, when the sand is sieved and examined in detail it is clear that much of it contains trace quantities of minerals that can be linked to the underlying or nearby rocks. So the inland sandplain may be the residual materials left after weathering of the inland granitic, metamorphic and sedimentary rocks, all the clays and other materials having blown away.

The Finke, Plenty and Hall Rivers, and the streams of the Channel Country in Queensland – the Diamantina and Georgina Rivers and Cooper Creek – flushed sand into the Simpson Desert. In Victoria, the Murray and Wimmera Rivers delivered the sand. In drier periods in the past even more sand was available for redistribution by the wind than now, since the vegetation along the river banks prevents erosion. River activity over the past 300 000 years was much greater than present, and large sand loads peaked around 110 000 years ago in central Australia.

SHRIMP dating of individual zircon grains in the sands, together with comparisons with ages of expected source rocks, shows that the sands can be derived locally or from distances up to 850 km, and that most have been brought in by rivers and then redistributed by the wind.

Many studies have shown that there were several cooler, windy periods during the last glacial cycle. For instance, thermoluminescence dating of quartz in sand dunes beside the Murrumbidgee River near Wagga Wagga show three periods of dune building: 120 000–80 000, 60 000–35 000 and 25 000–15 000 years ago. The sand was sourced from local river beds, when the rivers were carrying much greater sand loads than today. Included in the dune sediments is a minor clay component blown from regions to the west. Globally the period between 25 000 and 15 000 years ago has been recognised as being drier, cooler and windier than present.

Australia was at its most arid between 25 000–18 000 years ago, corresponding to the greatest lowering of sea levels during the Pleistocene ice ages (the Last Glacial Maximum). Fine quartz sand was apparently blown off eastern Australia to the north, and further east than New Zealand. Moisture sourced from the sea had receded farthest from the continental interior. Lake Eyre was completely dry. Large areas where the landscape consists of linear sand dunes, now heavily vegetated, – were active dune fields – on the Eyre and Yorke Peninsulas, and in northwestern Victoria. The Little Desert in Victoria extends nearly as far south as Horsham, so clearly there was a phase when the continent was drier than it is now.

The earliest dates for formation of the Australian deserts are not well established. Certainly periods of sand dune formation have been dated at around 256 000, 173 000, 112 000 and 56 000 years ago, with other phases less than

40 000 years ago. This does not mean there were no phases before 256 000 years ago. Firstly, sands are very difficult to date; and secondly, earlier systems could have been remobilised. We know from the fossil records at Riversleigh, for instance, that the continent was drying out rapidly by 5–10 Ma ago.

Cores taken in the sea off northeastern Australia have also been analysed to determine their pollen profiles. These profiles show a major change in the vegetation some 140 000 years ago, from moist climate, fire-sensitive rainforest communities to drier climate, fire-tolerant eucalypt communities, with increased amounts of fine charcoal in the cores above this point.

Salt lakes

Water that flows into the deserts collects in shallow depressions, forming claypans and salinas (salt lakes). These salinas vary in size from small ponds to large bodies such as Lakes Eyre, Torrens, Frome and Amadeus, each of which cover several thousand square kilometres.

Cores taken in the southern part of Lake Eyre show that it was originally full of water up to 25 m deep, though typically saline, from 130 000 to 90 000 years ago. Then it was intermittently saline and dry until about 60 000 years ago, when the modern pattern of a large salt lake periodically filled by monsoonal rains was established.

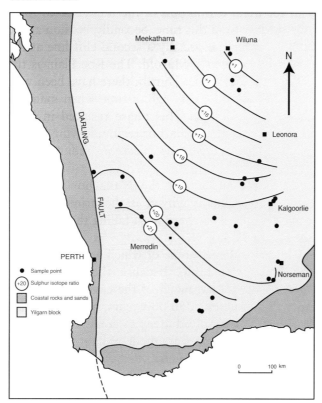

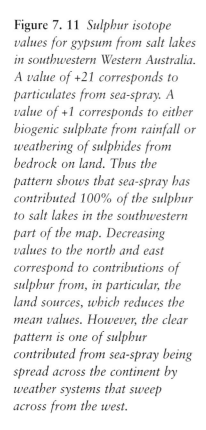

Figure 7.11 *Sulphur isotope values for gypsum from salt lakes in southwestern Western Australia. A value of +21 corresponds to particulates from sea-spray. A value of +1 corresponds to either biogenic sulphate from rainfall or weathering of sulphides from bedrock on land. Thus the pattern shows that sea-spray has contributed 100% of the sulphur to salt lakes in the southwestern part of the map. Decreasing values to the north and east correspond to contributions of sulphur from, in particular, the land sources, which reduces the mean values. However, the clear pattern is one of sulphur contributed from sea-spray being spread across the continent by weather systems that sweep across from the west.*

Where does the salt come from? Some comes from the leaching of soluble minerals and other weathering of rocks around the basins, and is transported in solution by the rivers before being deposited by evaporation in the desert salinas. However, isotope analyses show that much of the salts is derived from the sea, either as windblown material or dissolved in the rain. The proportion that is of marine origin is generally greater towards the coastline, but it is still observable in central Australia. A lesser proportion is derived from weathering of the inland rocks.

While rainwater seems as fresh as is imaginable, it does contain minute quantities of dissolved salts. In higher-rainfall regions this salt is always flushed through the soil and back into rivers running to the sea. But in drier inland areas, this salt gradually accumulates over thousands and millions of years as the water is evaporated or transpired by the plants. The result is a steadily increasingly salt content.

Nullarbor Plain

The Nullarbor Plain is a flat grassy plain formed mainly of limestone, with some sandstones closer to the older bedrocks in the west and north. An upper plateau covers most of the plain, and the lower Roe Plain borders the coast between Eucla and Israelite Bay.

The underlying rock sequence dates from 35 Ma ago, and was formed as Australia was drifting northwards after the separation from Antarctica. Limestone was first deposited 45–35 Ma, and another period of limestone deposition occurred from 25 to 15 Ma ago. At this point there was uplift and a global lowering of sea level, which left the surface of this 15-million-year-old limestone seabed exposed as the initial plateau.

The gradually drying climate since then has meant there has been little erosion, except for three events. Firstly, a wet phase 5–3 Ma ago led to some erosion at the edges of the plain, and probably to the main dissolution of the limestone to form caves. Most of the large caves probably date from this time. Secondly, erosion at the seaward edge formed cliffs typically 75 m high, especially a second cliff line at the back of the Roe Plain, which is probably around 3 Ma old. The Roe Plain is the uplifted bed of the sea, which cut that lower cliff line. Thirdly, there have been very dry, windy phases, especially one around 17 000 years ago, a time when many of the dunes around the salt lakes were also mobile. This phase resulted in huge quantities of dust being blown eastwards from the Nullarbor region, leaving behind the gibber plain north of the railway and spreading a fine calcareous sediment over South Australia.

Not much else has happened to the surface of the Nullarbor Plain for thousands and perhaps millions of years. That is why it is the prime place in Australia to search for meteorites and tektites. They have been falling there for ages and have been left undisturbed (see Figure 11.3a).

Underground there are extensive cave systems, some of which are shallow and some deep and extending below the water table. Because the present and geologically recent climate is so dry it is clear that much of the cave formation is very old. Some calcite deposits have been dated at 350 000 years, and these must have crystallised long after the limestone was dissolved away to create the space of the cave. The most recent crystals forming in the caves are made of halite – common salt. Halite precipitates only from very concentrated brines, which would be forming in the present arid climate. Other rare sulphate and phosphate minerals, typical of brine environments, can also be found. It is also clear that the salt crystals are corroding the rock walls and thus enlarging the caverns, and in some cases leading to collapse of the roof, forming openings to the surface.

So the Nullarbor caves have had two stages of formation: an older phase of calcium carbonate dissolution with precipitation of calcite stalagmites and stalactites during wetter climates perhaps over a million years ago, and the present phase of salt crystallisation and corrosion.

The caves contain much fossil bone and pollen material. These included two intact thylacine (Tasmanian Tiger) carcasses, an animal that has been absent from the mainland for 3000 years. One specimen dated at 4600 years had the hair, desiccated eyeball and tongue preserved. An analysis of the gut contents showed a similar vegetation to present plant communities.

Summary

Australia separated from Antarctica after 99 Ma ago, and initially both remained warm and humid, with rainforest vegetation and inland Australia a series of rivers and lakes with abundant life. Peatlands, which later changed to brown coals, and oilshales accumulated in the humid climates. Fossil birds, pythons, frogs and platypus are evident. Cooling after 30 Ma ago led to development of the Antarctic ice sheet after 15 Ma ago, and the increasing aridity of Australia.

Sand deserts and large inland salt lakes formed within the last 5 million years. Climatic oscillations in the Pleistocene during the last million years led to repeated phases of glaciation with lower sea levels and warmer periods with higher sea levels.

The Australian megafauna, with giant kangaroos and emus, plus extinct animals like the Diprotodon died out around 46 000 years ago, or perhaps more recently.

Figure 7.12 The southern edge of the Nullarbor Plain, showing the cliff eroded by the sea into Tertiary sedimentary rocks.
Photo: David Gillieson

SOURCES AND REFERENCES

Archer, M. & Clayton, G., 1984. *Vertebrate Zoogeography and Evolution in Australasia*. Hesperian Press. 1203 pp.

Archer, M., Hand, S. & Godthelp, H., 1991. *Australia's Lost World – Riversleigh, World Heritage Site*. Reed New Holland. 264 pp.

Archer, M., Hand, S.J. & Godthelp, H., 1995. 'Tertiary environmental and biotic change in Australia' in E.S.Verba, G.H. Denton, T.C. Partridge & L.H. Burckle (eds) *Paleoclimate and evolution, with emphasis on human origins*. Yale University Press, pp. 77–90.

Blunier, T., et al., 2004. 'What was the surface temperature in central Antarctica during the last glacial maximum?' *Earth and Planetary Science Letters*, 218: 379–388.

Brown, C.M. & Stephenson, A.E., 1991. *Geology of the Murray Basin, Southeastern Australia*. Bureau of Mineral Resources Australia, Bulletin 235. 430 pp.

Chivas, A.R. et al., 1991. 'Isotopic constraints on the origin of salts in Australian playas. 1. Sulphur'. *Palaeogeography, Palaeoclimatology, Palaeoecology*, 84: 309–332.

Clarke, J.D.A., 1994. 'Evolution of the Lefroy and Cowan palaeodrainage channels, Western Australia'. *Australian Journal of Earth Sciences*, 41: 55–68.

Coventry, R.J., 1978. 'Late Cainozoic geology, soils, and landscape evolution of the Torrens Creek area, north Queensland'. *Journal of the Geological Society of Australia*, 25: 415–427.

Dettmann, M.E. et al., 1992. 'Australian Cretaceous terrestrial faunas and floras: biostratigraphic and biogeographic implications'. *Cretaceous Research*, 13: 207–262.

Geological Society of Australia Inc., 1996. *Mesozoic Geology of the Eastern Australia Plate Conference, Extended Abstracts No. 43*. 575 pp.

Hand, S.J. & Laurie, J.R. (eds) 1998. *Riversleigh Symposium*. Association of Australasian Palaeontologists, Memoir 25. 154 pp.

Hou, B., Frakes, L.A., Alley, N.F. & Clarke, J.D.A., 2003. 'Characteristics and evolution of the Tertiary palaeovalleys in the northwest Gawler Craton, South Australia'. *Australian Journal of Earth Sciences*, 50: 215–230.

Kemp, E.M., 1978. 'Tertiary climatic evolution and vegetation history in the southeast Indian Ocean region'. *Palaeogeography, Palaeoclimatology, Palaeoecology*, 24: 169–208.

Kershaw, A.P., 1986. 'Climatic change and Aboriginal burning in north-east Australia during the last two glacial/interglacial cycles'. *Nature*, 322: 47–49.

Kershaw, A.P., 1994. 'Pleistocene vegetation of the humid tropics of northeastern Queensland, Australia'. *Palaeogeography, Palaeoclimatology, Palaeoecology*, 109: 399–412.

Kershaw, A.P. & Sluiter, I.R., 1982. 'The application of pollen analysis to the elucidation of Latrobe Valley brown coal depositional environments and stratigraphy'. *Australian Coal Geology*, 4: 169–183.

Langford, R.P. et al., 1995. *Palaeogeographic Atlas of Australia. Volume 10 – Cainozoic*. Australian Geological Survey Organisation, Canberra.

McLoughlin, S., 2001. 'The breakup history of Gondwana and its impact on pre-Cenozoic floristic provincialism'. *Australian Journal of Botany*, 49: 271–300.

McNally, G.H. & Wilson, I.R., 1996. 'Silcretes of the Mirackina Palaeochannel, South Australia'. *AGSO Journal of Australian Geology and Geophysics*, 16: 295–301.

Magee, J.W. et al., 1995. 'Stratigraphy, sedimentology, chronology and palaeohydrology of Quaternary lacustrine deposits at Madigan Gulf, Lake Eyre, South Australia'. *Palaeogeography, Palaeoclimatology, Palaeoecology*, 113: 3–42.

Martin, H.A., 1978. 'Evolution of the Australian flora and vegetation through the Tertiary: evidence from pollen'. *Alcheringa*, 2: 181–202.

Martin, H.A., 1987. 'Cainozoic history of the vegetation and climate of the Lachlan River region, New South Wales'. *Proceedings, Linnean Society of New South Wales*, 109: 213–257.

Nanson, G.C., Price, D.M. & Short, S.A., 1992. 'Wetting and drying of Australia over the past 300 ka'. *Geology*, 20: 791–794.

Nott, J., 1995. 'The antiquity of landscapes on the North Australian Craton and the implications for theories of long-term landscape evolution'. *Journal of Geology*, 103: 19–32.

Nott, J.F. & Owen, J.A.K., 1992. 'An Oligocene palynoflora from the middle Shoalhaven catchment N.S.W. and the Tertiary evolution of flora and climate in the southeast Australian highlands'. *Palaeogeography, Palaeoclimatology, Palaeoecology*, 95: 135–151.

Page, K.J. et al., 2001. 'TL chronology and stratigraphy of riverine source bordering sand dunes near Wagga Wagga, New South Wales, Australia'. *Quaternary International*, 83–85: 187–193.

Roberts, R.G. et al., 2001. 'New ages for the last Australian megafauna: continent-wide extinction about 46,000 years ago'. *Science*, 292: 1888–1892.

Salama, R.B., 1997. 'Geomorphology, geology and palaeohydrology of the broad alluvial valleys of the Salt River System, Western Australia'. *Australian Journal of Earth Sciences*, 44: 751–765.

Specht, R.L., Dettmann, M.E., & Jarzen, J.M., 1992. 'Community associations and structure in the Late Cretaceous vegetation of southeast Australasia and Antarctica'. *Palaeogeography, Palaeoclimatology, Palaeoecology*, 94: 283–309.

Stephenson, A.E. & Brown, C.M., 1989. 'The ancient Murray River system'. *BMR Journal of Australian Geology and Geophysics*, 11: 387–395.

Thiede, J., 1979. 'Wind regimes over the late Quaternary southwest Pacific Ocean'. *Geology*, 7: 259–262.

Veevers, J.J. (ed.) 1984. *Phanerozoic Earth History of Australia*. Clarendon Press. 418 pp

White, M.E., 1986. *The Greening of Gondwana*. Reed. 256 pp.

Williams, M.A.J., 2001. 'Quaternary climatic changes in Australia and their environmental effects' in V.A. Gostin (ed.) *Gondwana to Greenhouse*, Geological Society of Australia Inc., pp. 3–12.

Wroe, S., Field, J., Fullagar, R. & Jermin, L., 2004. 'Megafaunal extinction in the Late Quaternary and the global overkill hypothesis'. *Alcheringa* (in press).

Websites

Earth history maps
www.scotese.com [*especially the Cretaceous 94 Ma, showing Gondwana as it breaks up*]

Antarctic ice-sheet history
www-odp.tamu.edu/publications/178_IR/chap_02/ch2_htm2.htm

The Australian megafauna
wwwrses.anu.edu.au/environment/eePages/eeDating/extinct_oz_mega.html
www.museum.vic.gov.au/prehistoric/mammals/australia.html

Important Australian fossils, including Eric the pliosaur
www.amonline.net.au/webinabox/fossils/australian/index.htm

Illustrations

Figs 7.1, 7.10: redrawn and reproduced by permission of John Veevers after fig. 45 in Veevers, J.J. (ed.) 2000. *Billion Year Earth History of Australia Atlas*. GEMOC Press, Sydney, and fig 69 in Veevers, J.J. (ed.) 2001. *Billion Year Earth History of Australia*. GEMOC Press, Sydney; fig. 7.10 also with permission from Martin Williams.

Fig.7.2: by permission of the Geological Society of Australia after fig. 7 in Clarke, J.D.A., 1994. 'Evolution of the Lefroy and Cowan palaeodrainage channels, Western Australia'. *Australian Journal of Earth Sciences*, 41: 55–68

Fig.7.3a: modified and reproduced with permission of Geoscience Australia, Canberra from figs 1, 2 in McNally, G.H. & Wilson, I.R., 1996. 'Silcretes of the Mirackina Palaeochannel, South Australia'. *AGSO Journal of Australian Geology and Geophysics*, 16: 295–301. Crown copyright ©. All rights reserved. www.ga.gov.au

Fig. 7.4: redrawn and reproduced with permission of The University of Chicago Press from fig. 2 in Nott, J., 1995. 'The antiquity of landscapes on the North Australian Craton and the implications for theories of long-term landscape evolution'. *Journal of Geology*, 103: 19–32. © 1995 by the University of Chicago.

Fig. 7.7: redrawn and reproduced by permission of Oxford University Press Australia from fig. 4.1 in Young, A. *Environmental Change in Australia Since 1788*. © Oxford University Press, www.oup.com.au. The original data was from Martin, H.A., 1987. 'Evolution of the Australian flora and vegetation through the Tertiary: evidence from pollen'. *Alcheringa*, 2: 181–202.

Fig. 7.9: redrawn and used with permission of John Wiley & Sons Inc. after fig. 14.17, p. 378, in Skinner, B.J. & Porter, S.C., 1995. *The Blue Planet*. Copyright © 1995 John Wiley & Sons Inc. 493 pp.

Fig. 7.11: redrawn and reprinted from fig. 3 in Chivas, A.R. *et al.*, 1991. 'Isotopic constraints on the origin of salts in Australian playas. 1. Sulphur'. *Palaeogeography, Palaeoclimatology, Palaeoecology*, 84: 309–332. Copyright © 1991 with permission from Elsevier.

CHAPTER 8

EASTERN HIGHLANDS AND VOLCANOES BARELY EXTINCT

Australia is the only continent in the world without active volcanoes – but that was not so in the past. There has been regular volcanism throughout Australia's geological history, especially andesitic and rhyolitic volcanism along the eastern margin. The most recent basaltic phase has barely finished. Basalt eruptions started around 70 Ma ago and lasted until the most recent eruptions in South Australia 4600 years ago. There is a close relationship between these basalts and the Great Divide: why?

Rich soils derived from the weathered basalt on elevated country combined with a cooler and wetter climate form the basis for many of our finest agricultural and horticultural areas.

Volcanic provinces

Most of the volcanism in Australia has been of andesites and rhyolites, rocks formed mainly during compressive tectonics when there is considerable melting of continental crust. Three episodes of basaltic igneous activity are evident: late Precambrian (600–555 Ma) before the final assembly of Gondwana, mainly Jurassic (185–132 Ma) in southern and western Australia before and during the initial Gondwana break-up, and Tertiary–Recent (70 Ma to 4600 years ago) along eastern Australia.

This most recent episode started at least 70 Ma ago; there are dates on intrusive basaltic rocks at Mount Woolooma near Scone of 85 Ma, and there are reported dates on basalt dykes in New South Wales of 90 Ma. However, the main action on the surface seems to have started after 70 Ma; there is plenty of well-preserved basalt dating from 30 Ma onwards. The map in figure 8.1 shows that the volcanics are grouped in patches, generally 50–200 km across, with each patch being erupted over typically 5–10 million years. While these patches are what is left after erosion, careful mapping has shown that each contains separate volcanic edifices.

Two distinct types of volcanic province have been recognised – lava fields and central volcanoes.

Lava fields

In lava fields, the bulk of the lava comes from numerous vents or fissures, often covering an area 100 km across, and building broad shield volcanoes with layered basaltic lavas up to 1000 m thick. Most of the lava fields are situated on or east of the Great Divide. In some cases the weathered lavas have left plateaus of rich red-brown soils; in others they have been dissected by eroding valleys so that only remnants remain on the ridges. The lava fields were erupted mainly between 60 and 20 Ma ago, and extend from northern Queensland to southern Tasmania.

The most recent eruptions, within the last 1 million years, occur at both the northern and southern ends of the belt: the Murray Islands in Torres Strait and the Atherton Tablelands in far northern Queensland, and to the south in western Victoria and southeastern South Australia. The very youngest eruptions are 13 000 years near Charters Towers in northern Queensland, and 4600 years in South Australia.

Central volcanoes

The large central volcanoes are typically preserved as elevated domes with subsidiary vents around them. In many cases the volcanic cones have been eroded leaving plugs protruding as prominent landmarks. The central volcanoes produced basalts, and also more siliceous lavas – rhyolites and trachytes. Erosion has exposed

Figure 8.1 *Eastern Australia, showing basalts and the line of the Great Divide in red. Central volcanoes are marked green. To the east of the continent lie the Tasman and Lord Howe seamount chains. The lines of the central volcanoes and the seamounts are parallel with the oldest at the north and youngest at the south. They were formed by mantle penetrating the crust in successively southward positions as the Indo–Australian plate drifted northwards.*

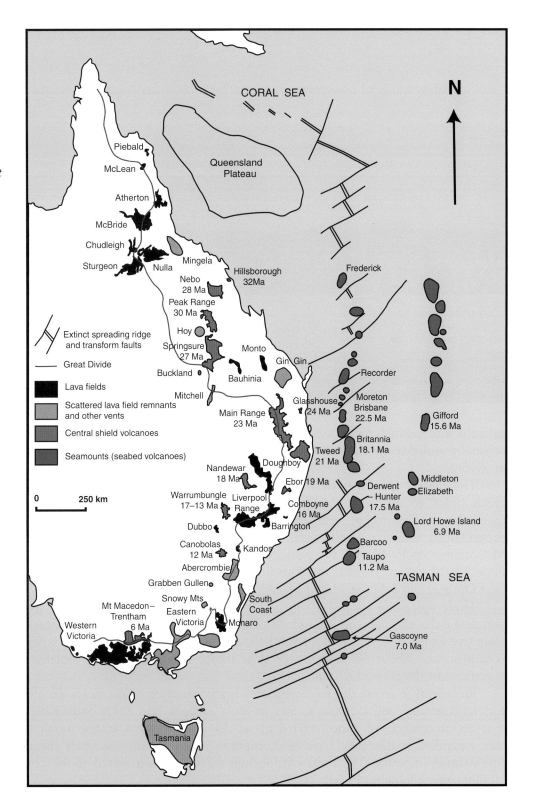

Figure 8.2 *Five of The Pinnacles (or Seven Sisters), a series of scoria cones on the Atherton Tablelands, between Atherton and Yungaburra, showing the rich brown soils formed on basalt.*
Photo: David Johnson

gabbros and related rocks which are the intrusive phases of this activity, and originally underlay the volcanic landscape. Many of the remnant plugs are trachytes or rhyolites.

The central volcanoes show a very interesting trend (Figure 8.1). They are much older to the north (up to 33 Ma at Cape Hillsborough in northern Queensland, 27–30 Ma at Nebo and Clermont in central Queensland) and become progressively younger southwards: 24 Ma at Toowoomba and the Glasshouse Mountains, 18–19 Ma for the Nandewar Volcano near Narrabri, 19–20 Ma at Ebor, 13–17 million years at the Warrumbungles, 10–12 Ma at Mount Canobolas near Orange, and 6 Ma at Mount Macedon north of Melbourne.

These volcanoes formed above a stationary mantle plume that successively punched holes through the lithosphere of the northward moving Indo-Australian plate (see Figure 8.9). The average southward migration rate of the central volcano volcanism was about 66 mm/yr, very close to the measured modern northward movement of the plate of 67 mm/yr.

In general it seems the volcanism was firstly producing lava fields all along eastern Australia, and that the activity at most of these lava fields paused before a series of central volcanoes started about 33 Ma ago, first in the north and then extending southwards. Lava field eruptions continued intermittently through the 70 million years. The total outpourings have been estimated to be about 20 000 cubic kilometres of basalts.

Northern Queensland

Basalts in northern Queensland have been dated as far back as 50 Ma, although much of the present outcrops range from around 8 Ma to as recent as 13 000 years old. Northern Queensland is one of three places in eastern Australia where basalts poured over the edge of the coastal escarpment, as they did down the ancient Russell River and Johnstone River valleys around 2 Ma ago. In the Johnstone system the drainage now forms two rivers, the North Johnstone and the South Johnstone, on either side of the blocked valley. The two rivers join on the coastal plain. This is a common pattern: paired lateral streams running either side of a lava flow that filled the ancestral valley.

Near Cooktown there are 32 vents (as basaltic hills and ash cones), to the north at the head of the Starke Valley and to the southwest in the McLean and Lakeland Downs areas.

In the Atherton basalt province eight major volcanoes form topographic highs up to 1100 m ASL, such as Hallorans Hill (Atherton), Bones Knob (Tolga), Windy Hill (Ravenshoe) and the Malanda and Jensenville volcanoes on the Kennedy Highway. These volcanoes delivered the bulk of the basalts that created the Tablelands between 3.9 and 1.6 Ma ago, and the basalts have weathered to rich red-brown soils. The scenic falls at Mungalli, Millaa Millaa and further inland at Millstream are all cut over basalts. In some cases the layering resulting from multiple flows can be seen. Commonly there is a columnar structure formed by cooling cracks as the lava solidified.

The basalts consist of a series of flows that filled old river valleys, and in some areas there are layers of sand and gravel left by streams that flowed between the eruption episodes. At Herberton, miners in the 1890s tunnelled under the basalt to recover alluvial tin from these ancient river deposits, called the Deep Leads.

There are also numerous scoria cones younger than around 200 000 years and many probably less than 20 000 years, such as Mount Quincan and The Pinnacles near Yungaburra. Some of the cones now contain deliciously cool waters, such as Lakes Eacham and Barrine. Craters such as Bromfield Swamp have a very broad crater with a low profile, and were probably maars. Maars are vents in which there was explosive interaction between rising lava and groundwater.

One unusual feature is the Mount Hypipamee crater east of Herberton, which is a vertical hole 60 m across and at least 140 m deep to water bottom. It is carved through solid granite, not basalt, and must have formed during an explosive event that blew a hole through the granite. In the surrounding forest there are granite blocks and basalt bombs containing granite fragments, which must represent the material blasted out to form this diatreme. If there was once a basaltic cone around the crater, it has long been eroded away.

This open pipe is the topmost section of a conduit that must extend some 35 km down through the Earth's crust to tap magma from the mantle. Such conduits, in the form of pipes or fractures, must underlie all basalt volcanoes.

Further south, across the Herbert River, are more lava fields, east of

Mount Surprise and Einasleigh. The lava fields form a broad dome centred at Undara crater, with lava flows radiating out and two flowing northwest, along older courses of the Einasleigh and Lynd Rivers. This region also hosts hot springs, at Innot Hot Springs and at Tallaroo, which may be due to residual heat in the crust from the geologically recent volcanism.

If you travel by train or drive on the long flat roads in central Queensland or west of Charters Towers, you will see extensive lava-topped mesas. Hughenden lies close to the Great Divide, so some of these basalts run northwestwards towards the Gulf and others run eastwards, one flowing into the Burdekin River and reaching almost to Charters Towers. Many were erupted from a volcano at Mount Desolation, north of Hughenden. Closer examination of the mesas between Torrens Creek and Hughenden show that the flat tops are basalt layers overlying a weathering surface on older rocks and sediments. By dating the basalt we can therefore estimate the age of the weathered surface, since it must be older than the basalt which has flowed over it. It turns out that the lowest mesas have basalts just under 1 Ma old, whereas the higher mesas are around 2.5 to 3.3 Ma old. The highest mesas have the oldest basalts, at 5.5 Ma. These basalts record the older age of the higher plateaus, and the successively younger age of the land surface closer to the present plain. It is clear from the dates that the land surfaces have been forming for at least 5 million years.

The original basalts flowed down valleys and onto plains. These basalts were very hard and resistant to weathering, so later erosion cut away the nearby hills, eventually cutting them down below the level of the original valley. Strange as it may seem, in many cases the original valley now sits on the hilltop – an inverted topography.

Figure 8.3 *Mesa capped by lateritised basalt at Boat Mountain, north of Murgon, Queensland.*
PHOTO: LIN SUTHERLAND, AUSTRALIAN MUSEUM

Eight of the lava flows in northern Queensland are notable because each flowed for more than 80 km. These are exceptional lengths. The Undara flow is the longest, flowing for 160 km from the source crater. The Undara volcano erupted 190 000 years ago, and its lavas cover 1550 km^2, with an estimated volume of 23 cubic kilometres.

Temperatures of erupting lava are around 1200–1300°C and a basalt normally spreads out and cools within a short distance, so that it slows down and stops. By the time the basalt temperature has dropped to 1100°C crystallisation is under way, and this will increase the viscosity of the flow, helping it slow. It seems that the very long flows were caused by continually high rates of eruption from the vent and the channelling of flows into river courses. During the younger flows the river courses would have been commonly dry for much of the year because of the semi-arid climate, and the flows would then have travelled unimpeded.

There is also considerable evidence of lava tubes, notably the spectacular examples at Undara itself, but there are many other places where there are remnants of tubes, or channels that are probably collapsed tubes. Lava tubes form where the surface of the lava solidifies, but the lava inside is still hot and keeps running downhill. These tubes can be seen in action in Hawaii today, where collapsed holes in the roof of a tube provide views of the red-hot basalt racing along inside the lava tube. Lava tubes insulate the flowing basalt, thereby maintaining a high temperature and allowing a long flow to develop. If the lava drains away then the tube is left, sometimes with 'tide-marks' showing the former levels of lava in the tube. These tide marks may be caused by an extra coating of basalt cooling against the wall, or to thermal erosion of the old wall by younger hot basalt.

Figure 8.4 *Lava tube (Taylor Cave) at Undara. These basalts are 190 000 years old.*
Photo: Jon Stephenson

Central and southern Queensland

Similar mesas and soil-covered lava plains occur in a belt running southwards from Cape Hillsborough past Nebo, Clermont and Springsure to Injune, with smaller outcrops to the east near Bauhinia and Monto. These occurrences form one of the main belts of central volcanoes, although there was activity as long ago as 55 Ma at Nebo and near Anakie, indicating that some of the basalts are remnants from previous lava fields.

The drive out from Seaforth to Cape Hillsborough winds past the massive domes of Mount Jukes and Mount Blackwood. These mountains, with their sheer rock faces, are massive intrusions and probably represent the solidified magma chambers underlying volcanoes that have since eroded away. The intrusions are the same age as the volcanic lavas and tuffs exposed along the coastal cliffs in the Cape Hillsborough National Park.

The area northwest of Anakie has more than 70 volcanic plugs in a circular area some 50 km across. Some of the plugs form sharp hills, others just low rises. The oldest known basalt is Policemans Knob, west of Rubyvale, which is 56 Ma old. This region is notable for its rich alluvial gem deposits of sapphire and zircon (see Box 8.1). The basalts at Buckland sit right on the line of the Great Divide, and most are the remnants of a large shield volcano with radial extensions that represent flows along old valleys.

In southern Queensland there are many basalt occurrences in the Bundaberg, Gayndah and Bunya areas and extending in an almost continuous strip through Kingaroy and Toowoomba to the New South Wales border. There are also scattered occurrences in the Ipswich–Brisbane area.

Figure 8.5 *Plug of trachyte remaining after erosion of the surrounding volcano, Belougery Spire, Warrumbungles.*
PHOTO: LIN SUTHERLAND, AUSTRALIAN MUSEUM

New South Wales

On the Queensland–New South Wales border, the Focal Point and Tweed volcanoes form overlapping major shield volcanoes each some 100 km across. The Focal Point volcano was the source for the basalts of the Lamington Plateau, now cloaked in dense rainforest vegetation. The Tweed volcano lies to the southeast and contains the prominent peak of Mount Warning. This is another sheer rock mass, most of which is actually intrusive – a solidified magma chamber rather than a volcanic neck.

The great arc of basalts across and inland of the New England area (see Figure 8.1) were erupted over a long period, 56–19 Ma ago. They extend towards the coast at Comboyne, and to the heights around the northern end of the Hunter Valley at

Figure 8.6 *a. The present extent of the Ebor central volcano. The Great Escarpment (heavy line) separates the edge of the New England plateau from the coastal plain of northern New South Wales. Erosion by coastal rivers formed the Great Escarpment, removing 90% of the original volcano. b. Reconstructions show that the original volcano was about 800 m high, and its summit was about 1900 m above the present sea level.*

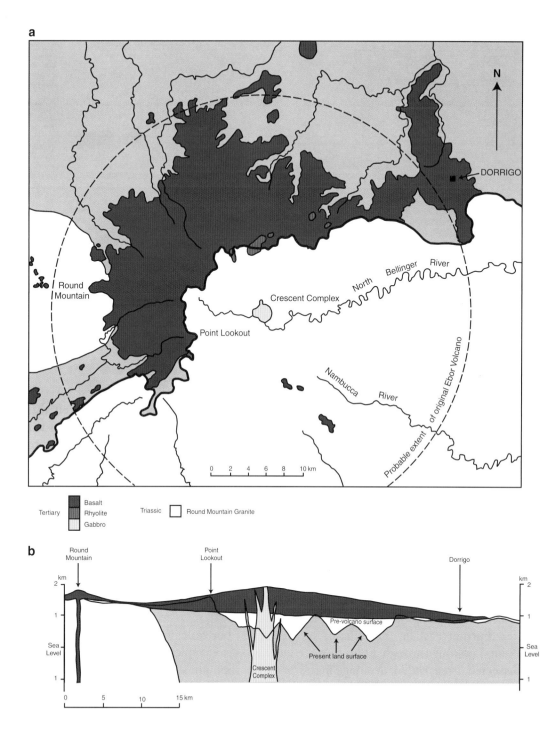

the Barrington Tops and the Liverpool Range. Inland are the basalts at Mount Kaputar, the Nandewars, the Warrumbungles and Dubbo. The nearby Nandewar volcanic area contains several erosional remnants, including Mount Kaputar (formed 17–18 Ma ago) and Killarney Gap (20–21 Ma).

Figure 8.7 *The Tweed central volcano. Despite considerable erosion the remnant core shape of the volcano is still evident on the skyline, with Mt Warning at the centre. The lava flows extended from this core past where the photograph was taken, some 20 km away.*
PHOTO: LIN SUTHERLAND, AUSTRALIAN MUSEUM

The Warrumbungle volcano covers a circular area 50 km across just west of Coonabarabran, a series of low hills rising above the Western Plains. Even this moderate height is enough to provide cooler air and lusher vegetation. Within the national park there is excellent scenery and bushwalking, with the opportunity to see tuffs exposed along the Grand High Track. Numerous plugs are obvious, and the Breadknife represents a dyke from which the wall rocks have been eroded. Parts of the original shield with layered lavas are visible in the erosional remnants at Mount Woorut, Mount Exmouth and Blackheath Mountain.

Lake sediments of grey and white diatomite were mined near Coonabarabran. Diatoms are microscopic algae that have an outer covering of opaline material. As they die and accumulate on the lake bed, a deposit of porous chalky material (diatomite) is produced. The Coonabarabran diatomites represent accumulations in lakes, perhaps in heated waters, and have yielded some fine fossils of fish, leaves, insects and birds.

Further east lie the remains of the large Ebor volcano, which extends from Point Lookout through Ebor to near Dorrigo. As is typical, the volcano has been dissected by a radial drainage pattern. An earlier phase around 48–45 Ma ago erupted the basalts capping the older (Triassic) granite Round Mountain to the west. Erosion had cut into granites and older rocks east towards the present site of Dorrigo. Then the basalts erupted at Ebor from 19–20 Ma ago. Reconstructions indicate that the Ebor volcano was over 45 km across and about 800 m high. Its centre must have been close to the present Crescent Complex in the upper North Bellinger River valley. However, the Crescent rocks are gabbro and other intrusives only about 18 Ma old, and were emplaced just after eruption of the Ebor basalts.

Erosion along the coastal escarpment by the North Bellinger, Nambucca and Macleay Rivers has removed 90% of the volcano. The lava flows that built the shield are now perched on top of the cliffs.

The most southerly central volcano in New South Wales is the Canobolas volcano near Orange, originally some 50 km across but now much reduced by erosion, especially on the southern side. The basaltic rocks are strewn around the edges, but the elevated central part is composed of rhyolitic and other volcanics. Flows exposed on the northern side are generally 2–3 m thick, commonly with soils or diatomite between them, indicating lengthy quiet periods between eruptions.

On the southern coast of New South Wales there are two places where basalts flowed into the coastal valleys. At Ulladulla near Lake Conjola there is 30 Ma old basalt near Lake Conjola and older basalts in the headwaters of the Towamba River valley near Bega.

Other lava field volcanics are scattered across the highlands in southeastern New South Wales, notably in the Canberra, Monaro and Snowy Mountains areas. The Monaro Plateau contains 65 volcanic plugs and extensive lava plains. The lavas extended to cover over 42 000 square kilometres, with a volume calculated to exceed 630 cubic kilometres, and range in age from 58 to 34 Ma.

The Monaro volcanic province sits squarely on the Great Divide, separating the headwaters of the Murrumbidgee (flowing inland to the northwest) from those of the Snowy River (flowing south to coastal Victoria). Outcrops show that the basalts filled deep valleys with up to 500 m relief. Clearly there was very major incision of the highlands early in their history, and the present erosion is only starting to uncover some of these older surfaces and deposits.

Box 8.1 Basalts as a source of gemstones

Basalts form in the Earth's mantle, and many basalts contain minerals or rock fragments that were plucked out of the surrounding mantle wall rocks as the magma ascended to the surface. Among these are a series of commercially valuable gemstones, including sapphires, rubies and diamonds, as well as stones that are less precious but interesting to amateur lapidaries: garnet, peridot (olivine) and zircon.

Detailed studies of the Barrington volcano, which contains rubies, sapphires and zircons, show that the volcano was active for a very long time – 55 million years. There were separate gem-bearing eruptions at 57 Ma, 43 Ma, 38 Ma, 28 Ma and 4–5 Ma ago.

Plenty of examples of these minerals have been found in the basalts, but commercial diggers and amateur collectors target the alluvial deposits around the basalts where creeks have concentrated the gems in layers within alluvial sands and gravels. In some cases it seems the gems were derived partly from weathered tuffs near the base of the basalts. The main areas of commercial mining and amateur fossicking are around Anakie in central Queensland and the New England and Crookwell–Oberon areas in New South Wales.

Sapphires in the Australian deposits are mainly dark blue; party sapphires, the pale yellow-green and blue variety, are more common around Anakie. Obvious crystal faces may be evident on both sapphires and zircons. Diamonds are generally hard to pick by the amateur. They are generally rounded, perhaps due to abrasion during transport, but most likely they came this way out of the volcanics.

The sapphires, rubies, garnets, peridots and diamonds in particular are normally formed at great depths under very high pressures and temperatures. Most of these minerals have probably come from the mantle at least 50–100 km below the surface, with the diamonds perhaps coming from over 200 km depth. Some of the basalts also contain large lumps of olivine-bearing rock (xenoliths), not just small mineral grains, which have also come from the mantle.

The traditional model for diamond formation exemplified by the South African occurrences requires the presence of very old, cold and thick crust to provide the high load pressures required to compress carbon into the diamond crystal structure. However the New South Wales diamonds have been formed under much younger, warmer, thinner crust, and seem to have been produced by dynamic pressure as a thick cold slab was subducted under eastern Australia, probably during the Palaeozoic. The only dated New South Wales diamond is about 300 Ma old, much younger than the 1100 Ma ages established for the South African gems.

These gemstones and other mantle-derived materials have been found in older basalts, such as the Jurassic diatremes in the Sydney region (see Chapter 6) and indicate that the volcanoes of eastern Australia have been tapping the mantle for an extended period, perhaps some 150 million years. These materials are rare in the Tertiary central volcano basalts but common in the lava field provinces.

Within lava fields the gems are most common in tuff cones and in the volcanic necks that represent the final plugging of vents, rather than in the extensive lava flows. It seems these basalts must have risen very quickly from the mantle and erupted at the surface or solidified in the vents. That is probably one reason why the gems are more common in fragmental tuff deposits than in the continuous lavas. The tuffs represent eruptions where the magma has risen rapidly through the crust. A quick ascent prevents gases from being liberated gradually, so they are released almost instantaneously at the surface. The magma explodes out of the vent, spraying ash and larger fragments rather than spilling over the brim as liquid lava.

The rapid release of steam or confining pressure in any situation can be very dramatic and dangerous. A comparable example is throwing a billy of water on a hot campfire, or accidentally lifting the lid on a pressure cooker – the whole contents explode, spraying hot matter in all directions, with potentially dangerous consequences.

Estimates have been made of the rate of ascent of magma from the mantle to the surface. The gems and mantle-derived materials are much denser than the basalt magma, so the basalt must be rising fast enough to prevent the gems and rocks descending under gravity. The estimates indicate that it takes between 5 and 50 hours for the travel, and the magma is moving at tens of metres per second.

Figure 8.8 *Mantle materials are transported to the surface by basalt volcanoes. This volcanic bomb from Mount Schank, South Australia, is a lump of olivine-rich mantle rock coated with basalt, which has been shaped by flight through the air.*
PHOTO: DAVID JOHNSON. SPECIMEN: PETER WHITEHEAD

Victoria and South Australia

Scattered remnants of basalts erupted 59–19 Ma ago occur south and east of Melbourne. There are both dissected lava fields preserved on uplifted highland areas and also valley fills around the margins of the highlands. Excellent exposures can be seen on the coast of the Mornington Peninsula. The remains of somewhat younger volcanoes around 7 to 4.6 Ma old form the elevated hills at Mount Macedon.

The youngest volcanic province erupted in the last 4.5 million years, and is known as the Western or Younger Volcanics. These form the extensive western plains of Victoria between Melbourne and Hamilton. They cover about 15 000 square kilometres and have nearly 400 known eruption points. Many of the features are very recent, less than 500 000 years old, and still preserve the original shape of cinder cones and craters. Mount Elephant is a scoria cone with the shallow crater still preserved, and Mount Rouse is another scoria cone showing the start of gully erosion where the vegetation has been cleared. In contrast, Mount Hamilton is the remnant of a lava volcano and is composed of much more solid basalt than the debris in the scoria cones. Many of these volcanoes display fine exposures of the basalts and scoria where they have been eroded.

Tower Hill near Warrnambool started erupting 20 000–18 000 years ago, and magma interacted with water in the underlying sediments producing explosions of steam and scoria. Eruption ceased about 7000 years ago, and a freshwater swamp formed in the crater.

Another vent is the Bridgewater volcano, exposed by coastal erosion near Portland. Cliff exposures show layers of ash dipping away from the central vent area, and thin basalt dykes cutting vertically through the tuff deposits. The Bridgewater volcano was a tuff cone or series of cones overlain by basalt flows forming the final lower profile shape of a shield volcano. At Byaduk and Mount Eccles are excellent examples of lava tubes as well as collapsed tubes forming lava canals, formed in a similar way to the Undara tube system in northern Queensland.

Across the border in South Australia are the volcanic vents at Mount Gambier, Mount Schank and Mount Burr. Mounts Schank and Gambier were active as recently as 4600 years ago – the most recent volcanic activity known in Australia. Mount Gambier contains a crater lake with walls that show layers where the volcano sprayed ash and larger bombs, building the sloping volcanic cone.

Tasmania

Basalts occur over most of Tasmania (except in the extreme southern and southwestern areas) and extend offshore as seamounts in depths of around 900 m of water. Onshore the basalts form lava fields and especially the fills of eroded valleys. The most extensive exposures lie between Waratah and Wynyard and there are many exposures along the coast, such as at Table Cape and at The Nut, where magma in a volcanic neck has intruded into bedded tuffs now exposed on the shore platform. Basalt remnants fill ancestral valleys of the Derwent River near Great Lake, former courses of the Macquarie and South Esk Rivers, and old valleys in the Tamar region.

Seamount chain offshore

The volcanism that extended down eastern Australia onshore is parallelled offshore by two lines of seamounts (see Figure 8.1). Individual seamounts are up to 60 km across at their bases, some of them 4000 m high, with flat tops eroded by the ocean waves before they subsided. These were first recognised as giant submarine volcanoes by Edgeworth David following a cable route survey by the SS *Britannia* in 1901. Further work mapped their outlines in the 1970s, and samples were dredged and dated carefully in the late 1980s.

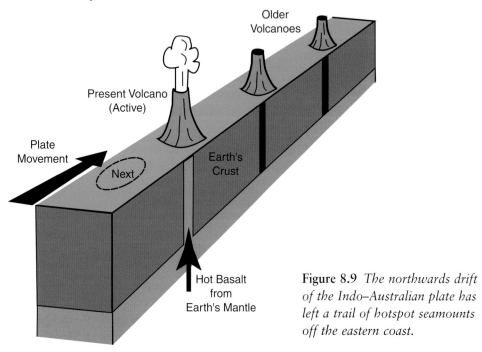

Figure 8.9 *The northwards drift of the Indo–Australian plate has left a trail of hotspot seamounts off the eastern coast.*

As the Australian continent and the Tasman Sea have drifted north towards Asia on the Indo–Australian plate, lines of volcanoes have been punched up through the crust. Three such lines extend down eastern Australia as central volcanoes, and two more down the Tasman Sea as the Tasman and Lord Howe seamount chains. The oldest volcano is at the northern end of each line and the youngest is at the south.

The western or Tasman chain extends some 1300 km north from the Gascoyne Seamount, which is in the middle of the Tasman Sea 500 km east of Bermagui. The volcanoes are youngest at the southern end; the Gascoyne Seamount is 6 Ma old, and the Queensland Guyot east of Brisbane is the oldest, at 21 Ma. If we divide the distance between these two by the age difference, we get an estimate of the rate at which the plate is being dragged northwards over the mantle plume. It turns out to be about 67 mm/yr, which is very close to the long-term spreading rate estimated from the magnetic stripes of around 70 mm/yr.

If the line of the Tasman seamounts is extended southwards we can estimate the present position of the mantle plume – near latitude 40°S, about opposite the northern coast of Tasmania. Whether we can expect the formation of another seabed volcano is a moot point. However, there is a cluster of seismic activity in the predicted region of the hot-spot, in an otherwise seismically inactive area.

The eastern chain extends north from Lord Howe Island. Lord Howe itself is 6.9 Ma old, and the Gifford seamount to the north is dated at 15.6 Ma. The Lord Howe volcano was built in over 2000 m of water to rise 875 m above sea level. The central part of the volcano collapsed to form a caldera now 900 m deep and 5 km by 2 km across, filled by younger basalts and surrounded by tuffs.

Origins of the volcanics and the Great Divide

The Great Divide and the Great Escarpment

The Great Divide forms a watershed running down eastern Australia from Cape York to Victoria, and separates the shorter rivers flowing to the eastern coast from the longer rivers draining westwards. The Divide itself is a line that can be drawn between the drainage basins, and it forms the central line within a much broader belt of uplands and plateaus some 400 km wide, sometimes called the Great Dividing Range or the Eastern Highlands.

The eastern edge of these uplands is dominated by a series of bluffs with narrow gorges running down towards the sea. This is the Great Escarpment. Roads climb with a series of steep turns out of the valleys to vantage points giving sweeping views of the coastal regions. Sublime Point north of Wollongong in New South Wales and Mount Spec near Townsville in northern Queensland are examples. To the west is generally a tableland or plateau typically several hundred metres in elevation, while below the escarpment to the east is a coastal plain less than 50 m above sea level. The escarpment lies closer to the coast than the Great Divide, and in central Queensland is 200–300 km closer.

Anyone lucky enough to fly along the edge of the escarpment can see how the coastal rivers are cutting back into the edge of the tableland. In some places the rivers have eroded the plateau edge so deeply that separate valleys have isolated a part of the plateau, leaving it stranded east of the escarpment; an example is the Comboyne plateau in central New South Wales (see Figure 8.6, page 171).

Figure 8.10 *Dissection along the edge of the Great Escarpment at Fitzroy Falls near Wollongong, where erosion has cut down steeply through the edge of the Illawarra Plateau.*
PHOTO: DAVID JOHNSON

The term 'Great Dividing Range' is not really appropriate, since in many areas it is only a line of very low hills. Even the highlands of the Divide are typically only 300–1600 m above sea level, which is low by world standards. The highest point is Mount Kosciusko at 2228 m above sea level, and the only extensive areas above 1000 m are the Snowy Mountains and Australian Alps and the New England areas in New South Wales. In contrast, the Divide is only 527 m above sea level about 10 km west of Mareeba in northern Queensland.

Typically the Great Divide lies 150 km inland from the eastern coast, and short steep river systems descend from it to the sea. This is true for the areas of far northern Cape York, and for areas in southern Queensland, New South Wales and Victoria. However, in central Queensland the Divide swings up to 400 km inland and consists of a much lower set of hills, with the very big drainage basins of the Burdekin and the Fitzroy rivers flowing eastwards.

The close association of the volcanics and the Great Divide indicate the two may have a similar origin. There is clear evidence that the hills and valleys were in existence during the Tertiary, because erosional remnants of basalts as old as Eocene age – over 40 Ma old – lie in the valleys. Clearly the valleys must be older than 40 Ma if basalts of this age flowed down them.

In three well-known places the basalts flowed over the scarp and down onto the coastal plain: at Innisfail in northern Queensland, where the basalt is still exposed at Clump Point; in New South Wales at Ulladulla and in the headwaters of the Towamba River south of Bega. The basalts on the southern coast of New South Wales have been dated at 51–40 Ma at Nerriga, 35–27 Ma at Moruya, and 30–29 Ma at Ulladulla. Thus these valleys and the coastal scarp must have been formed before 30–50 Ma ago. More recently, basalts reached the coast in the Warrnambool–Port Fairy region of the Victorian coast, where young maars also form parts of the coastal geology.

The Tasman Sea started to open about 84 Ma ago after some 10 million years of gradual cracking of the crust. Somewhere around 70 Ma ago very deep cracks had formed extending down through the crust to the mantle, and allowing basaltic magmas to ascend steadily to the surface, erupting as a series of lava provinces all down the eastern coast.

It is worth remembering that 70 Ma ago Australia was still positioned in the southern ocean, about where Kerguelen is now. The Tasman Sea kept opening for about 30 million years, probably maintaining the steady tension on the crust and keeping the cracks open. Off northeastern Australia the Coral Sea opened only at the end of this time, but for less than 10 million years, stopping about 48 Ma ago. For some reason, at present unknown, both the Tasman and Coral Seas stopped opening at the same time.

As the sea-floor spreading slowed, uplift of the continental margin ceased and the Earth's crust began to cool and subside. The result was a plateau that then began to flex down towards the east, like a sheet of hardboard supported only on one edge. Part still dipped inland, but the eastern part sloped down towards the Tasman Sea, and it is now about 1000 m high on the top of the New England area and 300–400 m high at the Great Escarpment.

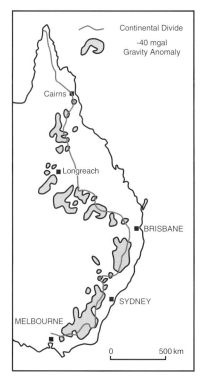

Figure 8.11 *The line of the Great Divide follows the trend of gravity anomalies, which probably indicate the presence of deep crustal roots of an old mountain belt.*

One view is that the highlands were formed by doming caused by the injection of basalt from the mantle into the lower crust, evident after 90 Ma ago. As the mountains eroded, the Earth's crust rose buoyantly, allowing further downcutting by streams. Alternatively, the uplift may have been caused by faulting associated with earlier parts of the rifting of eastern Australia over 100 Ma ago, before the basalts started to erupt. At these times of basaltic volcanism the landscape would have contained large areas bare of vegetation, with glassy surfaces of chilled lava, much as can be seen in Hawaii today (see Figure 2.4). When the climate became wetter the basalt would have weathered and become vegetated again.

In southeastern Australia and Tasmania, fission track studies show denudation rates around 10 m per million years from 300–100 Ma, rising sharply to 30–40 m per million years around 50 Ma ago, probably as a result of rapid uplift.

However, it has been pointed out that the line of the Divide follows the trend of major anomalies in the Earth's gravity field. These gravity anomalies are known to be caused by thicker than average crust, formed under mountain belts. So the Divide may represent the remnants of an old mountain chain dating from the Mesozoic or Palaeozoic, maybe 100–200 Ma ago.

Although there is evidence for the eastern highlands having very old roots, this does not mean that the old concept mooted by early geologists of a more recent 'Kosciusko Uplift' is wrong. Studies of Tertiary sediments in western Victoria and the Otway Ranges have shown an uplift of between 175 and 240 m occurred there, dating from around 1 Ma ago. Similarly, the Mount Lofty Ranges in Australia experienced uplift from around 5 Ma ago. So some of the high topography may be due to more recent crustal elevation, and is not just an erosional remnant of old, much higher mountains.

Figure 8.12 *Basalt overlying older bedrock in the gorge at Einasleigh, Queensland. The old landscape under the basalt can be dated by the age (0.26 Ma) of the basalt that flowed over it.*
Photo: David Johnson

Lateritic soils developed on the eroded bedrock and granites, as well as on many of the basalts. There are many places where these basalts can be used to determine the age of the underlying eroded landscape, and in some cases the lateritisation. For example, lavas from the Barrington volcano, which started erupting 59 Ma ago, overlie older folded rocks intruded by Permian granite. So the old granite landscape here is at least 60 Ma old in places, but has been modified by more recent erosion.

Estimated rates of denudation (that is, general lowering of the plateau lands by erosion), are around 5–7 metres per million years for Tasmania and eastern Victoria, 2–5 metres per million years in southern New South Wales, and 3 metres per million years for northern New South Wales and Queensland. The valleys tend to cut down much faster, at 5–30 metres per million years. A detailed study of the New England region has shown that the escarpment retreats along river valleys at about 2 kilometres per million years.

A similar rate of denudation (3.3 metres per million years) has been estimated for the Dugald River area in northwestern Queensland using radiometric (argon–argon) dating of weathering minerals. As is the case for the basalt-covered mesas near Hughenden (see p. 168) the oldest dates are on the highest parts of the landscape.

Summary

Although there is no active mainland volcanism in Australia, it stopped only 4600 years ago. A chain of mainly basaltic volcanoes extended along the line of the Great Divide from the Torres Strait to southeastern South Australia. Two styles of volcanism occurred, starting 70 Ma ago:

1. Lava fields, which erupted from fractures and spread over areas 100 km across.
2. Central volcanoes, which are younger towards the south. A similar trend is apparent in the lines of volcanic seamounts in the ocean east of Australia. These volcanoes represent periodic eruptions from underlying mantle plumes as the Australian plate drifted northwards.

The volcanoes carried precious and semiprecious gems, including sapphires, rubies, diamonds, zircons and peridots, from the mantle into the erupted lavas and ash. These gems have been weathered into river and creek sediments.

The elevated land along the Great Divide was gradually worn down by rivers, forming deep valleys carved into the edges of the ranges and tablelands. Since the basalts were erupted periodically over 70 million years, there were many times when the lavas poured down and filled valleys, sometimes blocking water courses to form lakes that accumulated diatomite.

Age determinations of basalt-covered hills and of weathered materials shows that the landscape has been lowered at typical rates of 2–7 metres per million years though individual valleys may cut down much faster.

SOURCES AND REFERENCES

Ashley, P.M., Duncan, R.A. & Feebrey, C.A., 1995. 'Ebor volcano and Crescent Complex: age and geological development'. *Australian Journal of Earth Sciences*, 42: 471–480.

Atkinson, A. & Atkinson, V., 1995. *Undara Volcano and its Lava Tubes*. Vernon and Anne Atkinson, P.O. Box 505, Ravenshoe, Australia 4872. 86 pp.

Coventry, R.J., Stephenson, P.J. & Webb, A.W. 1985. 'Chronology of landscape evolution and soil development in the upper Flinders area, Queensland, based on isotopic dating of Cainozoic basalts'. *Australian Journal of Earth Sciences*, 32: 433–447.

Duggan, M.B. & Knutson, J., 1993. *The Warrumbungle Volcano. A geological guide to the Warrumbungle National Park*. Australian Geological Survey Organisation. 51 pp.

Irving, A.J. & Green, D.H., 1976. 'Geochemistry and petrogenesis of the Newer Basalts of Victoria and South Australia'. *Journal of the Geological Society of Australia*, 23: 45–66.

Johnson, R.W., 1989. *Intraplate Volcanism in Eastern Australia and New Zealand*. Cambridge University Press. 408 pp.

Kohn, B.P., Gleadow, A.J.W., Brown, R.W., Gallagher, K., O'Sullivan, P.B. & Foster, D.A., 2002. 'Shaping the Australian crust over the last 300 million years: insights from fission track thermotectonic imaging and denudation studies of key terranes'. *Australian Journal of Earth Sciences*, 49: 697–717.

McDougall, I. & Duncan, R.A., 1988. 'Age progressive volcanism in the Tasmantid Seamounts'. *Earth and Planetary Science Letters*, 89: 207–220.

Ollier, C.D., 1982. 'The Great Escarpment of eastern Australia: tectonic and geomorphic significance'. *Journal of the Geological Society of Australia*, 29: 13–23.

O'Reilly, S.Y. & Zhang, M., 1995. 'Geochemical characteristics of lava-field basalts from eastern Australia and inferred sources'. *Contributions to Mineralogy and Petrology* 121: 148–170.

O'Sullivan, P.B., Orr, M., O'Sullivan, A.J. & Gleadow, A.J.W., 1999. 'Episodic Late Palaeozoic to Cenozoic denudation of the south-eastern highlands of Australia: evidence from the Bogong High Plains, Victoria'. *Australian Journal of Earth Sciences*, 46: 199–216.

Roach, I.C., McQueen, K.G. & Brown, M.C., 1994. 'Physical and petrological characteristics of basaltic eruption sites in the Monaro Volcanic Province, southeastern NSW'. *AGSO Journal of Australian Geology and Geophysics*, 15: 381–394.

Sandiford, M., 2003. 'Geomorphic constraints on the Late Neogene tectonics of the Otway Range, Victoria'. *Australian Journal of Earth Sciences*, 50: 69–80.

Stephenson, P.J. & Griffin, T.J., 1976. 'Some long basaltic lava flows in North Queensland' in R.W. Johnson (ed.) *Volcanism in Australasia*, Elsevier, p. 41–51.

Sutherland, F.L., 1995. *The Volcanic Earth*. UNSW Press. 248 pp.

Sutherland, F.L., 1998. 'Origin of north Queensland Cenozoic volcanism: Relationships to long lava flow basaltic fields, Australia'. *Journal of Geophysical Research*, 103: 27,347–27,358.

Sutherland, F.L. & Fanning, C.M., 2001. 'Gem-bearing basaltic volcanism, Barrington, New South Wales: Cenozoic evolution, based on basalt K–Ar ages and zircon fission track and U–Pb isotope dating'. *Australian Journal of Earth Sciences*, 48: 221–237.

Sutherland, Lin & Graham, Ian, 2003. *Geology of the Barrington Tops Plateau*. The Australian Museum Society, Sydney. 56 pp.

Vasconcelos, P.M. & Conroy, M., 2003. 'Geochronology of weathering and landscape evolution, Dugald River valley, NW Queensland, Australia'. *Geochimica Cosmochimica Acta*, 67: 2913–2930.

Wellman, P., 1987. 'Eastern Highlands of Australia; their uplift and erosion'. *BMR Journal of Australian Geology and Geophysics*, 10: 277–286.

Wellman, P. & McDougall, I., 1974. 'Cainozoic igneous activity in eastern Australia'. *Tectonophysics*, 23: 49–65.

Whitehead, P.W. (ed.) 1996. *Conference Abstracts: Long Lava Flows*. EGRU Contribution 56, Department of Earth Sciences, James Cook University.

Young, R.W., 1981. 'Denudational history of the South-Central Uplands of New South Wales'. *Australian Geographer*, 15: 77–88.

Young, R.W., 1989. 'Crustal constraints on the evolution of the continental divide of eastern Australia'. *Geology*, 17: 528–530.

Young, R.W. & McDougall, I., 1993. 'Long-term landscape evolution: Early Miocene and modern rivers in southern New South Wales, Australia'. *Journal of Geology*, 101: 35–49.

Websites

Australian general information
www.agso.gov.au/education/lava/
www.agso.gov.au/geohazards/grm/volcano/volcano.html

Volcanoes in the Australian region
www.volcano.und.edu/vwdocs/volc_images/australia/australia.html

Illustrations

Fig. 8.1: redrawn by permission of the New South Wales Department of Mineral Resources after map, p. 344, in Scheibner, E. & Basden, H. (eds) 1996. *Geology of New South Wales. Vol. 2: Geological Evolution*. Geological Survey of New South Wales Memoir 13

Fig. 8.6: by permission of the Geological Society of Australia after figs 2, 6 in Ashley, P.M., Duncan, R.A. & Feebrey, C.A., 1995. 'Ebor volcano and Crescent Complex: age and geological development'. *Australian Journal of Earth Sciences*, 42: 471–480.

Fig. 8.11: modified, redrawn and reproduced with permission of the publisher, the Geological Society of America, Boulder, Colorado, USA. © 1989 and 1999 Geological Society of America: after fig. 1 from Young, R.W., 1989. 'Crustal constraints on the evolution of the continental divide of eastern Australia'. *Geology*, 17: 528–530

Origin of the outline

Australia is surrounded by ocean. For most of its geological history it was embedded in the larger continents of Rodinia and Gondwana. Now Australia is floating free.

Australia has a mainland coastline of 35 877 km, with another 23 859 km of coastline contributed by the 8222 islands around the mainland. Fraser Island covers 1653 square kilometres and is the largest sand island in the world.

The main outline of Australia derives from the dismemberment of Gondwana. The curve of the Great Australian Bight mirrors the circular outline of Antarctica where it was formerly joined. The angular shape of the Western Australian coast, the general trends of the southeastern margin and the northwesterly trend of northeastern Queensland follow the orientation of the major lines of spreading that fractured the ancient land mass. The original outline after rifting, where Australia had been joined to other parts of Gondwana, lies closer to the outer edge of the continental shelf rather than the present coastline. The Gulf of Carpentaria is an embayment that is the last remnant of the subsidence which formed the Cretaceous inland sea.

Continental shelf and slopes

The continental shelf is a large terrace around the continent (see also Figure 2.25), with a steeper seaward slope passing into the deep sea. The shelf is 40 km wide off the central New South Wales, only 20–30 km wide off the southeastern coast, but up to 220 km off the Great Australian Bight and the North West Shelf and some 320 km wide in the Timor Sea. The outer edge is typically shallower than 200 m water depth.

The upper surface is relatively flat, with a gradient of 0.5–1.5 m/km. At the shelf break this increases to 10–50 m/km on the continental slope. In places the edge is an almost sheer cliff. The shelf break is at only 80–100 m water depth in the Great Barrier Reef, 100–160 m along New South Wales, but closer to 200 m around southern and western Australia. Off South Australia the shelf edge drops from 200 m to over 5000 m water depth in less than 40 km – a precipitous drop.

The origin of the continental shelf lies in two separate but related events: fractured basement that subsided after rifting, and later filling by sediments to a flat surface.

Rifting of Australia from Gondwana was accomplished by stretching of the continental crust until it fractured and broke apart along normal faults. These faults allowed pieces of crust tens of kilometres wide and hundreds of kilometres long to slip against each other and subside, as the crust was pulled apart. Generally each fault block was displaced downwards against the one to landwards, although some blocks were left elevated above the ones on either side. The vertical movements on these faults were up to 2000 m.

What happened to the other side of the rift? In some cases, substantial blocks of rock that were formerly joined to Australia now sit a short distance away. For instance, the Naturaliste, Wallaby and Exmouth Plateaus off Western Australia, the

CHAPTER 9

BUILDING THE CONTINENTAL SHELF AND COASTLINES

Why is Australia the shape it is? What determines the type of coast we see?

Sea level has been at its present level only for about 6500 years, and it rose and fell many times over the last million years in response to the climatic cycles of the ice ages. How has this affected our present shorelines?

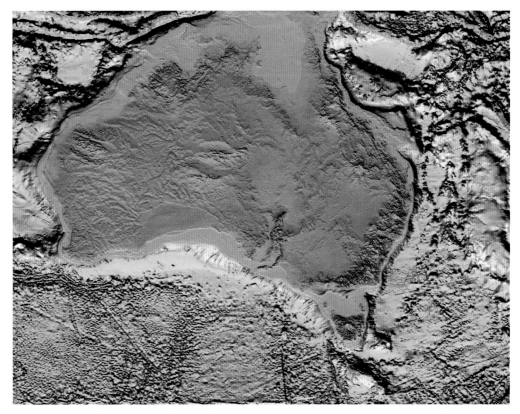

Figure 9.1 *Shaded topographic and bathymetric view of Australia and surrounding ocean regions, showing landmasses (brown) and shallow seas in yellow (< 2000 m water depth) tones, and deeper sea floors in blue (> 2000–4000 m) and green (> 4000 m). The extensive continental shelf joins the Australian mainland to Papua New Guinea in the north and to Tasmania in the south.*

During the Gondwanan break-up some parts of the continental mass did not drift far and are preserved as the surrounding plateaus, just off the shelves. East of Australia lies a sinuous strip of continental rocks, the Lord Howe Rise, which was part of Australia before the opening of the Tasman Sea. This continental mass extends southwards to join with the Campbell Plateau on which New Zealand is situated.

To the east lies the line of seamounts in the Tasman Sea that formed as the continent drifted northwards after rifting from Antarctica.

The east-west zone at the bottom of the diagram is elevated seabed marking the axis of sea-floor spreading between Australia and Antarctica.

The deformed terrains of Indonesia and Papua New Guinea, folded like ruckles on a blanket, are evident at the top of the map, and of New Caledonia and the Solomon Islands in the northeastern corner. The Timor Trough lies in a curve at the northwestern corner. Australia has so far remained unaffected by the considerable tectonic deformation suffered by these other regions.

IMAGE AND SOURCE DATA: COURTESY GEOSCIENCE AUSTRALIA, CANBERRA. CROWN COPYRIGHT ©. ALL RIGHTS RESERVED. WWW.GA.GOV.AU

Marion and Queensland Plateaus off the northeastern margin, and the South Tasman Rise which hangs off Tasmania, are all pieces of Australian continental crust still connected in the subsurface. All have subsided so that their upper surfaces are now 1000–2000 m below sea level, and in the case of Wallaby Plateau 3000 m below sea level.

Considering that these rocks could well have been at the same altitude as Mount Kosciusko when part of Gondwana, the total drop may be in the order of 3–5 km. This gives some idea of the amount of vertical movement that occurs regularly in the Earth's crust.

In other cases, a former piece of Australia has been moved by sea-floor spreading much further away. A good example is the Lord Howe Rise, which is now separated from Australia by the basaltic sea-floor of the Tasman Sea (Figure 9.3). Of course, the other side of the rift off southern Australia is the Antarctic continent, now separated by over 3000 km of sea-floor basalts.

The second process involved in forming the continental shelf is sedimentation, which fills and covers this faulted basement. Erosion of the landmass would have started during the uplift phase, before the Gondwana crust began to fracture. As the separation developed, continental mountains were further eroded and active volcanoes delivered large volumes of lava and ash into the developing rifts.

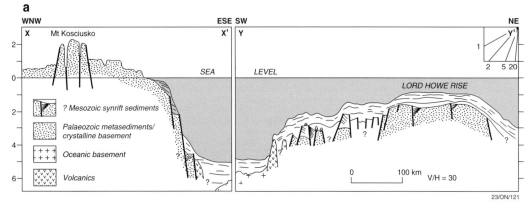

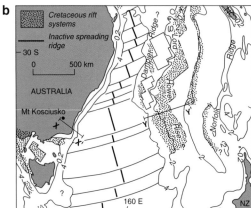

Figure 9.2 *a. Reconstructed schematic sections between southeastern Australia and the Lord Howe Rise, with about 800 km of oceanic crust removed: the left part (X–X') is across the Australian margin and the right part (Y–Y') is across the Lord Howe Rise. This section is based on mapping onshore and seismic work offshore. The sections have vertical scale exaggeration (30 times), so that the slopes appear much steeper than would really occur on the ground. The small box in the top right corner illustrates the real slope in degrees for the angle of a line drawn on the diagram. Note the faulted basement, which is exposed by erosion on Mount Kosciusko but is buried by sedimentary sequences offshore. b. This diagram shows both sides of the rift that opened to form the Tasman Sea. Water depths in kilometres. The thicker lines in segments down the centre of the Tasman Sea represent the line of seafloor spreading, which is now inactive. For much of the Australian margin we cannot clearly identify the other side because it has now drifted away.*

IMAGE AND SOURCE DATA: COURTESY GEOSCIENCE AUSTRALIA, CANBERRA. CROWN COPYRIGHT ©. ALL RIGHTS RESERVED. WWW.GA.GOV.AU

Figure 9.3 *a. A geological section based on AGSO seismic lines 199-08 and 199-11 across the southern Australian offshore continental margin. The units are colour coded to match the stratigraphic column in figure 9.3b, which shows the geological history for this section with a sea level curve.*

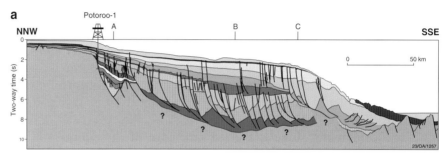

The continental shelf lies to the northwest (left) of the oil well Potoroo-1, and the deep-sea plain is the flat, aqua-coloured area on the right-hand side. Seaward of the shelf lies a terrace at 200–2000 m water depth. In the water column, 4 seconds of two-way travel time for the seismic wave corresponds to 3000 m water depth. In the rock sequence under the continental shelf the seismic speeds are faster and the sequence is actually thicker than it looks on the section.

Most of the continental shelf is constructed of Jurassic and Cretaceous sequences deposited in the sedimentary basin forming as Australia was separating from Antarctica. Note how the continental crust (pink) is cut by a series of normal faults that displace the crust successively downwards. The upper level of this crust was at one time the land mass of Gondwana and was exposed to the atmosphere. There was very extensive stretching off the continental crust in this region, meaning that the sequences extend laterally for long distances. The basal sequences, of mainly Jurassic age were formed in small rift basins (beige, magenta, brown units).

Break-up occurred between 99 and 86 Ma ago, when the units coloured dark blue, orange, pale blue and purple were deposited. Note how they too are heavily faulted because they were being stretched as the break-up continued. Continued subsidence allowed deposition of the green units, and finally the aqua units that form both the surface of the continental shelf and also the deep seabed.

All the sedimentary rocks forming these extensive units came from erosion of the Australian continent.

SOURCE DATA: COURTESY GEOSCIENCE AUSTRALIA, CANBERRA. CROWN COPYRIGHT ©. ALL RIGHTS RESERVED. WWW.GA.GOV.AU

The seismic sections show these sedimentary sequences are layered over the faulted basement. These sequences started forming during break-up, commonly in the Jurassic or Cretaceous. The upper sequences are of Tertiary and Quaternary age, with the surface layers generally formed during the Pleistocene ice ages. The details of sedimentation patterns vary around the continent, and depend on such factors as the local subsidence rate, the delivery of sediment from the land, and changing global sea levels.

The continental slope is the seaward edge of the continental shelf. In some places bare rock is exposed on this slope, but generally it is covered by a drape of sediments. Some sediments were delivered by rivers at times of lower sea level, though much is composed of oceanic plankton that has settled to the seabed. These plankton, mainly algae and foraminifers, have tiny skeletons of calcium carbonate that accumulate to form a pale grey or brown, calcareous pelagic ooze.

The slopes are incised by deep canyons that are conduits for sediment to the deep sea. Again it is thought that the upper parts of some of these were cut at sea level lows. However, the canyons extend down to water depths of 2000 m or more, way below even the lowest sea levels. So these must have been cut by downslope submarine erosion.

Box 9.1 Australia's Exclusive Economic Zone

Clearly, much of the offshore parts around Australia are of the same geology and are extensions of the land mass, though at the present time they are under the sea. And parts of the former continent have now broken away, such as the Lord Howe Rise. There are also nearby territories, such as Macquarie Island. So where are Australia's outer limits? How big is Australia really? Where is the limit of Australia's sovereignty and responsibility?

The land area of Australia is 7.7 million square kilometres, and there is a further 2.0 million square kilometres of continental shelf adjoining the main land mass. Under the provisions of the United Nations Convention on the Law of the Sea (UNCLOS), Australia is entitled to make a seabed claim out to a distance of 200 nautical miles (370 km) from the shoreline – the so called Exclusive Economic Zone, or EEZ. The EEZ includes the continental shelf but goes much further out, and is really a broader interpretation of the traditional 200 nautical mile fishing zone. The EEZ is the area over which a country has the right to explore and exploit living and non-living resources of the seabed, subsoil and superjacent waters. In terms of Australia's interests, the two obvious resources are fisheries and petroleum. Australia's EEZ, including that around territories such as Lord Howe, Norfolk Island in the Pacific Ocean, Macquarie and Heard–McDonald Islands in the Southern Ocean, and the Cocos–Keeling and Christmas Islands in the Indian Ocean, and the Australian Antarctic Territory, covers 11.1 million square kilometres.

Furthermore, under UNCLOS, Australia can claim areas that are geologically the same as the continental shelf beyond the 370 km line. Figure 9.4 shows that these include submerged areas such as the Naturaliste, Wallaby and Exmouth Plateaus and the Lord Howe and Tasman Rises. Geological and geophysical surveys have confirmed that these areas are the same geology as the mainland, and in the case of the plateaus are still connected to Australia in the subsurface. These potentially comprise a further 3.7 million square kilometres of EEZ. Thus the total offshore area to which Australia can lay claim is some 14.8 million square kilometres – almost twice the land area.

Where other countries have an interest within the 200 nautical mile radius, the boundary has to be negotiated. Such is the case in the Torres Strait, where Australia has a negotiated boundary with Papua New Guinea, and at Kerguelen and towards New Caledonia, where the boundary has been negotiated with France.

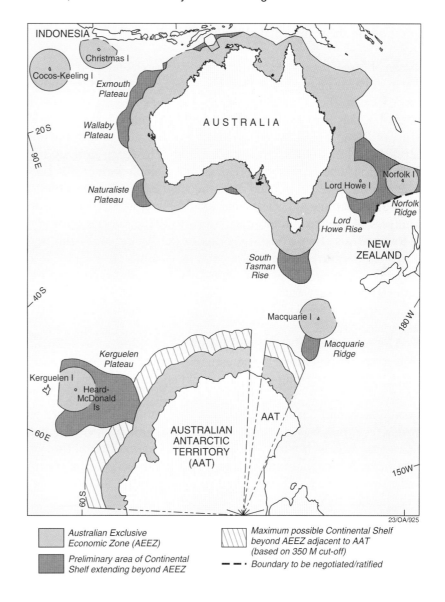

Figure 9.4 *Sketch map of the main marine jurisdictional zones around Australia and its territories. The Exclusive Economic Zone (green) and the legal continental shelf (magenta) beyond the 200 nautical mile limit are shown. Australia claims the oceanic zones around the Christmas, Cocos–Keeling, Lord Howe, Norfolk, Macquarie and Heard–McDonald Islands. The normal continental shelf is within the green zone, and the extra legal shelf may be claimed as a geological extension of the Australian continent.*

The hatched extension to the Antarctic EEZ is based on a formula allowing the continental shelf to be claimed up to the 350 nautical mile distance from the territorial sea baseline around the continent. These areas are subject to acceptance and ratification under the United Nations Convention on the Law of the Sea (UNCLOS). If granted, Australia will have responsibility not just for the zones connected to the mainland but for many beyond it.

SOURCE DATA: COURTESY GEOSCIENCE AUSTRALIA, CANBERRA. CROWN COPYRIGHT ©. ALL RIGHTS RESERVED. WWW.GA.GOV.AU

Sea levels

The sea level has risen and fallen in response to major climate changes (see Figure 7.9). When the Earth cooled, polar ice caps increased in size and ice sheets extended across Europe and North America. Small glaciers formed in the Tasmanian highlands at these times. The water for this ice was drawn from the oceans, so that during these cold periods sea levels fell, all around the globe. When ice sheets thawed, the sea rose. This has happened repeatedly over the last million years, so the continental shelf has been inundated as it is now, then dry when the sea level fell, then inundated again when sea level rose.

Oscillating sea levels during the Quaternary imposed two separate regimes of sedimentation on the continental shelf, as follows:

a   *Sea level lows during glacial phases*
    During glacial phases global sea levels were up to 130 m below present levels. Where the continental shelf break is at 200 m, such as off southern Australia, this low-level shoreline would still have been on the shelf. The Murray River would have flowed across the shelf to another shoreline, perhaps 50 km south of Kangaroo Island. However, in the Great Barrier Reef region where the shelf break is much shallower, at 80–100 m, the river would have debouched over the shelf edge, straight down the slope.

    Waters bathing the shelf edge would have been cooler, though not icy. Analyses of the strontium : calcium ratio (which varies with temperature) in fossil corals from tropical regions of the Pacific and Atlantic show that water temperatures were depressed 4–6°C some 10 000–14 000 years ago. Oxygen isotope analysis of microfossils in a core taken from the seabed under 1094 m of water south of Mount Gambier shows the ocean water temperature was 9–12°C 18 000 years ago, compared to 14–18°C over the last 14 000 years.

    The present continental shelf would have been a vast alluvial plain, with meandering channels and estuaries, mangroves and stranded old beach deposits marking the transitional shorelines. At this time the Australian coastline would have been much longer (see figure 7.10a). Tasmania would have been joined to the mainland, and there was a land bridge to Papua New Guinea. The last time this happened, up to 18 000 years ago, was well within the time of Aboriginal occupation of the continent. Aboriginal families would have camped on the continental shelf, and in successive generations had to camp further inland as the rising sea level pushed them back.

b   *Sea level highs during interglacial times*
    Water locked in the polar ice caps was released to the oceans as the ice sheets receded. While some research has assumed a steady return of

meltwater to the oceans, and hence a steady rise in the sea level, other evidence indicates the sea level rise was episodic. That is, there were short periods when the sea level rise paused and was stable or even dropped slightly, and periods of fast sea level rise to catch up between the pauses. The last time sea level was at its present level was around 120 000 years ago. Then for 100 000 years it was much lower, and of course Australia's dry landmass much greater.

Long-term tidal records confirm that the sea level is rising at around 1 mm per year. This has been attributed to thermal expansion of the ocean water body as the planet's atmosphere and ocean systems gradually warm. We have not seen, and do not expect to see, major sea level rises in the order of metres until the ice sheets melt, which will occur over a time span of hundreds of years.

The edge of Australia has been subjected to this oscillating sea level many times. If we had a time lapse movie made over a period of a million years we could watch it. We just happen to be living during an interglacial time, a time of high sea level. Imagine the chaos if we had started to build all the Australian coastal cities during a sea level low, out on the continental shelf, and then watched them all go under! Alternatively, the sea could recede, leaving our harbours high and dry. In a geological time frame, that will still happen one day.

Processes that form coastlines

Present shorelines result from both erosion of older rocks by the sea to form rocky headlands and wave-cut platforms, and sediment that accumulates to form coastal plains and beaches.

The sediment along Australia's coastline comes from five sources:

- river inputs from the mainland or locally off islands
- erosion of coastal rocks
- carbonate sediments resulting from marine animals, e.g. shells, corals, bryozoans
- possibly offshore sediment moved to the coast by tsunamis and other processes
- pumice from distant volcanoes.

Pumice floats, and is swept along by ocean currents. There are no active volcanoes in Australia; and this pumice comes from Tonga, Vanuatu and perhaps other Pacific islands. Pumice is swept westwards by the currents in the tropics, and then southwards by the East Australia Current along the edge of the continental shelf off eastern Australia. Geochemical analysis of some pumice can identify its source. For instance the eruption near Curacoa Reef, Tonga, on 14 July 1973 produced pumice that reached Townsville in North Queensland in February 1974, Brisbane in March 1974, and Seaspray in Victoria in June 1974.

Types of coasts

Coasts are classified as either erosional (drowned fluvial landscapes or eroding rocky coasts) or depositional (deltas or sandy or muddy shorelines).

Erosional coastlines

Drowned fluvial landscapes

The rising sea level flooded the low-level estuaries and eventually filled the valleys eroded in the land mass. Thus old river valleys became harbours and inlets.

Good examples are Port Jackson (Sydney Harbour), where the lower reaches of the ancestral Parramatta and Lane Cove Rivers have been flooded; and the Hawkesbury River system, Broken Bay and Pittwater to the north. Other examples are the protected inlet of Mourilyan in North Queensland, which is entered though a narrow inlet between a pair of rocky headlands, and the Kimberley coast where there are long indentations such as the entrance to the Prince Regent River, and where the Bonaparte Archipelago would have been a hilly landscape until it was surrounded by seawater. In Tasmania, the sea rose and flooded the Derwent River valley at Hobart and isolated Bruny Island offshore.

In all these areas the tide is now rising and falling daily against old rocky outcrops that were carved primarily by river erosion thousands of years ago. This is a drowned landscape. If the sea level was to fall again these rivers would run again, and the offshore islands would become hills on an alluvial plain.

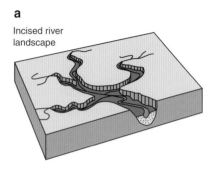

a
Incised river landscape

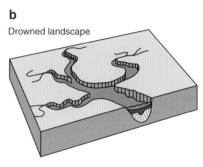

b
Drowned landscape

Figure 9.5 *A fluvial landscape becomes a drowned coastline as the sea level rises. a. The original river has cut a valley down through the pale brown rocks and is flowing across a fluvial plain (green). River deposits in the valley are coloured yellow. b. A rise in sea level has flooded the valley. The fluvial sediments are preserved in the base of the old valley and overlain by younger marine sediments (brown).*

Figure 9.6 *Sea level rise drowned the palaeo-valley of the Derwent River, where Hobart is now built, Tasmania.*
PHOTO: DAVID JOHNSON

However, not all these drowned landscapes present steep hillsides and angular skeletal outlines on maps. Broader, sweeping embayments, perhaps with some alluvial flats around the edges, form where large open valleys have been flooded. Commonly, alluvial deposits along the individual shorelines have been modified, forming narrow beaches around the estuaries.. Examples are Twofold Bay at Eden, Lake Macquarie in New South Wales, and Port Phillip Bay in Victoria.

But whether it is a narrow incision or a broad valley, the main outline has been inherited from a landscape fashioned by rivers and creeks thousands of years ago.

Eroding, rocky coastlines

Old landscapes facing the open sea are commonly eroded by the present ocean. Storm waves undercut cliffs and blocks of rock fall down, to be further broken up and the pieces swept away by tidal currents. Storm waves can be very powerful agents of erosion. Large boulders can be thrown up from low-tide platforms or even from below sea level and left stranded on headlands or low cliff tops. One well-documented example was the block lifted from the intertidal platform edge, 3 m up and 48 m backwards, at Ben Buckler near Bondi in Sydney on 14 or 15 July 1912. The block was 6 m long, and weighed about 235 tons (263 tonnes), and was turned completely upside down. This storm did considerable damage to jetties, breakwaters and retaining walls along the New South Wales coast.

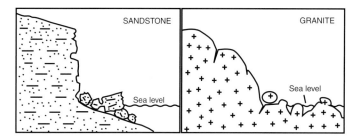

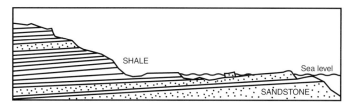

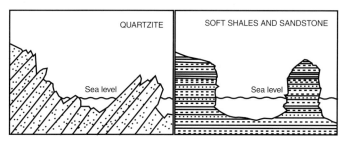

Figure 9.7 *Coastal geology determines shoreline type. Cliffs tend to have blocks of eroded rock at their bases, and these blocks are typically angular in the case of sandstones and rounded in the case of granites. Well-bedded sedimentary rocks commonly erode near sea level along weaker units, forming rock platforms. Steeply dipping resistant rocks such as quartzites or other metamorphic rocks create very jagged shorelines. Thickly bedded sedimentary sequences can be eroded by the sea, leaving remnant pillars, known as stacks, offshore.*

REPRODUCED AND REDRAWN WITH PERMISSION FROM HARPERCOLLINS PUBLISHERS AFTER FIGURE 24 FROM LASERSON, C., 1954 *The Face of Australia*.

Figure 9.8 *Rock platforms develop especially where flat-lying sedimentary sequences are eroded.*
PHOTO: DAVID JOHNSON

Figure 9.9 *Examples of eroding rocky shorelines. The Twelve Apostles on the Otway coast, Victoria, show sedimentary bedding (left). Massive rocks form rounded headlands at Bass Point near Wollongong, New South Wales (right).*
PHOTOS: DAVID JOHNSON

There is also strong evidence that the Australian coast has been subject to tsunamis. Beach sands and gravels deposited up to 30 m above sea level have been mapped on the cliff tops around Jervis Bay and Tuross (near Moruya) in southern New South Wales (see Box 9.2).

Good examples of an eroding rocky coastline are the Twelve Apostles (stacks) along the Otway coast of Victoria, the Sydney Heads, rock platforms north and south of Sydney, the cliffs around Kangaroo Island off Adelaide, and the limestone cliffs along the edge of the Nullarbor Plain west of Ceduna.

Where the rock is relatively soft and the ocean vigorous, erosion can be rapid, with changes apparent within a human lifetime. On some parts of the coast south of Sydney, cliff-top houses which were built in the middle of the last century are now uncomfortably close to the cliff edge as it is worn back (Figure 9.10). Where the rocks are very hard the erosion is not so rapid.

Figure 9.10 *Eroding cliffs at South Thirroul, near Wollongong in New South Wales, are now uncomfortably close to houses. The wedge of rock and soil at the base of the cliff was once at the top.*
Photo: David Johnson

Rock platforms

Wave erosion tends to cut away the rocks to about low-tide level. The actual level depends partly on the rock type, for the erosion will leave a particularly resistant layer, perhaps just above or just below low tide. As the cliff recedes this level remains as a rock platform extending out to sea and exposed at low spring tides. The surface of the rock platform can be cut by more deeply eroded fissures that follow joints in the rocks or that have been formed by the erosion of igneous dykes. The surface may also have deep circular potholes where loose rocks have been stirred round by wave turbulence, abrading the sides and bottom of a hole.

Box 9.2 Tsunamis

A tsunami is not a 'tidal wave' because it has nothing to do with the tides caused by the gravitational attraction of the Moon and the Sun. A tsunami is a wave generated by some massive disturbance of the ocean floor, such as faulting or a massive undersea landslide, especially during an earthquake. Other causes may be a volcanic eruption at sea or a meteorite impact in the ocean.

A tsunami dwarfs our usual experiences of waves. Normal waves approaching a surf beach have wave heights of 1–3 m and wavelengths of 100–300 m. The speed of a normal wave can be up to 100 km/h. A tsunami may have a wavelength of 1000 km. Since the speed of a wave in deep water is proportional to the wave length, tsunamis, with their long length, can travel at up to 700 km/h – about as fast as a passenger jet! This means that a tsunami wave could travel from Perth to Sydney in about 4 hours.

In mid-ocean the height of a tsunami is not great and may be imperceptible from a ship at sea. But as a wave approaches land it slows and the waveform steepens. Normally this is fine; it is why waves steepen and break on a beach, providing good surfing conditions. However, in a tsunami, the great speed and the great volume of water contained in that crest, stretched out at sea, becomes a monstrous wave when compressed at the coastline. The arrival of a tsunami is generally heralded by a short, sharp wave, and then an exceptional withdrawal of water as the first trough reaches the coast. Reefs become exposed, water drains out of bays, and boats are stranded at moorings. This is a great danger because people are likely to venture out to see all the seabed exposed.

Within minutes the giant wave arrives, and with immense power sweeps all before it, smashing boats and buildings, and carrying the debris and sediment inland. Large objects such as railway engines are carried like missiles. The wave crest may be 30–50 m above sea level, though 5–20 m is more common. Some have reached 115 m above sea level. The wave may extend many kilometres inland, depositing washovers of sand and debris, and there are records of ships being left stranded in river valleys.

Following the Chilean earthquake in the mid-morning of 22 May 1960, a tsunami rolled outward across the Pacific Ocean, reaching Hilo in Hawaii just after midnight. The lower township was wrecked within an hour. As an example of the power, large boulders from the sea-wall weighing up to 22 tonnes were carried nearly 300 m across an esplanade park, *without leaving noticeable grooves in the lawn.*

Other well-known tsunamis followed earthquakes in Lisbon in 1755, Suva in 1953, Alaska in 1964, Japan in 1983, Mexico in 1985, and Papua New Guinea in 1998. Tsunamis followed the volcanic eruptions at Santorini in the Mediterranean in 1490 BC, New Hebrides (Vanuatu) in 1878, and Krakatoa in Indonesia in 1883.

The Krakatoa tsunami reached the northwestern coast of Australia within 4 hours and was recorded at Geraldton, 2100 km from the volcano. In northern Western Australia the 1994 tsunami from Indonesia carried sand and coral boulders through a camp ground near Ningaloo.

Tsunami deposits have been identified around the northern cost of Australia and along the Great Barrier Reef, as well as down the eastern coast. These deposits have been dated in the range from less than 1000 to over 6500 years. The most recent may have been 250–400 years ago.

Tsunamis on the southeastern coast of Australia may have deposited beach sands and gravels high up on headlands. There is evidence that waves reached elevations 80 m above Jervis Bay and swept 10 km inland over the Shoalhaven delta. Stacked deposits of boulders and rock slabs on headlands, which are all dipping seawards, are also thought to be the result of tsunamis; major storms may not be capable of moving and emplacing such massive rock masses, especially where many boulders are piled together.

Depositional coastlines
Deltas

Deltas are the sedimentary deposits formed when a river meets the sea, or in some cases a large inland lake. The original name was coined by the Greek historian Herodotus to mean that land at the mouth of the Nile River in Egypt. The Nile divides into a series of channels radiating from a point about 120 km upstream from the mouth. These radiating channels encompass a triangular piece of land – the shape of the Greek capital 'D' or delta. Now we use the term to cover river deposits at the coast, whether triangular or not.

In general the sediment brought by rivers is dropped mainly near the mouth, so deltas on an open coast tend to form a bulge in the shoreline. The Burdekin delta in

Figure 9.11 *The Burdekin delta, Queensland. Deltas form where rivers bring sediment to the coast. The sediment then can build the coastline outwards at the river mouth, but much is generally transported alongshore by waves and currents to form spits and offshore sand bars. Successive sand bars extending from the river mouth become stabilised and grow above sea level, building the shoreline seawards.*

Photo: David Johnson

Queensland and the Gascoyne, De Grey and Ashburton deltas in Western Australia, are good examples. This bulge is evidence that the river is depositing sediment faster than waves and tides can carry it away.

Rivers that empty into deep coastal embayments do not form a coastal bulge. There is a delta of sediment at the river mouth, but the sediment gradually fills the embayment. In a million years or so, when the embayment is full, the sediment can then fashion a new coastline at the seaward end. Examples where the river delta is still at the head of the old valley system are the Fitzroy River where it empties into the head of King Sound near Derby, and the Ord River where it empties into Cambridge Gulf, both in Western Australia.

Long sandy or muddy shorelines
Sediment moves along shore from river mouths, transported by waves and tides along the beach face or in the shallow offshore surf zone, mainly in the direction of the prevailing wave systems.

In many places the beach builds outwards, and sand is blown up by onshore winds to form long sandhills behind the beach. As more sand is transported alongshore, another beach and ridge may develop until a series of ridges parallel the shoreline in a long curve. Thus a series of shorelines is preserved, building successively seawards. Each ridge represents the line of the shore at an earlier time. The inner edge of the ridges was where the initial shoreline stood. The Eighty Mile Beach that borders the Great Sandy Desert between the Pilbara and Kimberley areas of Western Australia, and the Ninety Mile Beach along eastern Victoria, are two very large examples.

At the mouths of many coastal streams a sand spit has grown across the mouth of a coastal embayment, forming a protected lagoon, generally with a tidal channel at one end. Where a spit grows across a river or creek mouth, the mouth is moved in the direction of the drift, leaving a long lagoon behind the beach with the mouth at one end. The river mouth may open during strong river flows but be sealed off by the beach at other times.

The Younghusband Peninsula in South Australia is a sand spit nearly 150 km long that has isolated The Coorong lagoon from the sea. At St Helens on the eastern coast of Tasmania there is a very large coastal lagoon shielded by a long spit. But many lagoons are not so large. Narrabeen Lake north of Sydney is a smaller example. Most creeks that empty onto an exposed beach have similar lagoons on a smaller scale.

The amount of sand moved along beaches and shallow offshore zones is prodigious. Calculations show that 500 000 cubic metres of sand are being transported northwards along the northern New South Wales coast and into Queensland. The interruption to this sand supply by the Tweed River training walls in the 1960s and early 1970s led to significant beach erosion in the Gold Coast of Queensland (see Box 9.3).

Figure 9.12 *The sandy beach and coastal dunefields at the western end of the Nullarbor. Such sandy shorelines can extend for many kilometres.*
PHOTO: RICHARD RUDD

Box 9.3 Coastal erosion problems

THE CAIRNS AND GOLD COAST BEACHES

Coastal erosion is a natural part of the geologic cycle, but it is also easily exacerbated by human activity. Sands washed down rivers to the coast and moved alongshore, primarily by waves, can accumulate to form beaches. It is typical for beaches to retreat and advance seasonally, so coastal development should always leave a buffer zone, typically around 50–150 m, to allow for this seasonal movement without causing damage to property and roads. In many cases of property damage the essential problem is not beach erosion, but that development has intruded into the zone of natural shoreline fluctuation.

Beaches are cut back on a short-term scale, such as by erosion in storms over a few days, and on a long-term scale because of changes in the sand supply budget over a period of years. In the long term, natural shifts of a river mouth over hundreds or thousands of years can interrupt the sand supply to a particular beach, or changing sea levels can modify transport patterns. Evidence of coastal erosion from such causes is common in the geologic record.

The following two examples show how human intervention can alter sediment supply to the coast and exacerbate beach erosion.

CAIRNS: NORTHERN BEACHES

Years of erosion of beaches north of Cairns since the 1950s led to investigations of the rates of sediment supply and movement alongshore.

The sole sediment supply to these beaches is the Barron River, which historically has had two outlets: Richters Creek and the present Barron River mouth. Sand transport northward along the coast is 10 000–13 000 cubic metres per year, depending on the particular beach aspect and processes. There is one groyne at Yorkeys Point, though the rocky headlands of Yorkeys, Taylors and Buchan Points provide natural anchors for beach formation. Sand tends to accumulate in river mouth bars and in the lee of the headlands.

Prior to 1939 the Barron River mouth was at Ellie Point, about 2 km south of its present position. After 1939 a northward shift of the mouth meant sand supply was interrupted for some years as the river built another bar at the new mouth. Meanwhile continuing northward longshore transport from beaches further north led naturally to significant erosion of Machans and Holloways Beaches after the 1950s. The calculated historical sand supply for the Barron River has varied greatly with the river flow conditions, from zero to 180 000 cubic metres per year, with a long-term average of 23 000 cubic metres per year. However, the demand for building and construction aggregate in the rapidly developing Cairns region from the 1960s to the 1980s was met by sand and gravel extraction from the Barron River bed. Sand extraction rates in the 1970s and 1980s reached 50 000–90 000 cubic metres per year.

This rate of sand removal from the river was at least twice the natural supply rate to the coast. But at the coast, alongshore transport of sand along the beaches by waves continued day in and day out. Thus the sand budget to the beach was severely imbalanced by human intervention, and beach erosion was inevitable.

GOLD COAST

Sand is transported northwards along the northern New South Wales coast and into southern Queensland at a rate of around 500 000 cubic metres per year. The Tweed River was originally trained between walls around 1900. However the strong sand transport rebuilt the bar across the entrance, and in 1962–64 the training walls were extended and the entrance dredged to allow boat movements. Monitoring of the beaches for the following 20 years showed the following effects:

- There was a beach sand build-up south of the Tweed River training wall, so that the coastline moved 250–300 m seawards and a new bar formed across the river entrance. The walls were damming the sand on the southern side.
- North of the training walls, beaches were eroded because longshore drift kept moving sand northwards, but the sand was not being replenished. The effect of the loss of sand was severe along the Gold Coast beaches from Point Danger to Bilinga. Erosion of 1–2 m depth of sand occurred in water originally 5–9 m deep.

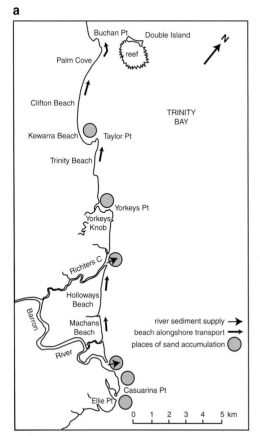

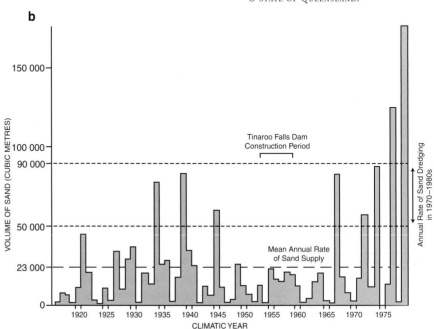

Figure 9. 13 *a. The beaches north of Cairns, showing the main sites of sand accumulation, and the northwards transport of sand along the coastline. b. Graph of the annual sand volume supplied by the Barron River to the coast. The average annual supply of 23 000 cubic metres was vastly exceeded by the dredging of 50 000–90 000 cubic metres of river sand per year in the 1970s and 1980s.*
REPRODUCED WITH PERMISSION OF THE ENVIRONMENTAL PROTECTION AGENCY, © STATE OF QUEENSLAND.

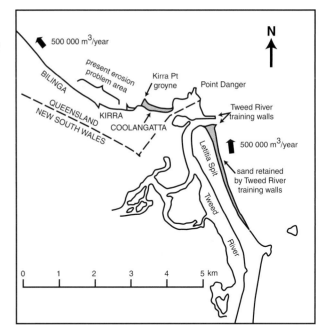

Figure 9.14 *The Tweed River mouth, showing the training walls damming sand on the southern side, depriving beaches to the north and west of sand. The rate of sand transport along this coast is 500 000 cubic metres per year.*
REPRODUCED WITH PERMISSION OF THE ENVIRONMENTAL PROTECTION AGENCY, © STATE OF QUEENSLAND.

- The groyne at Kirra Point did trap some sand on the eastern (updrift) side, mitigating the erosion problem on Coolangatta Beach. However, this exacerbated the situation downdrift at Kirra. Here massive erosion during storms in 1974 caused the collapse of a rockwall and the erosion of a campsite carpark, and the coastline was cut back 80 m.

Thus a major interruption to the sand supply had led inevitably to beach erosion.

How old is all this sand?

The sand on Australian beaches is millions of years old. In southern Queensland, SHRIMP dating of zircons in beach sand showed that the grains are 300–200 Ma old. This sand is derived from igneous rocks formed at that time in the New England area. On the southern New South Wales coast, the zircon ages are 450–350 Ma old, reflecting erosion of slightly older rocks of the highlands. In southern Western Australia, two sets of zircon dates are identified: most are 500–1300 Ma old, and some are as old as 3300 Ma. The SHRIMP ages tell us that these zircons, and probably much of the quartz, were originally derived from rocks of this age. The Western Australian beach sands, for instance, were originally eroded from rocks while Australia was still part of Gondwana and joined to Antarctica.

The sand has been eroded from the old granites and volcanic rocks, deposited in sedimentary basins that were then buried and uplifted, and then eroded again. It is quite possible that many of these sand grains have passed through more than one sedimentary cycle, perhaps being deposited in Permian systems, then re-eroded into

Figure 9.15 *Muddy shoreline, Cleveland Bay, Queensland. The low tidal flat has a mangrove fringe facing the ocean, with tidal salt flats and chenier ridges to landward.*
Photo: David Johnson

Cretaceous rivers, and then re-eroded into modern river and beach systems. This is recycling on a geological time scale!

In coastal regions that are protected from the ocean waves, muddy sediment can accumulate at the shoreline. This is particularly the case in northern Australia, where two factors assist the deposition of mud. Firstly, the weathering of rock and soil in northern tropical regions tends to cause a complete breakdown of component rock minerals, leading to fine muddy sediment. The Burdekin River, for instance, is typical of many tropical river catchments. The average Burdekin load is 69% fine suspended material (mainly mud) and 21% dissolved material; only 10% is the sand and gravel seen in the river bed.

The second reason is that these sheltered regions often support dense mangrove forests, kilometres wide, which trap the muddy sediment and protect it from erosion. Erosion does occur along the tidal channels, but there is a long-term accumulation of mud in the mangrove systems. The shores of Bowling Green Bay near Townsville, the Alligator River plain and much of the coast of the Gulf of Carpentaria, and the immense mangrove tidal flats in King Sound and near Port Keats, are good examples.

The Australian coastline

The southwestern coast

The coast of the Nullarbor Plain consists of cliffs 100 m high cut into Tertiary limestones. In the central part, from near Eucla west for 300 km, these cliffs have a sandplain along the seaward edge, forming a coastline of sandy beaches and dunes.

Further west, the eroding rocky coastline consists of Precambrian granites and gneisses, which have been deeply eroded forming protruding headlands with rocky bays, such as Point d'Entrecasteaux, Bald Head protecting King George Sound at Albany, or further east at Bremer Bay and Esperance Bay. The scale of erosion is evident from the scattered remnants of these rocks up to 50 km offshore in the Archipelago of the Recherche. Headlands cut in these old basement rocks of granite and gneiss form rounded, commonly bare surfaces with large boulders. Similarly the old, well-cemented sediments form prominent headlands.

Overlying these old rocks, south of the Stirling Ranges and extending across to the western side of Doubtful Island Bay, are much younger rocks: sandstones, limestones and some lignites that date from the Eocene around 40 Ma ago. Being softer they form coastlines with much lower profiles.

Many of the names here were given by the French scientific expedition of 1801–1802. The ships were the *Géographe* and the *Naturaliste*, with Nicholas Baudin the commander, Emmanuel Hamelin the other captain, Louis-Claude de Freycinet the cartographer and Francois Péron the naturalist. Leeuwin is derived from the Dutch ship of that name, which sailed the coast in 1622.

Sandplains of the western coast

The strip of headlands from Cape Leeuwin to Cape Naturaliste is the most westerly major outcrop of Precambrian rocks in southern Western Australia. The Darling Fault has ensured that Precambrian rocks are absent along the coastline between Cape Naturaliste and the Pilbara. All rocks east of this fault line are Precambrian; to the west, younger sediments overlie the downfaulted basement rocks. The coast reflects this lack of older, resistant rocks. There is a long sandplain from south of Perth to Geraldton, interrupted by the gorge and sea cliffs cut into red Silurian sandstones by the Murchison River mouth at Kalbarri.

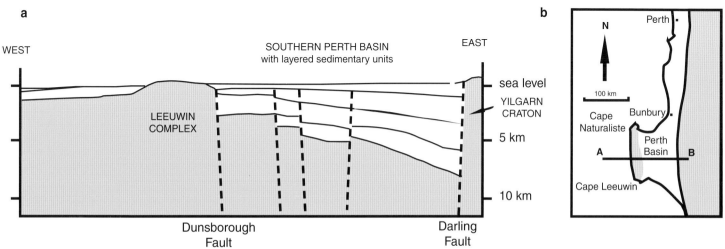

FIGURE 9.16 *Cross-section of the southern Perth Basin, a graben between the Archaean rocks of the Yilgarn craton to the east and Proterozoic rocks of the Leeuwin Complex to the west. The location of the cross-section is shown in the map at right.*

The shoreline itself is composed of extensive sand deposits, some of which formed in the Pleistocene and are partly cemented, such as the islands forming the seaward side of Cockburn Sound, Rottnest Island and small headlands north of Perth. These older cemented sands also exist offshore, forming north–south lines of shallow reefs on the inner continental shelf. They represent dune systems formed at Pleistocene shorelines during lower sea levels.

In the Nambung National Park near Cervantes, loose sand has been blown away from the land surface, leaving cemented columns in the Pinnacles desert. Individual pinnacles represent sand cemented around the deep tap roots of vegetation that grew on the top of the sand dunes.

Offshore from Geraldton lie the Houtman Abrolhos, a group of Pleistocene and modern coral reefs now at the southern extremity of reef-forming conditions. Seawater temperatures there can be less than 20°C for almost a third of the year, and the reefs are only maintained by the southwards flowing Leeuwin Current, which brings warm tropical water down the continental shelf.

Further north are cemented Pleistocene sandstones that form the steep Zuytdorp Cliffs south of Shark Bay, and Dirk Hartog, Dorre and Bernier Islands, which enclose the bay. These sandstones were formed as large, aeolian dune systems, and the two embayments of Shark Bay – Disappointment and Freycinet Reaches, separated by the Peron Peninsula – are eroded depressions between different dune systems. The north–south trend of the individual dunes can be seen in the elongated inlets in southern Freycinet Reach, and in the L'Haridon Bight south of Monkey Mia.

It was these cliffs and the reefs of the Houtman Abrolhos that caused so much grief to the early Dutch traders and explorers. After rounding the Cape of Good Hope they set sail eastwards, aiming for the East Indies (Indonesia). If they strayed south while crossing a huge Indian Ocean, without islands to judge their position and with deep water right up to the Australian coast, they could find themselves too close to these cliffs and reefs. Several ships were lost in this manner, smashed against the rocks. Among them were the *Zuytdorp* on the cliffs in 1712, and the *Batavia* in 1629 and *Zeewijk* in 1727 on reefs in the Houtman Abrolhos.

North of Carnarvon is a beach ridge system deposited by sands moved alongshore from the Gascoyne delta. The next section of uplifted Pleistocene and Tertiary rocks begins just south of Point Quobba and stretches to North West Cape at the tip of Exmouth Gulf.

The northwestern coast of the Pilbara and Kimberley

The Precambrian rocks of the Pilbara, which are so spectacular inland, form only a short coastline near Dampier. Most of the Pilbara coast is modern sediment brought down by the Ashburton and De Grey Rivers. Long sandy beaches alternate with muddy coastlines and mangroves.

The Great Sandy Desert reaches the coast between the Pilbara and the Kimberley. There are no rocks here to form headlands, and the sea has used the local product – desert sand. The sands have been reworked into beaches and dunes, including Eighty Mile Beach.

From the northern side of King Sound around to Joseph Bonaparte Gulf is one of the most isolated yet spectacular coasts in Australia. Precambrian sedimentary rocks have been carved into deep embayments, with stunning scenery and waterfalls at the heads of the inlets. Erosional remnants remain as hundreds of islands scattered offshore, forming the Bonaparte Archipelago. These old, hard rocks form a strong, resistant coastal topography.

The Top End coast

The eastern side of Joseph Bonaparte Gulf has an erosional remnant of Permian coal measures, with mangroves and tidal flats. This remnant is all that is left of the older sedimentary basin, some of which also exists under the seabed offshore.

From just southeast of Darwin to the western side of the Gulf of Carpentaria and Groote Eylandt, the coastal region consists of older Precambrian rocks that were deeply incised and have now been drowned. In much of this strip the actual coastline is formed of thin edges of Cretaceous sedimentary rock laid down during the time of the inland seas. These Cretaceous rocks also form the Cobourg Peninsula and underlie Melville Island.

East of Darwin, the Mary and Alligator Rivers have produced a coast with great alluvial plains passing near the sea into marshes, mangroves and tidal flats. Much of the southern and eastern side of the Gulf is an apron of muddy sediments and mangroves with sand and shell ridges thrown up by storms. Part of the eastern side consists of an extensive chenier plain. Cheniers are low sand ridges parallel to the shoreline, thrown up by storms or formed during times of shoreline erosion when coarser material accumulates at the shoreline. They are separated by broad mudflats and salt flats.

The northeastern coast

The eastern Queensland coast consists mainly of rocky promontories and headlands formed by the north-trending Palaeozoic rocks, with some granites. Cape York itself, like Prince of Wales and Thursday Islands, is composed of Carboniferous volcanics.

A succession of capes alternating with modern sedimentary coastlines extends southward down the coast: Cape Grenville, Cape Weymouth, Cape Melville, and Cape Grafton near Cairns. In between these capes are long areas of sandy beaches with tidal flats and mangroves in the lee of the headlands. Trinity Bay at Cairns has extensive mangrove systems at the southern end behind the protection of Cape Grafton, with clean sandy beaches north of Cairns.

The extensive dunefield at Cape Flattery, where silica sand is exported from Shelburne Bay, was formed mainly in the Pleistocene in three phases of sand dune development, dated so far at greater than 48 000 years, around 7000–8000 years, and within the last 1000 years.

To the south are Cape Cleveland at Townsville and Cape Upstart towards Bowen, so named by Captain Cook because it is surrounded by low land and, from the sea, rises or starts up at first sighting. Cape Hillsborough differs in that it is composed of much younger, Tertiary volcanic rocks. The rocks at the Cape itself are extrusive volcanics and sediments.

The Whitsunday Group is interesting because it is composed mainly of Cretaceous volcanics; that is, rocks that must have been erupted just before the severing of the northeastern continental margin of Australia in the early Tertiary. Presumably these volcanics were widespread along that margin but have now been eroded. The resulting sediment is thought to now form part of the continental shelf, but we know much also was washed inland from the old volcanic hills, into the Cretaceous inland sea stretching across central Queensland and up to the Gulf of Carpentaria.

The region south of Mackay is distinguished by a much higher tidal range compared to the coasts to the north and south. This higher tidal range means stronger currents offshore, and also more extensive tidal flats in Broad Sound, because the sea is able to flood a greater area of land each day.

The southern Queensland coastline is of a similar style, with the bedrock forming headlands near Gladstone, at Noosa, and Redcliffe near Brisbane. The intervening coastline is mainly eroded rocks of Devonian, Carboniferous and Permian age, with Broad Sound, Keppel Bay and Port Curtis as deeper embayments. The folding and faulting of these old rocks gives the coastline a very irregular shape.

South of Round Hill Head near Bundaberg the coast changes, and long sandy beaches extend from Hervey Bay to Cape Byron in northern New South Wales. This may be partly because this coast is beyond the southern end of the Great Barrier Reef, so that oceanic swells and waves can reach the coast, forming a higher-energy surf environment. It may also be because of a change in local rock type and the supply of sand from rivers further south.

Offshore lies Fraser Island, the largest sand island in the world, and then further south near Brisbane, Bribie Island, Moreton Island and North and South Stradbroke Islands. These massive islands attest to the large amount of sand that has accumulated on the shelf and coast over the past few hundred thousand years. Much of the sand is thought to have come from the New South Wales coast and been moved northwards.

The southeast

The northern coast of New South Wales consists of long sandy beaches with bedrock headlands protruding between them; for example, the beaches north of the Clarence River, extending to Evans Head and Ballina. In this region, erosion of the Jurassic and Cretaceous sedimentary sequences has provided an ample sand supply. So when walking on these beaches you can be aware that the sand under your feet was originally washed down rivers and onto coastlines over 100 Ma ago. It was buried and consolidated into rock, and is now being recycled onto the modern coastline. These sand grains have been around a long time!

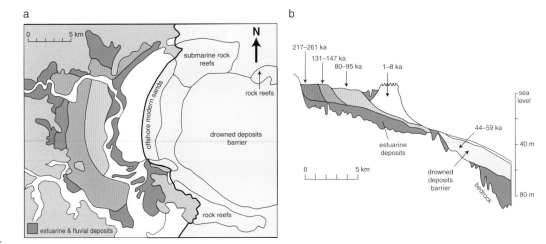

Figure 9.17 *Successive barriers in the Tuncurry area, north of Newcastle, New South Wales. a. The four main barriers, with a drowned barrier below sea level offshore. b. Cross-section showing the relative positions and ages of each barrier system.*

Detailed drilling and dating of the sand dunes has shown there are four episodes onshore and one drowned offshore. The innermost series of dunes was deposited 261 000–217 000 years ago during a sea level high, and the next two 147 000–131 000 and 95 000–80 000 years ago. The recovery of a coral sample during drilling under one of the inner dunes indicates that warmer waters prevailed at these times. The next deposits were deposited below sea level 59 000–45 000 years ago when the sea level was lower. The final deposits are the modern sand beaches and dunes at the coast, deposited since 8000 years ago.
Modified from Roy 1999, figure 4.

In the Tuncurry, Myall Lakes and Port Stephens area the coast is oriented across the strong wave systems, and a massive windblown dune field has developed. Northwards transport along the coast is a response to southerly storm and ocean swell waves. Seal Rocks, Cape Hawke and Hallidays Point are bedrock masses that extend eastwards as offshore reefs and form a northeast-trending coastline, trapping the sand eroded from the Hunter Valley. Again much of this sand has been recycled through the Permian and Triassic sandstones of the Sydney Basin. For those grains that came via the Permian marine sediments, their last outing was 260 Ma ago, when the water was a lot colder than it is now.

The New South Wales coast south from Newcastle to Cape Howe is mainly an eroding rocky coastline of cliffs and headlands separated by sandy beaches. Only the Shoalhaven River at Nowra has filled the ancestral valley and is now delivering sediment to the open coast.

From Newcastle to just north of Batemans Bay the rocks forming the coast are almost horizontal sedimentary rocks, and many headlands typically have long intertidal rock platforms where softer mudstones and coals have been eroded to sea level. Where massive sandstones occur at the coast there are cliffs, such as Sydney Heads. South of Batemans Bay are steeply dipping, deformed rocks as far as Bermagui, and rock platforms are rare on this coast. Angular headlands surrounded by ragged rocks are the norm. South of Bermagui, red Devonian sandstones form high cliffs with some rock platforms.

The Victorian coastline has a central rocky section from Wilsons Promontory to Warrnambool. In contrast, extensive sandy shorelines lie to the east along the Gippsland coast, and to the west in Discovery and Portland Bays. Cliffs and bluffs form 46% of the overall coastline, sandy shorelines 42%. Salt marshes and mangroves make up the remaining 12%, mainly in the protected embayments of Western Port, Port Phillip Bay and Corner Inlet.

The Gippsland coast has long sandy beaches with isolated granite bedrock outcrops, such as Point Hicks and Rame Head. Remnants of igneous bedrock also extend offshore, for example at Gabo Island. Most of the coast is the Ninety Mile

Beach and similar sandy beaches to the east, which have cut off estuaries of the coastal rivers, forming protected lagoons. The Gippsland Lakes, with a narrow access at Lakes Entrance, and similar estuaries further east at Lake Tyers and Sydenham, Tamboon and Mallacoota Inlets, are examples. Sand transport can be both to the east and to the west, depending on the local angles of wave approach. The Snowy River does reach the sea, and probably had a historically more open entrance which has been constricted due to the decreased river flow. The strong eastward shoreline transport at this point is evident in the spit covering the river mouth, which extends east of Marlo.

The Gippsland coast shows an inner barrier sand system deposited during the last sea level high, and the present outer barrier system. The outer barrier clearly has a long-term pattern of prograding seawards, but comparison of coastal surveys made from 1847 to 1849 with air photos in 1976 shows that the coast has been cut back 50–1100 m over that period. The early parts of the barrier had more calcareous sand, while the present sand is dominantly quartz, and appears to have been derived from the shallow offshore areas.

Wilsons Promontory is an isolated high outcrop of Devonian granite with adhering deposits of alluvium and coastal sands. These sands extend to the

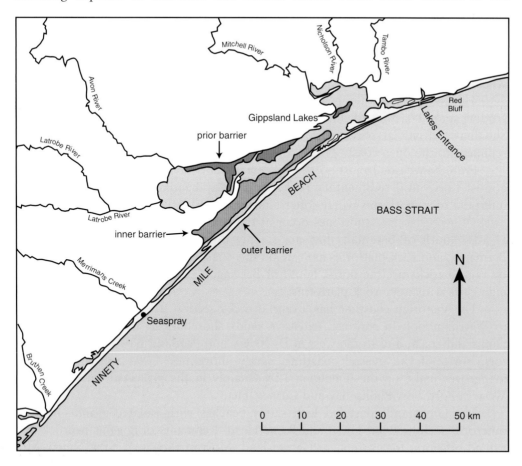

Figure 9.18 *The Gippsland coast of Victoria shows three barriers that have built the coast successively seaward.*

northeast as a series of barrier islands across Corner Inlet, and to the west as Venus Bay. The coastal geology of the central part of the coastline is affected by very recent faulting, which has produced sunken areas such as Western Port and Port Phillip Bay. Fragments of Cretaceous and Tertiary rocks form much of the shoreline, with patches of recent sands.

The Otway coast, including Apollo Bay and the Twelve Apostles, and west to Moonlight Head, consists of cliffs cut into Cretaceous sandstones. The cliffs on the coast westwards to Warrnambool are cut into Miocene sediments.

The coast of Two Mile Bay at Port Campbell preserves a line of bluffs inshore of, and parallel to, the main coastal cliffs. These bluffs are about 50 m high, with a steep slope descending to swampy lowlands. Underneath these swampy areas lies the horizontal surface of an ancient wave-cut platform, which sits about 3 m above the platform presently being eroded by the sea. The bluffs and this old platform were eroded during the last sea level high, some 120 000 years ago. So visitors now can see a remnant of the former Victorian coastline formed long ago. Of course Aboriginal people would have climbed down this slope during the sea level lows 50 000–20 000 years ago and then walked on towards Tasmania.

Box 9.4 Comparison of Sydney Harbour and Port Phillip Bay

Sydney Harbour and Port Phillip Bay represent two contrasting types of geology that form good harbours.

SYDNEY HARBOUR

Port Jackson is a deeply drowned river valley system. The bedrock of the original river valley was cut during the late Tertiary and Pleistocene. It now lies at 85 m below modern sea level at Sydney Heads, 40–45 m in the middle area near Fort Denison and 30 m at the Ryde Bridge on the Parramatta River.

Sediments partly fill this old valley, and it is clear that the rate of sedimentation has increased following European settlement and clearing of the surrounding country. The sediment fill is 25–50 m thick just seawards of the harbour bridge, and 20–35 m in channels and bays upstream. Some of this sediment fill is valley fill from a time when the valley had rivers flowing through it, some is estuarine material that accumulated as the harbour was flooded by the rising sea level, and some is sediment being moved by modern tidal currents.

A similar pattern exists at the mouth of the Hawkesbury River and Pittwater. Seismic profiling there has shown that the ancient Pittwater River ran south of Barrenjoey Head, then under the present Palm Beach before joining the ancestral Hawkesbury River seawards of the present coast.

PORT PHILLIP BAY

Port Phillip Bay is a broad embayment that represents the final phases of infilling of

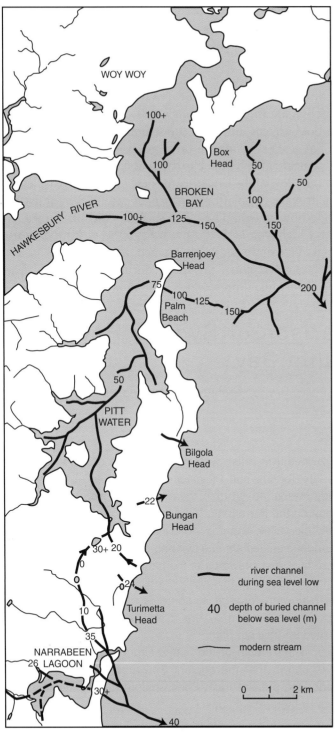

Figure 9.19 *Hawkesbury River and Pittwater areas north of Sydney, showing courses of ancestral, buried rivers.*

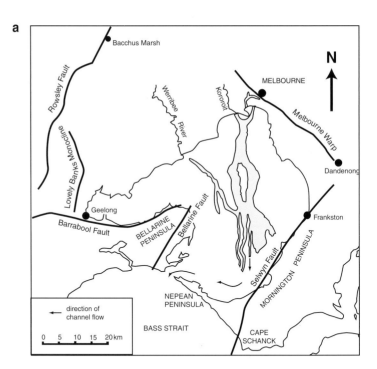

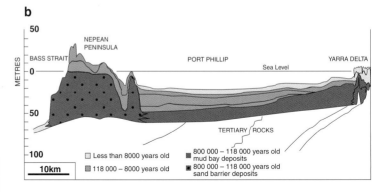

Figure 9.20 *a. Port Phillip Bay, showing the extent of old fluvial channels (grey) that flowed through the depression at lower sea levels. b. Cross-section showing the sedimentary units filling the Bay and the position of the sand barriers forming the Nepean Peninsula.*

what is termed a 'sunkland'. Large faults formed as Australia was separating from Antarctica, allowed a rhomboid-shaped block about 100 km across to settle downwards as the crust was being stretched apart and broken. The main boundary faults are the Selwyn Fault to the southeast, which runs through Frankston southwest to Cape Schanck, and the parallel Rowsley Fault running southwest through Bacchus Marsh. A sharp downwarp in the rocks along a northwest-trending line through Melbourne marks the northern side of the sunkland. Smaller faults mark the southern side near Geelong.

Drillhole evidence indicates that this depression became an open marine embayment, surrounded by rocky uplands, early in Tertiary times. Younger basalts have filled much of the area east of Bacchus Marsh to the shoreline of the Bay.

The present, enclosed form did not develop until the late Pleistocene, when calcareous dune sands accumulated east of Geelong and as the Nepean Peninsula. Only a small opening, 3.2 km wide at high tide, is left into the Bay, with strong tidal rip through this entrance. The modern outlines of the Bay were established around 6000 years ago as the sea flooded the embayment, forming the present beaches and tidal flats.

The present bay is generally 9–24 m deep, except for The Rip at the entrance where tidal scouring has eroded the bed to 60 m depth.

Seismic profiling has shown that the bay sediments overlie Tertiary sequences and a series of old buried river channels that flowed southwards from the mouths of the Yarra River and Kororoit and Werribee Creeks towards the Mornington Peninsula. These channels were active before 10 000 years ago, when the sea levels were much lower. Some of the sands from these rivers may have been blown onto the primitive Nepean Peninsula, because the age of sands recovered during drilling matches the age of these channels.

Most of the modern bay fill is up to 27 m thick, filling the old channels and overlying the old land surface. The sediments have accumulated at around 0.4 mm/yr, most of it in the last 6500 years.

The coastline between Warrnambool and Portland is the only place in Australia where recent volcanoes have strongly determined the coastal geology. Several volcanoes suffered central collapses after volcanism stopped, forming a caldera. The Tower Hill volcano has a lake-filled caldera with several final tuff cones in the middle. The volcano last erupted about 10 000 years ago. To the west, calcareous dune sands overlie basalts that extended seawards of the present shoreline and are exposed as reefs on the coast west of Port Fairy (see Figure 9.21).

The headlands southwest of Portland comprise four calderas that have had their southern sides eroded and breached by the sea, forming Grant Bay, Nelson Bay and Bridgewater Bay. Stony Hill between Bridgewater and Duquesne Bays is the remnant of a volcano that has been faulted, eroded and partly overlain by calcareous dune sands.

The coastline from Discovery Bay to Encounter Bay in South Australia is part of a series of old beach and dune systems that form long arcs, extending inland to Naracoorte. The ridges, with intervening flat corridors, are highest inland, and succeeding ridges lie downhill to the southwest towards the coast. The Younghusband

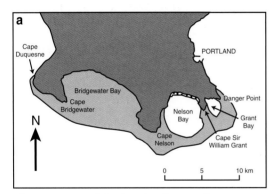

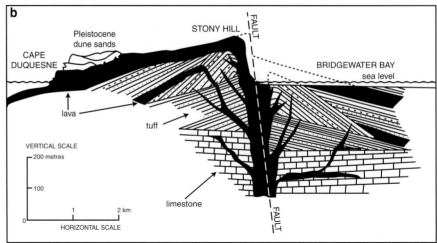

Figure 9.21 *a. The Portland coastal region, Victoria, showing calderas. b. Section through the peninsula southwest of Portland, showing the truncated Pleistocene volcano.*

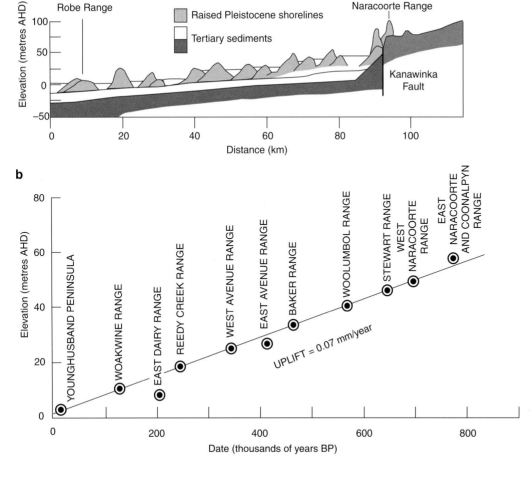

Figure 9.22 *a. Cross-section showing the successive barriers developed on the coastal region near Naracoorte, northwest of Mount Gambier, South Australia. b. Dating on several barriers confirms a long-term uplift rate of 0.07 mm/yr.*

Peninsula represents the latest phase of these ridges, and are still being deposited. Dating has shown the uplift has averaged 0.07 mm/yr over 800 000 years, so that each new shoreline was deposited to the southwest and downhill of the preceding shoreline.

Spencer Gulf and Gulf St Vincent are the seaward ends of downfaulted terranes that extend north through the inland to Lake Torrens and Lake Eyre. The coast of the Eyre Peninsula is composed of Proterozoic rocks, with an overlay of more recent sands.

Summary

The gross outline of Australia is the result of the fracture pattern that broke up Gondwana. Surrounding oceanic plateaus are pieces of the old continent that have only partly broken free.

On a more detailed scale, individual coastal shapes are the result of drowning or erosion of old rocky landscapes, or the smoother outlines of modern deposition in deltas or on long sandy coastlines. Much of the sand is millions of years old, and has been recycled through older sedimentary rocks before being eroded and delivered to modern beaches. Multiple barrier systems have been deposited at successive sea level highs in some areas, dating back over 200 000 years.

SOURCES AND REFERENCES

Beach Protection Authority Queensland, 1984. *Mulgrave Shire Northern Beaches*. 366 pp.

Bird, E.C.F., 1993. *The Coast of Victoria*. Melbourne University Press. 324 pp.

Bryant, E., 2001. *Tsunami. The Underrated Hazard*. Cambridge University Press. 320 pp.

Bryant, E.A. & Nott, J., 2001. 'Geological indicators of large tsunami in Australia'. *Natural Hazards*, 24: 231–249.

Bryant, E.A., Young, R.W. & Price, D.M., 1992. 'Evidence of tsunami sedimentation on the southeastern coast of Australia'. *Journal of Geology*, 100: 753–765.

Collins, L.B., *et al.*, 1993. 'Late Quaternary evolution of coral reefs on a cool-water carbonate margin: the Abrolhos Carbonate Platforms, southwest Australia'. *Marine Geology*, 110: 203-212.

Collwell, J.B., Coffin, M.F. & Spencer, R.A., 1993. 'Structure of the southern New South Wales continental margin, southeastern Australia'. *BMR Journal of Australian Geology and Geophysics*, 13: 333–343.

Etheridge, M.A., Symonds, P.A. & Lister, G.A., 1989. 'Application of the detachment model to reconstruction of conjugate passive margins' in A.J. Tankard & H.R. Balkwill (eds), *Extensional Tectonics and Stratigraphy of the North Atlantic margins*, AAPG Memoir 46: 23–40.

Gagan, M.K., Ayliffe, L.K., Beck, J.W., Cole, J.E., Druffel, E.M., Dunbar, R.B. & Schrag, D.P., 2000. 'New views of tropical paleo-climates from corals'. *Quaternary Science Reviews*, 19: 45–64.

Harvey, N., Belperio, A.P. & Bourman, R.P., 2001. 'Late Quaternary sea-levels, climate change and South Australian coastal geology'. in V.A. Gostin (ed.) *Gondwana to Greenhouse: Australian Environmental Geoscience*, Geological Society of Australia Special Publication 21: 201–214.

Herbert, C. (ed.) 1980. *Geology of the Sydney Basin* 1:100,000 Sheet 9130. Geological Survey of New South Wales. 91 pp.

Holdgate, G.R., *et al.*, 2001. 'Marine geology of Port Phillip, Victoria'. *Australian Journal of Earth Sciences*, 48: 439–455.

McNamara, K.J., 1995. *Pinnacles*. Western Australian Museum. Revised edition. 24 pp.

Nott, J., 2000. 'Records of prehistoric tsunamis from boulder deposits: evidence from northern Australia'. *International Journal of the Tsunami Society*, 18: 3–14.

Petkovic, P. & Buchanan, C., 2002. 'Australian bathymetry and topography grid', Data package, Geoscience Australia, Canberra.

Playford, P.E., Cockbain, A.E. & Low, G.H., 1976. *Geology of the Perth Basin*, Western Australia. Geological Survey of Western Australia Bulletin 124. 311 pp.

Roy, P.S., 1980. 'Quaternary Geology' in C. Herbert (ed.) *Geology of the Sydney Basin* 1:100,000 Sheet, Geological Survey of New South Wales.

Roy, P.S., 1999. 'Heavy mineral beach placers in southeastern Australia: their nature and genesis'. *Economic Geology*, 94: 567–588.

Roy, P.S., *et al.*, 1997. *Quaternary Geology of the Forster–Tuncurry coast and shelf, Southeast Australia*. Geological Survey of New South Wales Report GS 1992/201. 405 pp.

Sircombe, K.N., 1999. 'Tracing provenance through the isotope ages of littoral and sedimentary detrital zircon, eastern Australia'. *Sedimentary Geology*, 124: 47–67.

Sircombe, K.N. & Freeman, M.J., 1999. 'Provenance of detrital zircons on the Western Australia coastline – Implications for the geologic history of the Perth Basin and denudation of the Yilgarn craton'. *Geology*, 27: 879–882.

Struckmeyer, H.I.M., Totterdell, J.M., Blevin, J.E., Logan, G.A., Boreham, C.J., Deighton, I., Krassay, A.A. & Bradshaw, M.T., 2001. 'Character, maturity and distribution of potential Cretaceous oil source rocks in the Ceduna Sub-basin, Bight Basin, Great Australian Bight' in K.C. Hill & T. Bernecker (eds) *Eastern Australasian Basins Symposium. A Refocussed Energy Perspective for the Future*. Petroleum Exploration Society of Australia, Special Publication: 543–552.

Stagg, H.M.J., Willcox, J.B., Symonds, P.A., O'Brien, G.W., Colwell, J.B., Hill, P.J., Lee, C-S., Moore, A.M.G. & Struckmeyer, H.I.M., 1999. 'Architecture and evolution of the Australian continental margin'. *AGSO Journal of Australian Geology and Geophysics*, 17: 17–33.

Symonds, P.A., Murphy, B., Ramsay, D., Lockwood, K. & Borissova, I., 1998. 'The outer limits of Australia's resource jurisdiction off Western Australia' in P.G. Purcell and R.R. Purcell (eds) *The Sedimentary Basins of Western Australia 2*. Proceedings of the Petroleum Exploration Society of Australia Symposium, Perth, WA.

Symonds, P.A. & Willcox, J.B., 1989. 'Australia's petroleum potential in areas beyond an Exclusive Economic Zone'. *BMR Journal of Australian Geology and Geophysics*, 11: 11–36.

Thom, B.G., *et al.*, 1992. *Coastal Geomorphology and Quaternary Geology of the Port Stephens–Myall Lakes Area*. Department of Biogeography and Geomorphology, ANU, Monograph No 6. 407 pp.

Ward, W.T. & Little, I.P., 2000. 'Sea-rafted pumice on the Australian east coast: numerical classification and stratigraphy'. *Australian Journal of Earth Sciences*, 47: 95–109.

Wells, P. & Okada, H., 1996. 'Holocene and Pleistocene glacial palaeoceanography off southeastern Australia, based on foraminifers and nanofossils in Vema cored hole V18-222'. *Australian Journal of Earth Sciences*, 43: 509–523.

Young, R.W. *et al.*, 1997. Chronology of Holocene tsunamis on the southeastern coast of Australia. *Japanese Journal of Geography*, 8: 1–19.

Websites

Sea levels and tides
www.auslig.gov.au/geodesy/abslma/abslma.htm

Status of Australian estuaries
www.ozestuaries.org

Illustrations

Figs 9.1, 9.2a, b, 9.3a, b, 9.4 ; digital images provided by Geoscience Australia: fig 9.1 from Petkovic, P. & Buchanan, C., 2002. Australian bathymetry and topography grid (January 2002), Data package, Geoscience Australia, Canberra; figs 9.2a, b (Geoscience Australia, diagram 23/0N/121) from Etheridge, M.A., Symonds, P.A. & Lister, G.A., 1989. 'Application of the detachment model to reconstruction of conjugate passive margins' in A.J. Tankard & H.R. Balkwill (eds) *Extensional Tectonics and Stratigraphy of the North Atlantic margins*, AAPG Memoir 46: 23–40; fig. 9.3a (Geoscience Australia diagram 23/0A/125), fig. 9.3b (Geoscience Australia diagram 14/0A/1087) from Struckmeyer, H.I.M., *et al.* 2001. 'Character, maturity and distribution of potential Cretaceous oil source rocks in the Ceduna Sub-basin, Bight Basin, Great Australian Bight' in K.C. Hill & T. Bernecker (eds) *Eastern Australasian Basins Symposium. A Refocussed Energy Perspective for the Future*. Petroleum Exploration Society of Australia, Special Publication: 543–552; fig. 9.4 (Geoscience Australia diagram 23/0A/925) in Symonds, P.A., *et al.*, 1998. 'The outer limits of Australia's resource jurisdiction off Western Australia' in P.G. Purcell and R.R. Purcell (eds) *The Sedimentary Basins of Western Australia 2*. Proceedings of the Petroleum Exploration Society of Australia Symposium, Perth W.A. Crown copyright © All rights reserved. www.ga.gov.au

Figs 9.13 a, b, 9.14 are not official copies, and have been redrawn and are reproduced with the permission of the Environmental Protection Agency. Copyright the State of Queensland.

Figs 9.13a and 9.13b after figs 13.15 and 13.16 respectively from Beach Protection Authority Queensland, 1984. Mulgrave Shire Northern Beaches. 366pp. Fig. 9.14 after fig. 1 in *Beach Conservation* No. 58. The Newsletter of the Beach Protection Authority of Queensland, January 1985.

Fig. 9.16: redrawn and reproduced with permission from the Department of Industry and Resources of Western Australia from fig. 50, p. 227, Playford, P.E., Cockbain, A.E., and Low, G.H., 1976. *Geology of the Perth Basin*, Western Australia: Geological Survey of Western Australia, Bulletin 124, 311 pp.

Fig. 9.17: redrawn and reproduced with permission from the Society of Economic Geologists Inc. after fig. 4, p. 570, Roy, P.S., 1999. 'Heavy Mineral placers in Southeastern Australia: Their nature and genesis'. *Economic Geology*, 94(4):567-588.

Figs 9.18, 9.21: redrawn and reproduced with permission of Eric Bird after figs 179, 18 in Bird, E.C.F., 1993. *The Coast of Victoria*. Melbourne University Press. 324 pp.

Fig. 9.19 : redrawn by permission of the New South Wales Department of Mineral Resources after fig. 14 in Roy, P.S., 1980. 'Quaternary Geology' in C. Herbert (ed.) *Geology of the Sydney Basin* 1:100,000 sheet, Geological Survey of New South Wales.

Fig. 9.20: redrawn with permission of the Geological Society of Australia after figs 6, 7 in Holdgate, G.R., *et al.*, 2001. 'Marine geology of Port Phillip, Victoria'. *Australian Journal of Earth Sciences*, 48: 439-455.

Figs 9.22a, b redrawn with the permission of the Geological Society of Australia after figs 17.2, 17.5 in Harvey, N., Belperio, A.P. & Bourman, R.P., 2001. 'Late Quaternary sea-levels, climate change and South Australian coastal geology' in V.A. Gostin (ed.) *Gondwana to Greenhouse: Australian Environmental Geoscience*, Geological Society of Australia Special Publication 21: 201–214.

CHAPTER 10 *Introduction to reefs*

GREAT BARRIER REEF

The Great Barrier Reef is one of Australia's prime assets. How old is it, and what is it built upon? Can we understand the major processes that built and maintain it?

To what extent is human activity on land affecting the reef?

The coral reefs were a great hazard to the early seafarers. Even today, with highly accurate navigation equipment and powerful engines, boats, and even ships, still run onto the reefs of the Great Barrier Reef.

How much more difficult must it have been for the early explorers like James Cook and Matthew Flinders, who had to command sailing vessels that were at the vagaries of the winds and tidal currents, in navigating these waters. On 1 June 1770 Cook's vessel, HM Bark *Endeavour*, was firmly fixed on a reef, yet he was able to refloat it, make hasty repairs, sail to the mainland, and there careen it on a river bank at present-day Cooktown. With the repairs complete he sailed a short distance before climbing the 300-metre-high granite hill on Lizard Island to search for a way free of the reefs. From this vantage point he took a bearing on a channel to the safety of the open sea. Flinders directed steerage of the *Investigator* from the masthead. Both Cook and Flinders and their vessels survived their encounters with the reef.

Not so fortunate were those on the *Pandora*, which was carrying mutineers from the *Bounty* home to England for trial. In 1791 she struck Pandora Reef at the entrance to Torres Strait, and was dragged across it through the night, before being tipped into 34 m of water the next morning. Those who survived took to four small boats and rowed to Kupang in Timor, and then to Batavia, repeating part of the journey that Bligh had done when his crew mutinied two years earlier.

A coral reef is particularly dangerous because it rises precipitously and unexpectedly from much deeper, safer waters. Sometimes the growl of breakers on the reef can be heard, or white foam and the paler green colour of the shallower water can be seen in the distance. But in rain, or when the Sun is low and the colour change in the water not obvious, or at night, reefs are perilous places for ships.

Corals grow in a very wide range of conditions, although they are most prolific in shallow, warm, clear waters. Individual coral species can be found right around the Australian coastline, including several species found in the cold waters off Busselton in southern Western Australia. However, large reef systems built by corals are a feature of tropical waters (see Figure 10.1). Average sea surface temperatures greater than 24°C seem to be the prime determinant of tropical coral growth. Although corals do grow in waters with a temperature of 20–24°C, they do not form large reefs. The cooler waters reduce coral growth during a longer winter, and lower the metabolic rate of symbiotic algae (Zooxanthellae) living in the coral skeleton.

Being a feature of tropical regions, coral reefs can also be devastated by periodic cyclones (or hurricanes, as they are known in the Americas).

Coral reefs are among the most diverse ecosystems on Earth, rivalling the tropical rainforests for their range of organisms. Most of this life is soft and is not preserved, but the corals, calcareous algae and shells have hard skeletons which remain after all the colourful flesh has decayed.

Australian coral reefs

The Great Barrier Reef is the largest and best-studied reef system in Australia. It extends for some 2000 km from near Bundaberg at 24°S latitude to Torres Strait and Papua New Guinea in the north. It contains around 2500 reefs and reefal shoals. Small patches of coral also grow further south. When John Oxley first entered the Brisbane River there was coral at its mouth, and until 1937 there were luxuriant corals at Peel Island in Moreton Bay.

Considering that the continental shelf around Australia was dry land from 120 000 to 18 000 years ago, the modern reefs are relatively young. However, there is evidence that reefs were also present at previous sea-level highs. For instance, the southernmost reef development is at the Houtman Abrolhos, off Western Australia, at 28–29°S. The Abrolhos reefs are growing on top of an old reef that accumulated between 130 000 and 120 000 years ago. This old reef was heavily weathered during the sea-level low, forming thick calcrete. The modern reef deposits have formed over 10 000 years during the last sea-level rise, and are now up to 26 m thick.

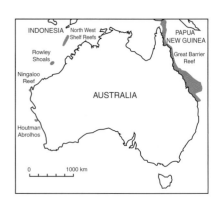

Figure 10.1 *The distribution of coral reefs around Australia. The Great Barrier Reef along the northeast edge is the most extensive, but there are also major reefs on the western coast: around the North West Shelf, and the Ningaloo and Houtman Abrolhos reefs.*

Box 10.1 Effects of cyclones on the Great Barrier Reef

In northern Australia cyclones occur mainly in the months of December to March, and generally start when the monsoonal trough moves southwards over New Guinea and Indonesia during the Southern Hemisphere summer. A low-pressure cell separates from the monsoonal trough and moves southwards, and then eastwards or westwards. While passing over warm tropical waters, the central pressure of the system is lowered and the wind systems begin to revolve around the centre, spiralling inwards in a clockwise direction. So on the southern side the winds are travelling westwards and on the northern side, to the east. When the pressure reaches less than 1000 hPa it is named as a cyclone. Severe cyclones have central pressures less than 960 hPa, and Category 5 cyclones can have central pressures less than 920 hPa.

Cyclones are categorised into five levels of intensity by the strongest wind gust:

Category 1 Strongest gust < 125 km/h
Category 2 Strongest gust 125–170 km/h, e.g. Cyclone Steve, Cairns, 2000
Category 3 Strongest gust 170–225 km/h, e.g. Cyclone Winifred, Innisfail, 1986
Category 4 Strongest gust 225–280 km/h, e.g. Cyclone Tracy, Darwin, 1974
Category 5 Strongest gust > 280 km/h, e.g. Cyclone Orson, North West Shelf, 1989

The last Category 5 cyclone affecting the Great Barrier Reef was in 1899 in Princess Charlotte Bay. Studies on storm ridges formed along the coast and on islands show such 'super cyclones' have a recurrence interval of two to three centuries.

The effects on the reefs, continental shelf and coasts are twofold:

1. Severe erosion by waves and storm currents as the cyclone moves towards the coast. At landfall on the eastern coast, damage is generally most severe on the southern side where the winds impact the coast from offshore. On the northern side the winds blow offshore at landfall. As the eye of the cyclone passes there is a lull, and then the cyclonic winds resume from the opposite direction. The cyclone creates a storm surge that pushes the sea level 1–3 metres above normal tide levels. This storm surge can cause extensive erosion and also deposit storm ridges of rock and coral gravel above normal tide levels.
2. Extensive flooding and discharge of sediment ensues from the coastal rivers and streams as the cyclone moves inland and very heavy rain falls. Even low-intensity cyclones that do minimal damage offshore can still precipitate large rainfall onshore.

A study in February 1986 immediately after Cyclone Winifred showed erosion of the continental shelf down to water depths of at least 45 m, with winnowed sands overlying muddy sands on the mid shelf and fresh mud from the rivers overlying muddy sands on the inner shelf. Coastal beaches and rock walls were eroded. Onshore the heavy rainfall produced widespread flooding, and turbid water extended well out to sea.

Coral reefs are damaged more on the windward side, where broken corals and the reef front are abraded by sand-laden water. Older coral colonies, especially branching forms, are more vulnerable to cyclone damage.

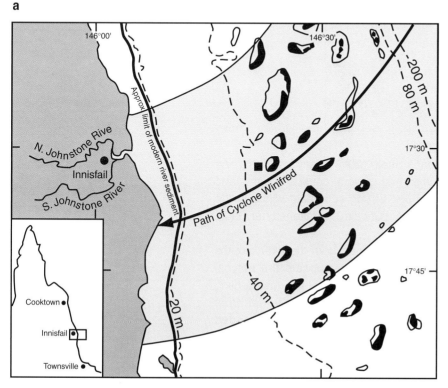

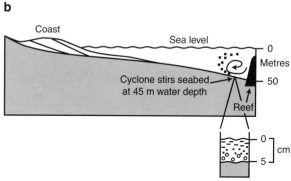

Figure 10.2 *a. Path of Cyclone Winifred across the Great Barrier Reef on 1 February 1986, where the cyclone eye crossed the coast. The eye was 50 km wide at landfall. Wind speeds reached 36 m/s (130 km/h) just before landfall, and dropped rapidly as the cyclone moved inland and degenerated into a rain depression. b. Offshore Cyclone Winifred stirred sediment to at least 45 m water depth, leaving a graded bed as the sediment re-settled.*

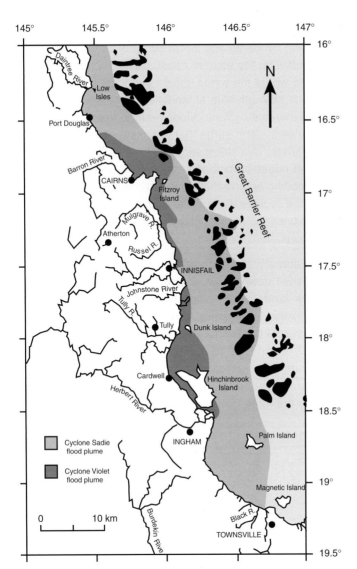

Figure 10.3 *Extent of flood plumes after cyclones Sadie (29–31 January 1994) and Violet (3–7 March 1995).*

Aerial surveys by GBRMPA following cyclones Sadie (29–31 January 1994) and Violet (3–7 March 1995) showed that flood plumes from several coastal streams and rivers amalgamated to form a single plume offshore. Strong southeast winds in the days after Violet had passed kept the plumes inshore, but calm conditions after Sadie allowed turbid water well out onto the mid shelf.

The turbid plumes of freshwater and sediment discharged from the river mouths can stress or kill corals in three ways: the low-salinity water upsets the polyps' metabolism, the turbidity decreases the light needed by symbiotic algae, and settling of the sediment can smother the corals. Records kept by the Great Barrier Reef Marine Park Authority (GBRMPA) show that low-salinity waters and suspended sediments, both of which can harm reefs, may extend for up to 25 km offshore and persist for three weeks. The damage done by these events will be greatest close to the shore, where the water is fresher and the amount of suspended sediment is greater.

Reef types

Two main types of reef are recognised around Australia: fringing and shelf reefs. **Fringing reefs** form close to land, typically around offshore islands. **Shelf reefs** lie well offshore; some are solitary and others form clusters. A third type of reef – atolls, which are ring-like reefs developed around the top of subsiding submarine mountains – are common in the Pacific Ocean, but do not occur around Australia.

Fringing reefs grow mainly around nearshore islands. Some examples are the reefs at Hayman Island in the Whitsundays, Magnetic Island near Townsville, Dunk Island off Mission Beach, and Green Island off Cairns. Fringing reefs are best developed along the protected, leeward sides of the islands. Most fringing reefs have grown under the influence of muddy water discharged from coastal rivers and creeks. On these reefs you can walk directly from the island, across the beach and out onto the reef flat at low tide. Commonly the intertidal surface consists of loose sand with mangrove trees in clumps near the creek mouths or scattered in protected areas of the flat. Live coral is generally at the outer edge and on the subtidal slope along the reef front.

There are also some fringing reefs attached to the mainland, such as the Alexandra Reef near Mossman and the reefs at Cape Tribulation. It is not clear why these mainland fringing reefs occur mainly north of Cairns. It may be that the water is warmer there than farther south, or that more protection is provided by a narrower continental shelf and more shelf reefs offshore. It may also be that the larger rivers further south supply too much fresh water and sediment for coral reef development nearshore. Although there are small coral patches growing close to shore north of Townsville at Paluma Shoals, and off Forrest Beach south of Ingham, inshore environments have not produced the major reef masses that have developed offshore.

The mix of coral species in reef communities on the inner shelf fringing reefs is different to that in the communities in clearer waters on the shelf reefs offshore. Fringing reef coral communities seem better adapted to growing in the more turbid conditions. However, we do not know how close these coral communities are to their environmental limits – an important issue because of the extra stresses placed on them by increased sediment, nitrogen and phosphorus outputs from the land after settlement and clearing. Both desktop studies and field data indicate that short-term sedimentation rates over the past 100 years are greater than long-term rates measured by radiocarbon over a 5000 year time-scale.

Figure 10.4 *Landsat ETM image of Hinchinbrook Island and the coast from Ingham to Innisfail. The turbid water of the inner shelf is clear. Offshore lie shelf reefs with strongly built windward edges on the southeastern side. The white dots in the northeastern corner are clouds.*
COPYRIGHT © COMMONWEALTH OF AUSTRALIA (1999) LANDSAT ETM SATELLITE IMAGE ACQUIRED BY THE AUSTRALIAN CENTRE FOR REMOTE SENSING, GEOSCIENCE AUSTRALIA

Drilling data shows that the present fringing reefs have grown mainly since sea level reached its present position around 6500 years ago (see Chapter 9). The growth history of one fringing reef, in the lee of Fantome Island off Ingham, has been delineated by coring and radiocarbon dating. The reef has an upper layer of sand and gravel formed from calcareous reef organisms, especially corals, shells and algae. Underneath this is a fine-grained unit containing mud washed out to sea from the nearby Herbert River mouth. So in deeper water the mud forms a foundation on which the reef is growing. A similar pattern has been found in the reefs off Cape Tribulation, north of the Daintree River. Cores of fringing reefs show that 60% of sediment between the large coral heads is derived from the land.

Radiocarbon dating shows that the fringing reefs at Fantome Island started growing close to the island just as sea level reached the present level, and then grew outwards. There would have been small corals at the shoreline, perhaps starting on large boulders, and then in the shallow waters offshore. As the corals and other reef organisms grew more thickly, some of the debris was washed back toward the island by waves, forming a small beach. Terrigenous (land-derived) sediment accumulated in slightly deeper water offshore from the island. Generally the coral communities colonised the upper part of this slope, with the debris extending from the beach to form a reef flat. With time the reef prograded seawards. At each stage of its growth there was a shallow, coral-rich unit overlying the deeper muddy unit deposited on the reef slope.

Shelf reefs occur well offshore. They form large underwater hills with their tops near sea level and their bases at 40–100 m water depth. The southeastern, windward edge is generally a steep, solid wall. The leeward side of the reef is commonly more open, with gaps between smaller patch reefs, and commonly there is a central lagoon. In general it seems that the reef edges grow and coalesce, the internal lagoon becomes filled with reefal debris, and eventually a flat platform top is developed.

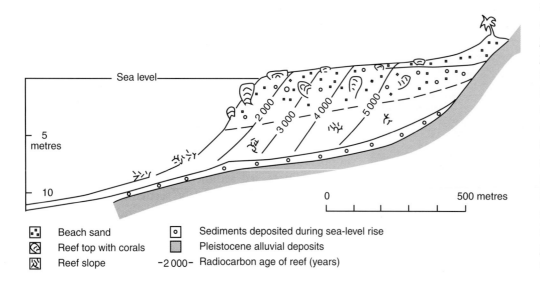

Figure 10.5 *Cross section of the Fantome Island fringing reef, showing the internal structure and radiocarbon ages of corals recovered by coring. The reef has prograded seawards, with the coral-rich upper reef unit overlying a muddy reef slope unit deposited on the reef slope. The entire reef has been built over the past 5500 years. The innermost parts formed when the sea level was just higher than present, because many of the coral heads on the surface near the beach are elevated up to 0.5 m above modern sea level, and are now dead. The reef has always grown in turbid conditions, as evidenced by the high content of terrigenous mud in the cores. In reality this mud forms the foundation on which the coral reef has grown.*

Figure 10.6 *A shelf reef, showing the pronounced windward edge and reef flat with lagoon behind.*
Photo: David Johnson

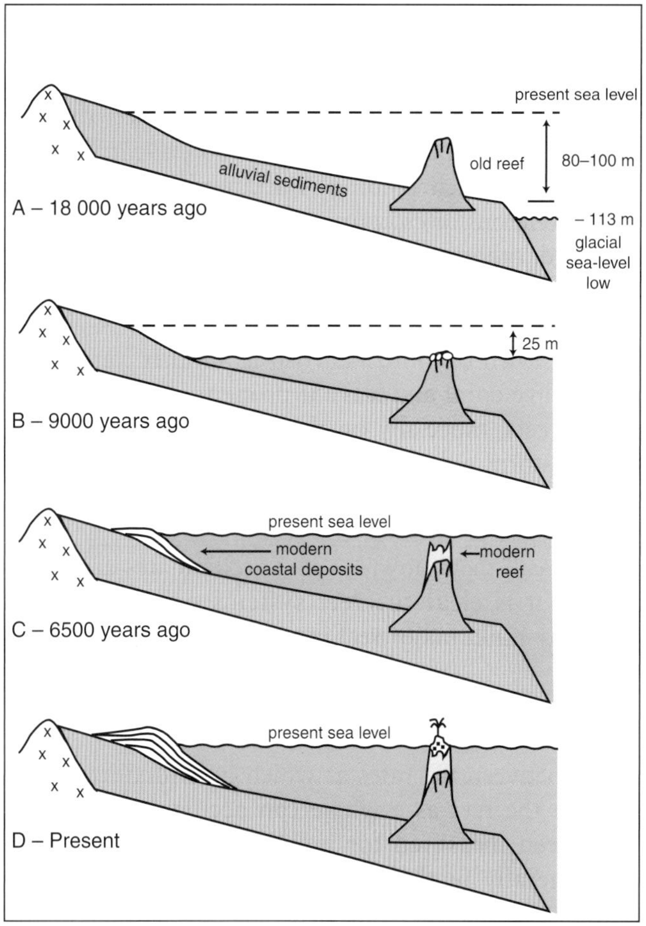

Figure 10.7 *Shelf reef growth. a. Old reefal hills standing 50–80 m above the plain during a period of low sea level. b. Modern reef growth starts about 8000–9000 years ago when sea level reaches the hilltops. c. Sea level rises to its present level, and the reef catches up. d. The reef fills in the lagoon to form a platform reef.*

In the more juvenile reefs that are still growing upwards, the sediment produced by the reef organisms is filling in lagoons or joining the large coral bomboras that form the reef framework. In the mature reefs, where the lagoons are full, this sediment is either shed sideways off the reef or starts to accumulate as a sandbank or cay on top of the reef, as at Wheeler Reef off Townsville. Eventually this cay may become large enough to support a rain-fed groundwater reserve of freshwater, and subsequently vegetation, such as at Heron Island reef.

The shelf reefs we see today are growing on top of old reefal hills. When the sea level was lower, during the ice ages, previous reefs stood as hills above the surrounding plain. The modern reef started growing when the sea level reached these hilltops, which are typically about 20 m below present sea level. Myrmidon Reef, the thickest modern reef drilled in the Great Barrier Reef, has its base 29 m below sea level. Radiocarbon dates show the base of most modern reefs is 8000–9000 years old. On thinner reefs such as One Tree Reef, where the modern reef is only 8–12 m thick, growth started around 6000–7000 years ago.

Coral reef growth

The reefs are not just coral, though it is the corals with all their colours and elegant shapes that often attract visitors. There are also exotic fish, and the giant clams with their iridescent flesh lining the cavernous openings. Then there are hundreds of species of animals and marine plants that are not so eye-catching but help build the reef and bind it together. Much binding is accomplished by particular calcareous algae, which secrete skeletons of calcium carbonate.

On the other hand, other organisms are busily eating and decomposing the reef materials – sponges and worms are boring into coral and shells, weakening them so that waves and currents more easily break and erode them. Parrot fish chew the corals for food. The Crown-of-thorns starfish is another natural predator that can destroy large areas of live coral and leave it prone to erosion. Cyclones periodically devastate sections of reefs, especially on the sides facing the high winds where the reef is sandblasted by storm waves carrying sand and gravel.

So the reef is growing as a balance between organisms forming hard materials and cementing them together to form the reef framework, and on the other hand processes tending to break them down. A reef accumulates as long as this balance is positive. Historically it is clear the reef systems can cope with the destructive processes and gradually build major reef masses.

How fast does a reef grow?

There is a difference between the rates at which individual coral colonies can grow and the rate at which the reef as a whole can accumulate. Growth rates of up to 100 mm/year have been measured for individual corals. Long-term rates of reef accumulation are generally in the range 5–8 mm per year, with some examples under ideal shallow water conditions up to 15 mm per year. When the sea level rises faster than 10 mm/year because of global warming the growth of coral reefs will almost certainly lag behind, and the reef will be drowned.

Reef deposits

The **reef wall** is composed of hard, almost rock-like material formed from corals and other hard organisms, bound together by calcareous algae. The corals are typically zoned, with different species (and especially different shapes) of coral in different environments. Compact head corals and branching staghorn corals are more common in the shallow turbulent waters, while platy corals are more common at depth where the plate-like form allows all polyps in the colony to face upwards and so maximise light interception. The analysis of cores drilled in reefs shows that natural crystalline cements also form. As seawater is pumped through the reef framework by big waves, crystals of calcium carbonate form in the voids, gradually helping to cement the framework together.

The **reef lagoon** is accumulating unconsolidated sediments, much of it fine sand and mud. Some of this sediment is composed of fragments washed into the lagoon from the reef wall by waves and tides. However, most of the lagoon sediment consists of organisms living in the lagoon: corals, foraminifers, shelly molluscs and calcareous algae. While the reef wall forms the vertical coating on the outside of the reef, drilling has shown that the inside of many reefs is mainly unconsolidated sediment, and this soft centre may in fact comprise most of the volume of the reef.

The **reef flat** forms in the intertidal zone on the reef crest. On juvenile reefs this may only be a rim around the top of the wall with the deeper lagoon behind it. On mature reefs the entire reef mass may be capped by a reef flat. Initially this reef flat consists of small corals bunched closely together. Some of these corals may show

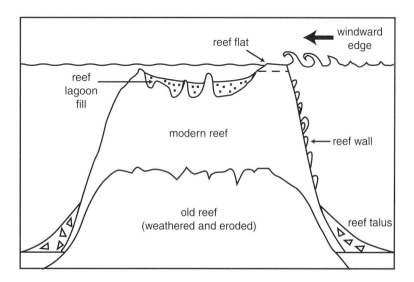

Figure 10.8 *Four main deposits forming in a reef: reef wall, reef lagoon, reef flat, and off-reef talus.*

Figure 10.9 *Photos of coral reef depositional environments: a. reef wall, b. reef lagoon, c. reef flat, d. micro-atoll.*
PHOTOS: DAVID JOHNSON

micro-atoll forms; that is, the colony is circular and tends to grow outwards, since it cannot grow upwards. The living coral then is restricted to a rim around the edge, and eventually the centre parts die, leaving a shape similar to the larger coral atolls. These micro-atolls are important indicators of sea-level. Eventually calcareous algae overgrow the corals, forming a solid pavement.

The **off-reef talus** forms around the base of the reef from material broken off by major storms or material swept off by tidal currents or waves. It can therefore vary greatly in size. Generally it forms a deposit that is thickest near the base of the reef, becoming thinner on the nearby shelf.

Formation of the Great Barrier Reef

The formation of the Great Barrier Reef has its origins in events millions of years ago, and its continued growth in the conditions of today. The essential requirement for massive reef growth is warm water (preferably free of turbidity from coastal sediments) and a largely stable geological situation. Tectonic instability can elevate reefs above sea level or drown them in deep water.

Four separate geological factors have contributed to what we see on the Great Barrier Reef today:

- stable continental margin at moderate depth
- northward continental drift into warm, tropical waters
- sea level changes
- sediment influx from the land.

Stable continental margin

The original rifting of the northeastern margin of the continent started when Australia was much closer to Antarctica, 60–55 Ma ago. At around 50 Ma the rifting appears to have stopped. By then a chunk of continental crust had separated eastwards, and is now preserved as the Queensland Plateau. This plateau subsided rapidly to a water depth of 1000 m, and only a few high points such as the Osprey, Flinders and Lihou Reefs were able to maintain reef growth. If the Plateau had been more stable and remained a shallow platform, we would have had another Bahamas off northeastern Australia.

The continental shelf also subsided, perhaps more slowly. Sediment influx from the Australian land mass, especially when the sea level was lower, has continued to build the shelf substrate upwards, and so help compensate for the subsidence. The width of the present continental shelf varies from less than 20 km off Cape Melville to over 400 km in the southern region. The gradient of the shelf is also steeper in the north: 1.5 m/km off Cooktown and only 0.7 m/km off Townsville.

Northward continental drift

After Australia separated from Antarctica at the break-up of Gondwana, water temperatures in the southern ocean became cold and the Antarctic ice sheet developed (see Chapter 7). As northern Australia passed into warmer tropical waters, conditions became suitable for prolific coral growth (Figure 10.10). With further northward continental drift, progressively more of the Australian landmass entered the warm, coral growing conditions. Consequently the Reef systems started growing at the northern end, the first part to enter the tropical conditions, and then with time expanded southwards.

If Australia continues its northwards drift the Great Barrier Reef will eventually extend down to the New South Wales coast and even around to Victoria. At present

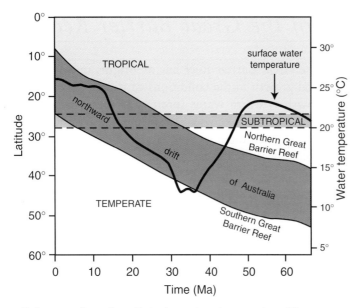

Figure 10.10 Northwards drift of northeastern Australia into tropical waters. The northern Great Barrier Reef line corresponds to Anchor Cay, now at 9°30'S in the Gulf of Papua, and the southern line to Heron Island (now at 24°S). Each position shows a steady northwards migration into more tropical waters, as Australia drifted northwards following the break-up of Gondwana.

However, this did not mean the waters got steadily warmer. The solid line shows the surface seawater temperature for the northern edge of the reef province. The temperatures were warm 45–60 Ma ago when Australia was still very close to Antarctica. The palaeoclimate and vegetation of this time were discussed in Chapter 7. Then the circumpolar cold current developed and the oceans off northeastern Australia were much cooler for over 25 million years. By around 15 Ma ago (even though the Antarctic ice-sheet was developing) northeastern Australia had moved sufficiently far north that the warmer water encouraged the formation of tropical carbonate sediments. However, the reef as we know it today did not start to form until much more recently, maybe less than 500 000 years ago.

rates there will be coral reefs off Sydney in about 14 million years. Seismic and drilling studies show deposition of tropical carbonate sediments started in the Miocene around 60 Ma ago, although real coral reef development may not have started until less than 500 000 years ago. Rapid fluctuations in carbonate production occurred over short time frames because of changes in ocean water temperatures.

The history of seawater temperature in the western Coral Sea and Great Barrier Reef region is known from the analysis of oxygen isotopes in sediments from a core (ODP 811) drilled in the deep-sea on the Queensland Plateau. The temperatures were 20–24°C for most of the last 10 million years, a range suitable for restricted coral growth. However, temperatures since 700 000 years ago have been consistently above 24°C, which allowed rapid growth of corals and consequently development of the reef.

Sea level change

The development of the modern Great Barrier Reef can be understood in terms of the changing sea level. Recent drilling on Ribbon Reef Number 5, down to 210 m below sea-floor and extending back some 700 000 years, has shown that massive reef-building corals first appeared well above the base of the hole. So the reef corals must be less than 700 000 years old. The upper part of the cores has pronounced erosion surfaces with rootlet horizons at 16 m, 28 m and 36 m below the sea-floor. These represent times of lower sea levels when the shelf and reefs were exposed to rain and river runoff. Thus the cores show cycles of alternating sea levels. Carbonate sediments accumulate when the sea level is high, but there is erosion, dissolution of limestone and plant growth when the sea levels are low.

Sea levels fluctuated between 120 000 and 18 000 years ago (see Chapter 9, pp. 188-9), when the shorelines were on the mid and outer shelves. The rivers must have ended as deltas on the present shelf, or fed into the heads of canyons incised into the shelf edge. The details of these patterns are yet to be investigated.

During the sea level lows, rivers extended across the present shelf, depositing fluvial sediments. There would have been vegetation, kangaroos and other animals on the exposed shelf, and Aboriginal people would have used this land. Much of the shelf was a large floodplain with channels extending to the shelf break. Since the sea level lows (over 100 m lower than at present) were below the shelf break (approximately 80–100 m lower than at present), the rivers debouched directly onto the shelf slope, through canyons incised at the shelf break. Dotted across the shelf were hills. On the inner shelf these hills are composed of granite, volcanics and other basement rocks similar to those on the present mainland. These represent remnant hills from Tertiary erosion that are now partly encased in more recent sedimentary sequences. As the mainland erodes and more sediment is delivered to the coast and shelf, the shorelines prograde, enveloping these offshore hills. We can expect that the coast will eventually envelop Magnetic Island off Townsville, the Frankland Islands off Innisfail, and Green Island off Cairns, in the same way that Cape Cleveland south of Townsville is now part of the mainland.

On the outer shelf the hills are made of limestone deposited as reefs and other carbonate bodies during previous sea level highs. These older reefs were then weathered and eroded by rain and wind during the sea level lows. The weathering has altered the nature of the limestones, dissolving some materials and changing the original aragonite of the coral and mollusc grains to calcite. On some reefs there are deep holes known as dolines, which probably formed by freshwater dissolution or cave collapse. At these times Aboriginal people would have walked across the floodplains, climbed the hills to admire the view, and fished in the estuaries.

We have a reasonably good idea of the sequence of events in the most recent cycle. Sea level was high around 120 000 years ago but dropped to a low of −113 m 18 000 years ago, before rising and reaching its present level about 6500 years ago.

As the sea level rose the continental shelf became inundated progressively from the outer edge, so that the shoreline moved across the shelf until it reached its present position. Coral larvae were swept in by the currents and began to colonise suitable substrates, especially older limestone hills. The sea level continued to rise and the coral communities accumulated reefal materials, building up the substrate.

Parts of this shoreline movement can be traced. If we drill the surface of the shelf, we find old mangrove mud deposits under 20–30 m of ocean water. Clearly the sea level was lower at the time these muds were deposited. The rate of sea level rise was reasonably high; the long-term average rate from the shelf edge to the present level was nearly 9 mm per year, or 9 metres per thousand years. Over slopes as gentle as the continental shelf off Townsville this sea level rise would have caused a very rapid, lateral advance of the shoreline, at least 10 metres per year.

Sea level rise was episodic and included periods when the rate of rise exceeded 10 mm per year. During these periods the shoreline change would have been greater – about the width of an average house block in 1 to 2 years.

In fact the sea level eventually rose slightly higher. The data also show that the sea level passed through its modern level about 6500 years ago, rising to 1–2 m above the present level at around 5700 years ago. Recent evidence from fossil oyster beds indicates that the peak in sea level was maintained until 3700 years ago, rather than falling smoothly to the present level over a 5000-year period.

Why did the sea level fall again? Two theories are current. Firstly, the weight of water on the shelf may have caused the outer shelf to sink slightly and the inner shelf to rise, hinging around the mid shelf. Such a small rise in the land would appear to be a sea level fall. Secondly, recent work in the Antarctic has shown that ice accumulation rates increased between 4000 and 2500 years ago, and this could have led to a fall in sea level of at least 0.7 m.

This fall and rise of sea level is only the latest in a series that have occurred over the past million years. So, the reefs grew at times of high sea level, and were then exposed and eroded at times of low sea level. During each period of high sea level another reef cap was added to the previous reef.

Coastal sediment influx

At the coast, rivers bringing sediment to the sea are depositing sand along the coastal beaches and bars, and some is blown inland to form coastal dunes. The mud is carried in suspension from the river mouths out to sea, where it settles in deeper, calmer water or is transported into protected estuaries and mangrove tidal flats. Sediment is moved mainly northwards along the coast in response to the prevailing southeasterly trade winds.

Little terrigenous sediment is carried out onto the Great Barrier Reef shelf. The sediment is trapped in north-facing embayments, and these areas are the likely long-term repositories of terrestrial sediment and contaminants. The thickest accumulations of sediment reach 30 m thick, and occur in the lee of headlands such as Edgecumbe Bay near Bowen, Cleveland Bay at Townsville and Trinity Bay at Cairns. Though visually conspicuous, muddy flood plumes from coastal rivers introduce only very minor amounts of sediment to the mid shelf and outer shelf. Most of the sediment contributed to the Great Barrier Reef is delivered by three river systems: the Fitzroy, the Burdekin, and the Normanby River on Cape York. Both the Fitzroy and the Burdekin Rivers have large catchments that extend well inland of the coastal ranges.

The modern continental shelf sediments form three zones: a coastal inner shelf zone with fringing reefs, where river influx is deposited mainly because of strong coastward seabed transport; a mid shelf starved of terrigenous and carbonate sediment; and an outer shelf dominated by carbonate sediment and reefal deposits. This pattern is as follows:

Zone	Water depth (m)	Carbonate % of shelf sediment
Inner shelf	0–20	> 30%
Mid shelf	20–40	30–80%
Outer shelf	40–80	> 80%

The general pattern that has emerged is that terrigenous influx is large around many inner shelf fringing reefs, and the coral communities of those reefs are adapted to high-water turbidities. Very little terrigenous sediment reaches the reef tract that starts 30–50 km offshore.

Box 10.2 Extent of terrigenous sediment in the Great Barrier Reef

Five lines of data confirm that the bulk of modern terrigenous (land-derived) sediment influx extends only 10–15 km offshore.

1 **Oceanographic studies**

Seawater analyses after the 1981 Burdekin flood, in which fresh river water diluted the ocean water, showed clearly that low-salinity water, less than 30 parts per thousand,* was held close to the coast. Ocean water is about 35 parts per thousand salinity, while fresh river water is almost 0 parts per thousand. Water with a salinity of less than 30 parts per thousand has a large admixture of river water. A similar pattern was found off the Johnstone River catchment after floods associated with Cyclone Winifred in 1986. Turbid water was noted well offshore from ships and reported from planes. However, the concentration of the particulate matter in the water was extremely low. Furthermore, the discoloration can be caused by a phytoplankton bloom, or to the resuspension of shelf sediment by storm waves associated with the cyclone.

2 **Seismic and core studies**

Seismic and core studies have shown that the terrigenous sediment bodies near the coast become thinner seawards, and are absent more than 15 km offshore. Seismic profiles and cores also confirm that the mid shelf has rarely more than 1 m of sediment cover. Most terrigenous sediment is being deposited within 10–15 km of the coast.

3 **Carbon isotope studies**

Organic materials in shelf sediment off Innisfail in 1986 were studied before and after the floods following Cyclone Winifred. Stable carbon isotope studies showed that most of the terrestrial organics reached only about 15 km offshore, a distance similar to the general distribution of inner shelf sand and mud deposits.

4 **Coral fluorescence**

Long cores drilled into big *Porites* coral heads contain bands that show enhanced yellow-green fluorescence under UV light in the laboratory, and these bands can be matched to historical records of coastal river flooding (see Figure 10.11). This enhanced fluorescence was originally attributed to organic acids delivered by river waters from the land and then incorporated in the coral

* 30 parts per thousand is equivalent to 30 000 parts per million (30 000 ppm)

skeleton. Recent investigations have shown the fluorescence is associated with an altered microstructure of the coral skeletons, one which has more holes and indentations. These narrow, low-density bands presumably form during periods when the corals are bathed in lower salinity waters containing organic acids derived from the land. Cores from nearshore reefs strongly display the fluorescence; while it is absent or weak in coral cores from offshore shelf reefs.

5 **Reef matrix**
Nearshore reefs can consist of 50–60% terrigenous material mixed with the coral and other carbonate debris, whereas offshore shelf reefs contain only traces. Clearly the modern influx from the land is not inundating the shelf reefs to be incorporated in their deposits.

Is the sediment runoff from the land destroying the Great Barrier Reef?
It is clear rivers are discharging much more sediment and nutrients since clearing of the onshore catchments and farming began. The 2001 report by GBRMPA shows that sediment, nitrogen and phosphorus exports may be up to eight times those of 1850. The health of the reefs may depend on how we manage the nearby land and river catchments. Certainly the sediment and freshwater influxes can have a great effect on the inshore fringing reefs, probably causing coral bleaching and death (although there is evidence of recovery within a few years), with changes in species composition of the coral communities. There is no evidence that influx from the land is affecting the shelf reefs offshore. However, around the globe there is evidence that bacteria and viruses carried by humans are spreading diseases that are killing corals. Increased river flows may have a similar effect by spreading soil bacteria into reef systems.

History of an individual reef: Britomart Reef

We can trace the growth history of a reef in detail by drilling holes from the surface down through the reef mass, such as the one at Britomart Reef, off Ingham.

The modern Britomart Reef began growing about 9000 years ago, when the first corals attached themselves to an older mound of limestone that had been exposed and weathered during a sea level low. The sea kept rising quite rapidly, because it reached present level 6500 years ago. The average rate of sea level rise over that time was about 10 mm/year.

Could the reef keep up with this rise? Evidently not, because the diagram shows that when sea level reached the present level 6500 years ago the reef deposits were still 14 m under water. By 3000 years ago the reef had reached sea level and had begun to grow sideways. This story is not uncommon for reefs; other drilling has found similar patterns in many reefs. Many reefs are still growing up to present sea level and have deep lagoons behind the reef wall; or the modern reef is just part of a larger shoal, most of which is still 5–20 m under water. Clearly it will take thousands of years for the whole shoal to build up to modern sea level. However, some reefs already form a platform at present sea level, with much of the platform exposed at low tide.

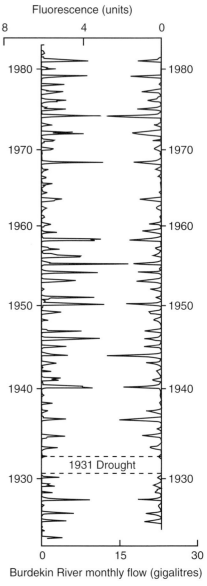

Figure 10.11 *Fluorescence data on corals is matched by river discharge from the floodwaters of the Burdekin River.*

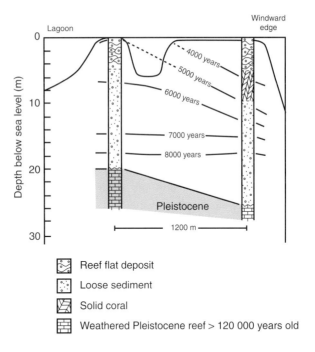

Figure 10.12 *A drilled section of Britomart Reef, showing modern reef deposits overlying the eroded older reef foundation, which is 20–25 m below modern sea level. The reef surface is shown at different ages based on radiocarbon dating of corals from the cores. Note that 6500 years ago, when sea level reached its present level, the reef surface was still at least 10 m below sea level, and that since the sea level stabilised the reef has caught up and steepened the windward edge.*

Continental slope and trough seaward of the Great Barrier Reef

East of the southern and central parts of the Great Barrier Reef lie the ocean plateaus – the Queensland and Marion Plateaus. These then pass into the deep sea of the Pacific Ocean, 300–500 km offshore. Extending east of Townsville is a deep-water embayment – the Townsville Trough.

Seawards of the northern part of the continental shelf and reefs is the Queensland Trough, which deepens northwards, being 1000 m deep off Townsville and 3000 m deep off Cape York. This trough is the submerged rift that formed from 60 Ma ago when the Coral Sea was opening. The trough is now filling with pelagic sediment from the Coral Sea waters, and also sediment derived from the Australian mainland and continental shelf that is swept into the trough, especially when the sea level is low. The main valley in this trough is over 400 km long, and 80–100 km across, bounded to the east by the steep edge of the Queensland Plateau and to the west by the continental slope of the Great Barrier Reef. The continental slope has a mantle of pelagic sediment but is marked by large submarine landslides.

In the southern part of the continental slope, off Townsville, there is little downslope movement. On the slope between Ingham and Cooktown there are small failures up to 30 km wide and 50 km long. The section north of Cooktown has a deeply incised shelf edge where deep canyons cut the slope and deliver sediment to the deep sea. Sediment was also delivered to the slope and trough during lower sea levels, and cores in deep water show alternating layers of carbonate and terrigenous sediment supplied by rivers and estuaries that cut across the present shelf.

Summary

Modern reef growth is localised on previous highs such as shallow hills of weathered limestone, or around bedrock islands. The Great Barrier Reef has developed as a result of four main effects: a stable continental margin at moderate depth, northward continental drift into warm tropical waters, sea level change that has allowed periodic erosion of the reef during sea level lows with further reef growth during sea level highs, and sediment influx from the land.

On the shelf, reef growth rates are 5–8 mm/year, rarely up to 15–18 mm/year. The reef can match sea-level rise only under conditions that are optimal for reef growth, but would be drowned by higher rates during episodic sea level rise. Typically the shelf reefs started growing some 9000 years ago, lagged behind sea level rise for a while, but were sufficiently close to the present sea level not to be drowned and to be able to catch up the rest of the height within a few thousand years.

Fringing reefs around the inshore islands show a different story to the shelf reefs. These fringing reefs are growing closer to the coast and within the zone where floods in the main rivers discharge muddy waters. Even in normal conditions the wave and current activity stirs up the muddy seabed, creating turbid waters. These corals are commonly different species to those on the shelf reefs because they have to tolerate muddy rather than clear water conditions. In a sense they are already stressed by the natural environment so it is important not to load too much further stress from human activities.

SOURCES AND REFERENCES

Barnes, D.J. & Taylor, R.B., 1998. *On the nature of luminescence in coral skeletons*. Technical Report No. 22. CRC Reef Research Centre Ltd, Townsville, 38 pp.

Belperio, A.P., 1983. 'Terrigenous sedimentation in the central Great Barrier Reef lagoon: a model from the Burdekin region'. *BMR Journal of Australian Geology & Geophysics*, 8: 179–190.

Boto, K. & Isdale, P., 1985. 'Fluorescent bands in massive corals result from terrestrial fulvic acid inputs to nearshore zone'. *Nature*, 315: 396–397.

Carter, R.M. & Johnson, D.P., 1986. 'Sea-level controls on the post-glacial development of the Great Barrier Reef, Queensland'. *Marine Geology*, 71: 137–164.

Chappell, J. *et al.*, 1983. 'Holocene palaeo-environmental changes, central to north Great Barrier Reef inner zone'. *BMR Journal of Australian Geology & Geophysics*, 8: 223–235.

Chin, A. & Ayling, T., 2000. 'Disturbance and recovery cycles: long-term monitoring on 'unlucky' inshore fringing reefs, Cairns section of the GBRMP'. *GBRMPA Reef Research*, 10 (1): 1–6.

Collins, L.B. *et al.*, 1993. 'Late Quaternary evolution of coral reefs on a cool-water carbonate margin: the Abrolhos Carbonate Platforms, southwest Australia'. *Marine Geology*, 110: 203–212.

Davies, P.J. & McKenzie, J.A., 1993. 'Controls on Pliocene-Pleistocene evolution of the northeastern Australian continental margin'. *Proceedings of the ODP, Scientific Results*, 133: 755–762.

Davies, P.J., Symonds, P.A., Feary, D.A. & Pigram, C.J., 1989. 'The evolution of carbonate platforms of northeast Australia' in *Controls on carbonate platform and basin development. Society of Economic Mineralogists and Paleontologists*, Special Publication No. 44. 233–58.

Done, T., 1982. 'Patterns in the distribution of corals communities across the central Great Barrier Reef'. *Coral Reefs*, 1: 95–107.

Gagan, M.K., Johnson, D.P. & Carter, R.M., 1988. 'The Cyclone Winifred storm bed, central Great Barrier Reef shelf, Australia'. *Journal of Sedimentary Petrology*, 58: 845–856.

Gagan, M.K., Sandstrom, M.W. & Chivas, A.R., 1987. 'Restricted terrestrial carbon input to the continental shelf during Cyclone Winifred: implications for terrestrial runoff to the Great Barrier Reef province'. *Coral Reefs*, 6: 113–119.

Goodwin, I. D., 1998. 'Did changes in Antarctic ice volume influence Late Holocene sea-level lowering?' *Quaternary Science Reviews*, 17: 319–332.

Hopley, D., 1982. *The geomorphology of the Great Barrier Reef: Quaternary development of coral reefs*. Wiley-Interscience. 453 pp.

International Consortium for Great Barrier Reef Drilling, 2001. 'New results on the origin of the Australian Great Barrier Reef: Results from an international project of deep coring'. *Geology*, 29: 483–486.

Isdale, P., 1984. 'Fluorescent bands in massive corals reflect centuries of coastal rainfall'. *Nature*, 310: 578–579.

Isern, A.R., McKenzie, J.A. & Feary, D.A., 1996. 'The role of sea-surface temperature as a control on carbonate development in the western Coral Sea'. *Palaeogeography, Palaeoclimatology, Palaeoecology*, 124: 247–272.

Johnson, D.P., Cuff, C. & Rhodes, E.G., 1984. 'Holocene reef sequences and geochemistry, Britomart Reef, central Great Barrier Reef, Australia'. *Sedimentology*, 31: 515–529.

Johnson, D.P. & Risk, M.J., 1987. 'Fringing reef growth on a terrigenous mud foundation, Fantome Island, central Great Barrier Reef, Australia'. *Sedimentology*, 34: 275–287.

Johnson, D.P. & Searle, D.E., 1984. 'Post-glacial seismic stratigraphy, central Great Barrier Reef, Australia'. *Sedimentology*, 31: 335–352.

Larcombe, P. *et al.*, 1995. 'The nature of post-glacial sea-level change, central Great Barrier Reef, Australia'. *Marine Geology*, 127: 1–44.

Larcombe, P., Woolfe, K.W. & Purdon, R. (eds) 1996. *Great Barrier Reef: Terrigenous Sediment Flux and Human Impacts*. 2nd Edition. CRC Reef Research Centre Current Research Townsville Australia. 174 pp.

Maxwell, W.G.H., 1968. *Atlas of the Great Barrier Reef*. Elsevier. 258 pp.

Nott, J. & Hayne, M., 2001. 'High frequency of 'super-cyclones' along the Great Barrier Reef over the past 5,000 years'. *Nature*, 413: 508–512.

Smith, S.E. & Kinsey, D.W., 1976. 'Calcium carbonate production, coral reef growth, and sea-level change'. *Science*, 194: 937–939.

Websites

Great Barrier Reef Marine Park Authority
[*Access to a wide range of data on the reef and its management*]
www.gbrmpa.gov.au

CRC Reef Research Centre
www.reef.crc.org.au

Australian Institute of Marine Science
www.aims.gov.au

James Cook University
[*Specialises in tropical marine and reef studies in Earth sciences, engineering and marine biology*]
www.jcu.edu.au

Slide show 'Wonders of the Great Barrier Reef'
www.jcu.edu.au/school/mbiolaq/mbiol/mbslides.gif

Cyclone Information
www.bom.gov.au/info/cyclone/

Illustrations

Figs 10.2, 10.10 redrawn and reproduced with permission of SEPM (the Society for Sedimentary Geology): fig. 10.2 after figs 1, 4 in Gagan, M.K., Johnson, D.P. & Carter, R.M., 1988. 'The Cyclone Winifred storm bed, central Great Barrier Reef shelf, Australia'. *Journal of Sedimentary Petrology*, 58: 845–856; fig. 10.10 after figs 15, 16 in Davies, P.J., Symonds, P.A., Feary, D.A. & Pigram, C.J., 1989. 'The evolution of carbonate platforms of northeast Australia' in *Controls on carbonate platform and basin development*. Society of Economic Mineralogists and Paleontologists, Special Publication No. 44. 233–258.

Fig. 10.3: redrawn with permission from the CRC Reef and Jon Brodie after fig. 1 in Brodie, J., 'River flood plumes in the Great Barrier Reef lagoon', in P. Larcombe, K. W. Woolfe & R. Purdon (eds). 1996. *Great Barrier Reef: Terrigenous Sediment Flux and Human Impacts*. 2nd Edition. CRC Reef Research Centre Current Research Townsville Australia. 174 pp.

Fig. 10.4: satellite image provided by the Australian Centre for Remote Sensing (ACRES), GeoScience Australia. Crown copyright © All rights reserved. www.ga.gov.au

Fig. 10.11: redrawn with permission of Macmillan Publishers Ltd after fig. 1 in Isdale, P., 1984. 'Fluorescent bands in massive corals reflect centuries of coastal rainfall'. *Nature*, 310: 578–579 with permission of Peter Isdale.

CHAPTER 11

PLANETS, MOONS, METEORITES AND IMPACT CRATERS

Earth is only one of a very large number of rocky spheres orbiting the Sun, and presumably only one of a vastly greater number in the Milky Way and galaxies beyond. In what ways is Earth unique, and do the other planets tell us something of our past and of our possible future evolution?

The Earth's natural satellite, our Moon, is visibly pock-marked with craters, evidence of massive bombardment by meteorites. When did it happen? How many meteorite impacts are preserved in Australia, and when did they strike? What role have meteorite impacts played in the geological and biological evolution of Earth?

Earth in context

As a result of exploration by US and Russian space probes and missions to the Moon and nearby planets, we know much more about our neighbours than ever before. This chapter summarises some aspects of planetary and stellar geology, in order to place the Earth in its planetary context within the Solar System and describe the range of impact features that are so well preserved in the arid outback of Australia.

The Sun

The mechanism that maintains the enormous heat output of the Sun was a puzzle until the 20th century. The prevailing theory for hundreds of years was that the Sun was simply a huge fire, burning by chemical combustion. Then in the mid 19th century Lord Kelvin calculated that if the Sun were made of coal it would burn up in 5000 years – far too short a time considering evidence in the fossil record on Earth that the Sun has been heating our planet for hundreds of millions of years.

In 1854 German physicist Hermann von Helmholtz proposed that the outer layers of the Sun may be collapsing inwards due to the great gravitational attraction, and that some of this energy was released as heat. His calculations showed that this could have sustained solar radiation for the past 22 million years – still too short a time given the length of the fossil record.

The solution had to wait until the 20th century, when the first clue was given by Einstein's famous equation of 1905, $E = mc^2$. That is, a small amount of matter can be converted to a huge amount of energy. In the early 1920s Arthur Eddington had calculated that the temperature at the centre of the Sun, where the gases are under enormous pressure, must be around 10 million degrees, a temperature barely conceivable. He proposed that at this temperature individual hydrogen nuclei, separated from their electrons (a state called a plasma), might collide and join to form helium nuclei. The specific reaction was not verified until 1938, by Hans Bethe working at Cornell University in the USA. The fusion of two hydrogen nuclei to form a single helium nucleus is a continuing chain reaction in the Sun, and releases an immense amount of energy.

The Sun forms the dominant body of the Solar System and makes up 99.9% of the mass. Its huge size and gravitational field hold the planets in their orbits. It is the source of most of the heat in the system, with a surface temperature of around 6000 K. An idea of the Sun's heat output can be grasped by noting that here on Earth, at a distance of 150 million kilometres, our face receives only a minuscule fraction of the ultraviolet and heat radiation from the Sun, and yet this radiation can burn skin and blind eyes. Another example of its reach is the solar wind, a stream of ionised particles spiralling away under the influence of the Sun's rotating magnetic field. During solar flares this wind can cause havoc to electronic and other communications systems on Earth.

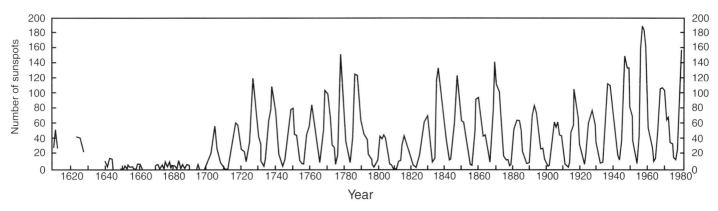

Figure 11.1 *Mean annual number of sunspots from 1610 to 1980. For every second cycle the magnetic polarity is reversed. Note the minimal number of sunspots measured in the years 1640–1710 (the Maunder Minimum), which corresponds to the Little Ice Age in Europe. Note also that there are periods of 60–80 years when the solar activity is high (e.g. 1730–1790) and others when the activity is much lower.*

REDRAWN AND REPRODUCED WITH PERMISSION OF CAMBRIDGE UNIVERSITY PRESS FROM FIGURE 47, GIOVANELLI, R.G., 1984, *Secrets of the Sun*.

The Sun has many other strange properties. One of particular interest to us is the regular cycle of sunspots. The sunspot cycle has a regular 11-year period, during which sunspots start forming about 32° north and south of the solar equator. Over 11 years they migrate closer to the equator and then disappear. However, the polarity of each 11-year sequence is the reverse of the previous sequence, so it is in fact a 22-year cycle for the Sun's polarity to return to the same state. Recent work shows that each 22-year cycle in fact starts as a set of ephemeral sunspots further from the solar equator towards the end of the previous 11-year period. That is, the solar magnetic reversal cycle is 22 years, but the next cycle starts about halfway through the present one.

The sunspot cycles have been mapped in detail since 1850, and less detailed observations go back to 1610. The sunspot cycles vary in intensity, sometimes alternating, and with a trend for recent cycles to be more intense. A lengthy interval from about 1640 to 1710 when there were very few spots – the so-called Maunder Minimum –coincides with the long period of cold weather in Europe called the Little Ice Age.

There is plenty of evidence for similar reversals of the Earth's magnetic field, although on a much longer and more irregular time scale. Presumably the processes are similar and both are thought to derive from changes in the circulation and convection of the inner parts of each body. However, we cannot yet explain why these changes occur, and in the Earth's case we have no way at all of predicting them.

The planets

Eight of the nine planets in the Solar System can be classified into two groups: the inner four terrestrial or rocky planets (Mercury, Venus, Earth and Mars), and the outer four giant, gaseous planets (Jupiter, Saturn, Uranus and Neptune), which are immensely larger than the rocky planets. A general theory for this division is that during the formation of the Sun only dense materials could accrete into planets in the zone close to the Sun, while in the farther, colder parts of the Solar System ices and gases were retained. The ninth and remote outermost planet, Pluto, is also rocky and its origin is unclear, although it is probably similar to Neptune's satellite Triton.

Feature	Mercury	Venus	Earth	Mars	Jupiter	Saturn	Uranus	Neptune	Pluto
Mean distance from Sun (million km)	58	108	150	228	778	1427	2870	4497	5900
Period of revolution around Sun	88 days	225 days	365 days	687 days	11.86 years	29.46 years	84.01 years	164.8 years	247.7 years
Rotation period	59 days	243 days*	23h 56m 4s	24h 37m 23s	9h 50m 30s	10h 14m	17h*	17h 50m	6.4d
Equatorial diameter (km)	4880	12 104	12 756	6787	142 800	120 000	50 100	49 500	2200
Mass (Earth = 1)	0.055	0.815	1.00	0.108	317.9	95.2	14.6	17.2	?0.1
Volume (Earth = 1)	0.06	0.88	1.00	0.15	1316	755	67	57	?0.1
Density (kg/m^3 x 10^3)	5.4	5.2	5.5	3.9	1.3	0.7	1.2	1.7	?

* Venus and Uranus have retrograde rotations (i.e. in the opposite direction to the other planets).

Table 11.1 *Solar and planetary data.*

Five planets can be seen regularly with the naked eye: Venus, Mars, Jupiter, Saturn, and less commonly Mercury.

Table 11.1 summarises some facts about the planets. The similar nature of the inner planets can be gauged from their similar densities. Jupiter and Saturn dominate the planets in terms of size and mass, despite their lower densities.

The inner terrestrial planets are relatively dense (3000–6000 kg/m^3), have few or no satellites and no rings. All have solid surfaces and rotate relatively slowly. The giant gaseous planets have low mean densities (less than 2000 kg/m^3), have multiple satellites and rings, are composed mainly of hydrogen and helium, probably with denser central cores, and rotate rapidly.

There is also a zone of asteroids, which are small rocky masses orbiting between Mars and Jupiter, about 500 million kilometres from the Sun. The first asteroid discovered, 1 Ceres, was found on 1 January 1801 by an Italian astronomer Piazzi. Ceres is about 1000 km across – not planet-sized but certainly a sizable rock. Over 2000 asteroids with well-documented orbits are named, and there are another 1000 lesser-known asteroids with temporary numbers. Most stay in the orbit between Mars and Jupiter, well away from Earth. However, one asteroid (433 Eros) was observed in detail in 2000 when it came close to Earth. The Near Earth Asteroid Rendezvous (NEAR) probe *Shoemaker* circled Eros for a year until it crashed into the asteroid's surface on Valentine's Day, 14 February 2001, taking its last picture 120 m from the surface. Eros is a 33 km long potato-shaped rock with a surface composed of ejecta from several impact events, so that boulders and dust partially cover older craters.

The farthest objects known in our Solar System lie in the Kuiper Belt, an elliptical zone of icy objects 10.8 billion kilometres from Earth, far beyond Pluto. The Hubble Space Telescope has confirmed the existence there of a large ball of ice and rock first detected by land telescope, a frozen mass about 1280 km across, now named Quaoar.

Mercury is closest to the Sun, completing an orbit every three months, which makes its movement seem very fast and hence its naming by the ancients as Mercury, the wing-footed messenger of the gods. The first clear images were returned by *Mariner 10* in 1974, and revealed a barren surface of broad plains, with some long fault scarps up to 2500 m high, covered with multiple craters but lacking the large maria of the Earth's Moon. The magnetic field is very weak and the atmosphere almost absent, only traces of argon and hydrogen being detected. Temperatures fluctuate wildly, from up to 330°C during the day to minus 174°C at night.

Venus is the second-closest planet to the Sun and is shrouded in clouds, so early observers could see little detail. The *Mariner 2* flyby in 1962 revealed the first reliable data. It found that Venus rotates backwards compared to Earth, that there is virtually no magnetic field, and that the surface temperature is extremely hot, almost a constant 500°C.

In 1990 the *Magellan* spacecraft undertook radar mapping, which showed that the surface is relatively featureless, although there are two very large mountainous uplands rising some 9000 m above the general surface level. Radar mapping has confirmed that the Venusian surface lacks the heavily cratered uplands characteristic of the Moon and Mars. The density of craters is lower on the volcanic uplands than elsewhere on the planet, indicating burial of older features by more recent and very large volcanic eruptions. Photographs from the Russian *Venera* lander, which functioned for just 20 minutes on the surface, showed a landscape strewn with rocks. The lander also showed that the clouds exist mainly 50–70 km above the surface, with a haze below them but a clear atmosphere for 20–30 km above the surface.

The heavy atmosphere is Venus's most obvious characteristic. It is composed of 96% carbon dioxide and 3% nitrogen, with traces of water vapour and sulphuric acid droplets. The atmospheric pressure on the surface is almost 100 times that of Earth. Venus's atmosphere creates an extreme greenhouse effect, by which the dense atmosphere traps and recycles heat.

Mars is a truly intriguing companion to Earth. Though slightly smaller it has similar rotation period and tilt, which means its days and months are comparable to those on Earth. The magnetic field is extremely weak. The gravity is less because of the smaller size, and so less atmosphere has been retained. The atmosphere is mainly carbon dioxide (95%), with minor nitrogen, argon and oxygen contents. Temperatures in polar regions are generally below minus 100°C, but summer temperatures at the equator a comfortable 27°C. The polar ices consist of water ice, but there is permanently frozen carbon dioxide (dry ice) at the southern pole.

The Martian landscape is very variable, with huge volcanic calderas, impact craters, long rift valleys and many patterns like big rivers which have long since dried up. The *Viking* lander images show that the ground surface is barren and strewn with rocks and sand, and the winds cause immense dust storms. Mars has the largest landscape feature yet mapped in the Solar System – Olympus Mons, a volcanic mountain 26 km high and 600 km across surmounted by huge calderas. This is truly a giant feature.

Water probably once flowed rapidly over the Martian surface, but has either now all dried up or is frozen in the ground. High resolution images from the *Mars Orbiter* camera taken in 1999 show gullies and rills on the insides of craters and edges of valleys, indicating groundwater seepage and surface runoff, though no evidence of precipitation. The Mars Explorer Rover Mission, which was launched in June 2003 and landed in January 2004, found bedded rocks and possible evaporites, which would be evidence of previous water on the planet.

Jupiter and Saturn are huge gas balls, very different to the rocky planets. Jupiter is the second largest body in the Solar System after the Sun, and like the Sun is composed mainly of hydrogen and helium. It may well be a failed star, a mass of gas too small to initiate the self-sustaining nuclear chain reactions. Even so it still radiates almost twice the energy that it receives from the Sun. It is thought the bulk of Jupiter is a mass of metallic hydrogen, condensed by the immense pressures, and surrounded by a swirling atmosphere of water and ammonia. The red coloration may well be due to red phosphorus. The atmosphere is rotating as a series of latitudinal belts at high speeds with enormous eddies revealed by the *Voyager* and *Cassini* flyby cameras. Alternating belts flow eastwards and westwards beside each other. There are at least 16 moons, of which the largest four – Io, Callisto, Europa and Ganymede – were observed by Galileo, and can be seen with good binoculars as small bright dots, except when they are obscured by the planet.

Although Jupiter is much larger than Earth, with a diameter of 143 000 km compared to Earth's 13 000 km, its rotation is much faster: once every 9 hours, 50 minutes and 30 seconds! The magnetic field is extraordinarily powerful, some 4000 times stronger than Earth, set at an angle and just offset from the rotational axis. The Jovian magnetosphere is some 30 million kilometres wide.

Saturn is the farthest planet from the Sun that is visible to the naked eye, and is similar to Jupiter. The planet is 120 000 km in diameter, and composed primarily of hydrogen and helium, with traces of ammonia, methane and phosphine. Saturn also has a banded atmosphere caused by winds swirling around the planet at speeds up to 1400 km per hour. The rotational period is also short, about 10 hours. Saturn's magnetic field is strong, about 1000 times that of Earth, although much weaker than Jupiter's field.

The most outstanding feature of Saturn is the complex ring system. The rings are composed of silica rocks, iron oxide and water ice particles, ranging in size from dust particles to lumps the size of a car. There are at least 17 moons.

Earth's Moon

The Moon is the only other body in the Solar System that we have sampled directly, and being Earth's closest neighbour it is one that can help us understand our own planet's history. We have known for a long while that it is a very different place, and

Figure 11.2 *The Moon, showing craters, uplands and maria. Note the uplifted peaks in the centre of many of the impact craters. The formation of these peaks is shown in figure 11.4.*

Reproduced from Kuiper, G. *et al.*, 1967, *Consolidated Lunar Atlas*, Lunar and Planetary Laboratory, University of Arizona; digital version by E. Douglass and M.S. O'Dell, 2003, Lunar and Planetary Institute, Houston.

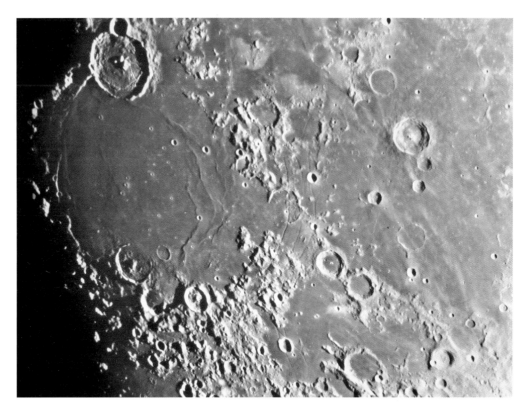

inhospitable to life as we know it. The lunar atmosphere is non-existent – a hard vacuum – and the lunar gravity about one sixth that of Earth. The surface temperature ranges from –158°C to over 130°C, depending on whether or not it is illuminated by the Sun.

The Moon is generally thought to have formed after a collision between the Earth and another object, perhaps when the Earth was only half or just formed. Such a collision must have been off-centre, so that mostly crust and mantle (but little of the metallic core) was removed. This is because the Moon has an unusually small core: less than 3% of its mass, compared to 30% for the Earth.

Since the earliest telescopic observations and drawings of the Moon were published by Galileo in 1610, the Moon has continually occupied the interest of amateur and professional scientists. The apparently barren rocky surface covered by craters clearly had a vastly different history to the Earth.

There are two distinctly different landscapes on the Moon: uplands, which appear bright and are heavily cratered; and the large patches of darker ground, which are depressions generally without craters. By analogy with the continental uplands and ocean basins on Earth, these two areas were called the highlands (terrae) and seas (maria). The maria are quite circular and very flat, some not varying more than 150 m in elevation over distances of 200–600 km. They lie below the average level of the Moon surface – up to 4 km lower in places. In contrast, the highlands have a rugged topography, with mountains and valleys of up to 4 km relief.

Two main theories were advanced for the origin of craters: that they were due to meteorite impacts, analogous to the Meteor Crater in Arizona, or that they were formed by huge volcanic eruptions. The difference between these theories is important: widespread volcanic activity implies a molten interior for the Moon, and major plate movements to allow the release of heat and magma from the interior. If none of the younger craters are volcanic, the Moon must have long ago lost its heat and is now a cold, rocky sphere.

During the 1960s a series of detailed maps of the Moon, prepared from photographs, showed that some craters are very old and some are much younger, with fresh debris visible around them. However, it was the close observations of the *Ranger*, *Surveyor* and *Orbiter* spacecraft, and finally the *Apollo* landings that recovered samples for analysis on Earth and left instruments on the Moon surface, that have resolved the debate. The craters are primarily of impact origin with later modification by volcanism, perhaps induced by lava eruptions triggered by fracturing due to the impact. The lavas flooded the crater floors and in places spilled over the edges.

Most of the Moon's surface is covered by a blanket of granular materials – the lunar soil – ranging in size from grains less than a millimetre to huge boulders several metres across. On Earth the soil is formed by the physical breakdown and chemical alteration of rocks, and is governed especially by the availability of water and aided by plant activity. On the Moon such processes are absent. The lunar soil has been formed by a continual bombardment of meteorites, spraying debris outwards from each impact. This blanket of ejecta is estimated to be around 2–8 m thick.

The composition of all Moon rocks recovered is basaltic and contains the same minerals as those found in the basalts on Earth. The highland rocks have slightly greater feldspar contents, but none of the samples recovered either from the highlands or from the maria are in any way granitic like the continents on Earth. The feldspars are less dense than the other components in the basalts, so the highlands probably represent the floating of early-formed lighter crystals to higher levels in the Moon crust, while the heavier crystals settled at depth.

What age is the Moon?

Before discussing the age of the Moon rocks it is important to note that there is an inherent bias. The Moon rocks had to be collected from the surface; there was no way a drilling rig could be transported to the Moon to recover rocks from depth. So the rocks are presumably the youngest, since they lie on top of others buried more deeply. The sample size is also small, and could have missed the oldest or indeed the very youngest. Nevertheless, the dates so far of the rocks collected by the *Apollo* astronauts tell a surprisingly consistent story. The highland crust samples are 4.4–4.6 billion years old, and the impact craters appear to have formed over 4 billion years ago. There was a cataclysmic episode in the Moon's history around 3.6–4.1 billion years ago when the circular maria were formed, and basalt flowed from the Moon's interior to fill them in the period 3.9–3.2 billion years ago. Since then sporadic meteorite impacts have formed scattered craters on the otherwise smooth surfaces of the maria. The large crater Tycho is estimated to be around 100 million years old.

Instruments left behind on the lunar surface have measured numerous, very low intensity moonquakes, about 1800 per year. Some tremors measured are the result of meteorite impacts. Even the largest moonquakes measure less than 2 on the Richter scale; they are measurable by instruments but would not be felt by a person standing on the surface. Compared to Earth, the Moon is virtually aseismic. If it were not for the low background activity of moonquakes and the high sensitivity of the seismometers, it might seem that the Moon is completely dead. The moonquakes are not random. They tend to occur monthly and have a common focal depth of about 800–1000 km, about half way to the Moon's centre. The monthly cycle is caused by the gravitational attraction of the Earth pulling on the Moon's interior. The common focal depth indicates that movement occurs at a major structural weakness.

That the Moon has a flexible interior is supported by recent analyses of the distance between the Earth and the Moon. The *Apollo* astronauts placed laser reflectors on the Moon in 1969, and for three decades this distance has been monitored. It takes about 2.5 seconds for the laser beam to flash from Earth to the Moon and back. The measurement precision of this round-trip time has improved and is now equivalent to less than a couple of centimetres. The records show that the Moon's surface moves in and out by about 100 mm in response to the monthly variation in the gravitational pull of the Earth. This movement is evidence of the elastic nature of the Moon's interior.

Because there are several seismic stations on the Moon we can estimate the velocities at which shock waves travel through the interior, and so deduce the structure of the Moon. The present model is of a lunar crust about 25 km thick, with a very fractured and fragmented outer layer up to a few kilometres deep. There is a transitional zone from 25–65 km, interpreted as gabbro (a coarser-grained equivalent of basalt), and below this the lunar mantle. The elastic interior of the Moon indicated by laser distancing indicates that the core and part of the lower mantle may still be molten.

In summary, the Moon appears to have been originally molten, but cooling left a solid crust and a molten interior. During the early formation of the Solar System the Moon was bombarded with meteorites, forming the heavily cratered uplands. In addition to impacts that formed the multitude of small craters, several huge impacts formed maria surrounded by circular mountain ranges. By this time the intense meteorite bombardment was finished. Then, either because of the heating induced by the impacts or as a result of radiogenic heat, or both, there was an immense phase of basalt outpouring that filled the biggest and deepest impact basins. For a period the maria were indeed seas of lava.

Small scale volcanic activity continued for about 1.3 billion years. Since then major meteorite impacts have been relatively few though smaller meteorite impacts are common, and a 1 tonne impact occurred in 1972.

The lack of an atmosphere on the Moon has two effects. Modern meteorites are much more common than on Earth because the incoming meteors do not burn up as shooting stars. Secondly, the impact structures are preserved, not destroyed by the wind and water erosion that happens on Earth. Like Mercury, the Moon records the early history of the Solar System, a history shared by the Earth but whose record has in most cases been weathered away on our planet.

Meteorites

Meteorites are lumps of rock from outer space. All meteorites show a glazed or flow-marked surface where the object has had its outer edges melted during passage through the atmosphere. Accounts of stones falling from the sky go back to ancient Chinese, Greek and Roman writers. In some cases they were attributed to stones blown off mountains by high winds! Other ideas were that the meteorites were the result of catastrophic volcanic eruptions that blasted rocks into orbit early in the Earth's history, or that they came from the Moon or other planets. Despite the large numbers studied we have not been able to positively identify their sources, though we are sure they are extraterrestrial. One theory is that they derive largely from the asteroid belt between Mars and Jupiter, and it has been suggested that some originated from Mars or the Moon. The ages of meteorites are all around 4.5 billion years. This is the same age attributed to the Moon, the Earth and probably for the whole Solar System. So meteorites date from the very origin of the planets as we know them. The main bombardment in the Solar System was between 4400 and 3800 million years ago.

Meteorite types

There are three types of meteorite: stony, iron and stony-iron meteorites. The stony meteorites are composed of basaltic rocks with various mixtures of mineral crystals and glassy material, indicating an origin from planetary material similar to the mantle of the Earth. Some are carbonaceous, with organic compounds including amino-acids – the basic building blocks of proteins and essential to life as we know it. The prevailing opinion is that these organic materials do not indicate life elsewhere in space, but indicate that there is an organic chemistry story in outer space, just like the inorganic rock chemistry story.

The iron meteorites are composed mainly of iron, with 5–15% nickel and minor amounts of other elements such as cobalt, chromium, sulphur, carbon and copper. Such a composition is similar to the supposed core of the Earth. Iron meteorites can be very large. Many museums have specimens up to 1 tonne, and there are records of meteorites of up to 100 tonnes still lying in remote areas where they fell. Fragments of iron meteorites have been found around several of the main recent impact craters on Earth, such as the Meteor and Odessa craters in the USA, and the Wolfe Creek crater in Western Australia. Iron materials oxidise rapidly at the Earth's surface, so if iron meteorites formed some of the older craters the evidence would have long since disappeared.

Figure 11.3 *a. The Mundrabilla meteorite (3.5 tonnes mass) found by Mr A.J. Carlisle on the Nullarbor Plain in 1988. b. A tektite 25 mm across.*
PHOTOS: TRUSTEES OF THE WESTERN AUSTRALIAN MUSEUM

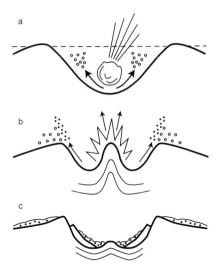

Figure 11.4 *Effects of a meteorite impact. First, a shock wave produces almost instantaneous deformation, compressing the ground and crushing and melting rocks. A decompression follows, with jets of melted meteorite and ground materials exploding from the interface between the meteorite and the ground. The decompression wave also sucks up material in the centre of the impact, forming a central uplift. The resulting explosion sprays ejecta outwards. Finally, the walls collapse and ring-shaped faults form around the impact site, with later erosion of the crater walls.*

MODIFIED, REDRAWN AND REPRODUCED WITH PERMISSION OF THE PUBLISHER, THE GEOLOGICAL SOCIETY OF AMERICA, BOULDER, COLORADO, USA. © 1989 AND 1999 GEOLOGICAL SOCIETY OF AMERICA. AFTER FIGURE 1 FROM MORGAN, J. & WARNER, M. 'CHICXULUB: THE THIRD DIMENSION OF A MULTI-RING IMPACT BASIN', *Geology*, 247: 407–10, WITH ADDITIONAL INFORMATION FROM GLASS, B.P., 1982, *Introduction to Planetary Geology*, CAMBRIDGE PLANETARY SERIES 2.

The stony iron meteorites are mixtures of metals and rock and are relatively rare, making up only 4% of the known meteorites.

The extreme energy of meteorites is a result of their very high velocity. The velocity of the meteorite on entering the atmosphere is many times the speed of sound, and sonic booms have been produced by meteorites or fireballs. The speed of sound is around 300 m/s. An Olympic sprinter contesting the 100 m is moving at 11 m/s. In contrast, a meteorite is travelling at around 15 km/s – over 1300 times faster.

Now the kinetic energy of a moving body is given by the equation:

kinetic energy = ½ × mass × velocity^2

The kinetic energy of an iron meteorite 20 m in diameter with a mass of 32 000 tonnes and travelling at a velocity of 10 km/s will be roughly the equivalent of 19 Hiroshima atomic bombs.

The melting and the glow of a meteorite in the night sky is caused mainly by the heat of friction as the meteorite passes through the Earth's lower atmosphere. Some of the heat also comes from compression of the air in front of the object, in a similar way to the heat of a bicycle pump during compression.

Impact craters

The examination of aerial and satellite photos has revealed over 150 impact craters on the Earth's surface, ranging in age from Precambrian to very recent. The craters have evidently been made by the impact of meteorites or comets. The famous historical example is the body, probably a comet, that exploded as an airburst above the remote Tunguska region of Siberia on 30 June 1908. The huge explosion flattened trees in a radial direction over an area of 2150 square kilometres.

A fresh impact crater is a circular structure with a rim dipping inwards, and the inner slope of the rim is typically steeper than the outside. These inner walls display collapse structures and other features indicating that the material has been pushed outwards and then collapsed inwards. There is commonly a central zone within the main ring, where the rocks have been uplifted by elastic rebound and suction as the meteorite or comet disintegrated.

Since meteorites could easily strike the surface at a low angle and thus create an elliptical hole, why are craters invariably circular? The answer lies in the mechanism causing the crater. As the meteorite approaches the surface it is travelling greater than the speed of sound, compressing the air in front. On impact there is compression of the surface rocks, and the high kinetic energy of the object is turned into heat. This combination of hot object and rocks explodes, deforming the surrounding rocks. The explosion is circular and obliterates any elliptical evidence of the crater impact.

A characteristic effect of the blast is to fragment the rocks, forming shatter cones, which are commonly found around impact craters but absent in the same rock units elsewhere. The area around and within the crater consists of a breccia of broken and shattered rock debris. Fused particles and small spherules of molten rock that have cooled are common. In some of the older impact craters, such as Gosse Bluff in the Northern Territory, the present structure is only the central uplift. The much larger outer ring has been eroded away.

Australian meteorites and impact structures

Until 1996, fragments from 474 distinct meteorites had been recovered in Australia, most from the Nullarbor region. There had been 13 observed falls where the object was recovered, including one at 10.10 am on 30 September 1984 that fell onto Binningup Beach, 143 km south of Perth. A brilliant fireball lasting a few seconds burst into four or five luminous fragments, and two loud bangs were heard. Two women sunbaking on the beach were startled by a whistling sound and a loud thud about 4 m from where they lay on the sand. A small stony meteorite was found shallowly buried in a small depression.

Another recent example was at Dunbogan, 30 kilometres from Port Macquarie in New South Wales on 14 December 1999, when a small meteorite fell through a house, leaving a 300 mm hole in the ceiling before breaking into hundreds of fragments.

The oldest evidence of extraterrestrial impacts in Australia is not a crater, but a layer of spherules embedded in Precambrian rocks. These spherules are similar to those strewn across the countryside by later impacts. They result from instantaneous melting of surrounding rocks and soil. Droplets of molten rock are splashed out,

Figure 11.5 *The Wolfe Creek impact structure: an aerial view taken with infrared film.*
Photo: Richard Rudd

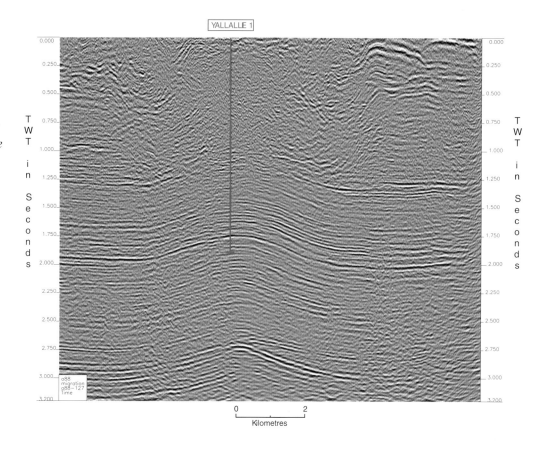

Figure 11.6 *Seismic section of buried Yallalie structure. Vertical scale is in seconds of seismic travel time. The red line under 'Yallalie 1' marks location of the drillhole which reached a total depth of 3322 metres. The upper layers are very broken, due to the impact. Lower down, the 500–700 m of uplift in the bedding caused by suction after impact is clearly visible. The Yallalie structure probably formed from an impact in the Pliocene, but may be as old as the early Cretaceous.*
IMAGE: ORIGIN ENERGY

and these cool to form the spherules. Layers of spherules have been identified in the Pilbara (aged 450 Ma) and in the Hamersley area (aged 2630 Ma).

Several impact structures buried hundreds of metres underground have been identified with 3D seismic profiling during petroleum exploration in Australia.

Among these is the Yallalie structure, about 30 km northwest of Moora in Western Australia. The Yallalie structure is circular, about 12 km across with a 3–4 km wide uplift zone in the centre, and radial structures extending as far as 40 km from the centre point. Deformed rocks can be identified over a depth of 1500 m depth. Others are the Tookoonooka structure, formed about 128 million years ago in southwestern Queensland, and the Puffin structure in the Ashmore region of the North West Shelf.

The old rocks of arid, outback Western Australia contain five major craters, and five more occur in the Lawn Hill area spanning the Queensland – Northern Territory border. Seventeen well-known craters in Australia are listed in Table 11.2. They range in age from 1600 million years or more in age to only 1000–2000 years old. Australia has one of the best collections of impact craters in the world, and it is possible that others are buried beneath the vast desert sands or have been removed by erosion.

Name	Diameter	Impact age
Young craters		
Boxhole, NT	175 m	25 000 years
Dalgaranga, WA	25 m	?1000–2000 years
Henbury, NT (13 craters)	up to 180 m	4000–5000 years
Mt Darwin, Tas.	1000 m	c. 740 000 years
Veevers, WA	70 m	c. 20 000 years
Wolfe Creek, WA	880 m	c. 300 000 years
Old craters		
Goat Paddock, WA	5 km	50 Ma
Gosse Bluff, NT	22 km	142 Ma
Lawn Hill, Qld	18 km	515 Ma
Liverpool, NT	1.6 km	150 Ma
Strangways, NT	25 km	646 Ma
Other (mainly Precambrian) craters		
Kelly West, NT	8–20 km	over 550 Ma
Acraman, SA	35–40 km	c. 590 Ma
Shoemaker, WA	30 km	c. 1630 Ma
Foelsche, NT	6 km	Proterozoic
Spider, WA	11 km	Proterozoic
Goyder, NT	> 3 km	pre-Cretaceous

Table 11.2 *Some well-known impact craters in Australia.*

The younger craters have associated meteorite fragments, while the older ones are only erosional remnants of originally much larger structures.

The oldest known structure is the Shoemaker Impact Structure, which is 130 km northeast of Wiluna. Upon its discovery in 1974 this structure was named the Teague Ring, but was renamed the Shoemaker Structure in 1998 in honour of Eugene and Carolyn Shoemaker, who made major contributions to the study of Australian impact craters. Eugene was killed in a car accident in outback Western Australia. It lies in Proterozoic sediments, with the central uplift zone exposing Archaean granites.

Two of the most spectacular are Gosse Bluff, west of Alice Springs, and the Wolfe Creek crater (see Figure 11.5) in northern Western Australia. Both have almost complete circular rings rising above the flat plains.

Perhaps the most intriguing ancient discovery is the Acraman structure in the Gawler craton in South Australia, which has been correlated with layers of ejecta debris in nearby sedimentary sequences. This is the only ancient structure in the world where the impact structure has been related to its ejecta. The debris layers have been dated at around 590 million years old.

Compared to the Moon and planets such as Mercury and Mars, Earth has preserved relatively few craters. Weathering by wind and water has worn most of them away. Also, the continuing effects of plate tectonics have buried most of them, in contrast to the stable crusts on the Moon where craters have been preserved for 4 billion years.

Tektites

Tektites are small glassy objects found scattered on the ground. In Australia most have been recovered from the outback where they are not buried by soil and plant growth. The Nullarbor Plain has been a fruitful searching ground. Three other major areas strewn with tektites (and their ages) are North America (35 Ma), Czech Republic (15 Ma), and the Ivory Coast (1 Ma).

Numerous tektites were recovered around the Henbury craters in central Australia. Two very extensive tektite fields lie adjacent to large impact sites: the European field near the Ries Crater in Germany, and the Ivory Coast field near the Bosumtwi Crater in Ghana. The North American tektite field originated from a buried crater under Chesapeake Bay. So far no impact crater has been identified as the source of the Nullarbor and South East Asian tektites, though a prime suspect is a crater-like lake 100 km long, the Tonle Sap in Cambodia.

Fission-track dating has shown the Australian tektites and those found across South East Asia formed around 790 000 years ago, and that both sets may be derived from one impact event. The tektites and impact debris are found over some 10% of the Earth's surface, including much of Australia and the surrounding oceans.

Tektites are made of a high-silica glass with rare mineral inclusions. Typically they are a centimetre or so across, though they range from a few mm in diameter (microtektites) to larger masses 100–200 mm across and weighing up to 12 kg. Their shape is usually round or teardrop-shaped, though a particular button shape is common in Australian tektites.

Two origins for tektites have been proposed. The first is that they represent ejecta from volcanoes on the Moon, and secondly that they are fragments of impact glass generated by meteorites hitting the Earth. An Earthly rather than lunar origin is more likely. The silica-rich composition of tektites is very different to the basaltic composition of the Moon, though very similar to granitic and sedimentary rocks on Earth. Also, the Moon's craters are primarily formed by meteorite impact, not volcanism. The close association of tektites with Henbury and other craters, together with their siliceous composition, has for long indicated that tektites form as droplets of Earthly rock melted and splashed out during a meteorite impact.

This theory is conclusive now. Detailed radiometric age analyses using neodymium and samarium isotopes show the glass was melted from rocks of the following ages: North America 650 Ma, Europe 900 Ma, Ivory Coast 1900 Ma and Australasia 1150 Ma. In other words, all the glass came from Precambrian rocks. And all these rocks are much younger than the rocks on the Moon.

Box 11.1 Large meteorite impacts: Eltanin and Chicxulub

ELTANIN

Because 70% of the Earth is covered by water it is probable that most of the impacts have occurred in the oceans. A major impact has been confirmed in the ocean between South America and Antarctica. Named the Eltanin asteroid, this body is estimated to have been 1–4 km across, and splashed down 2.15 million years ago.

Seismic profiles of the seabed show a 20–24 m thick layer of disturbed sediment extending hundreds of kilometres across the ocean floor. Cores taken of the abyssal seafloor at water depths of 3965 m and 4961 m show four sedimentary units in the impact deposit, and these overlie older undisturbed sediments. From the base (figure 11.7b) these are:

Unit IV – Chaotic mixture of mixed lithologies and ages of older sediments, with larger fragments (including manganese nodules) in the lower part and finer material in the upper part. This unit contains material ripped up from the seafloor during the impact and redistributed over a wide area. Coarser, heavier grains settled first, with the finer material settling later.

Unit III – Laminated, well-sorted sand with some basalt fragments, becoming finer upwards. This unit was sediment deposited from dense, turbid water flowing back and along the seabed, allowing finer material to settle out after the impact threw large quantities of sand and mud into the water column. The preserved laminae confirm that there was nothing living to mix the sediment.

Unit II – Fine sediment with up to 20% ejecta material in the lower part. Chromium and iridium levels correspond to layers with a high proportion of meteorite fragments. This unit records settling of fine material and meteorite ejecta, most of which was probably airborne and settled through the atmosphere and then the ocean.

Unit I – Late Pliocene to modern sediment, showing mixing by animals.

One core, taken on top of a seamount at 2707 m water depth, contains fragments of clays from the deep ocean floor which were displaced up onto the seamount standing over 1000 m higher.

Clearly the asteroid plunged into the ocean, causing massive erosion of the deep-sea floor, followed by the settling out of this sediment and the fragments of the meteorite, which disintegrated on impact.

This impact would have generated a major tsunami, estimated to involve 20–40 m waves over deep water approaching South America and Antarctica. Run-up heights could have been 10–25 times higher on continental margins. One possible result of such a tsunami is the unexplained sedimentary deposits on the Peruvian coast, which are of similar age, and which contain a mixture of complete skeletons of both marine and land mammals. Could these be the death assemblage from that tsunami?

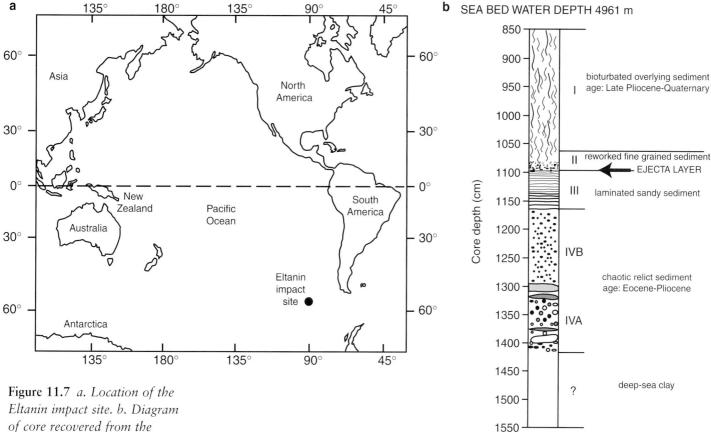

Figure 11.7 *a. Location of the Eltanin impact site. b. Diagram of core recovered from the abyssal sea bed showing the four units associated with the impact event 2.15 million years ago.*

REDRAWN WITH PERMISSION OF MACMILLAN PUBLISHERS LTD AFTER FIGURES FROM *Nature*: FIGURES 1A AND 3 RESPECTIVELY FROM GERSONDE, R. *et al.*, 1997, 'GEOLOGICAL RECORD AND RECONSTRUCTION OF THE LATE PLIOCENE IMPACT OF THE ELTANIN ASTEROID IN THE SOUTHERN OCEAN', *Nature*, 390: 357–63, AND WITH PERMISSION OF RAINER GERSONDE.

On a statistical basis it is probable that some 500 extraterrestrial bodies of similar size to the Eltanin asteroid have hit the oceans over the last 540 million years.

CHICXULUB

The Chicxulub meteorite or comet impact coincided with the extinction of the dinosaurs and much other life around 65 Ma ago. We have known for decades that this extinction event occurred, but its probable cause has only recently been established. The 251 Ma extinction event may also have been due to an impact (see Box 6.1).

The first signs of the impact were detected by very precise geochemical analyses of clays at the Cretaceous–Tertiary boundary in sedimentary sequences in Italy. The clays show unusually high contents of iridium, an element more concentrated in meteorites than rocks on Earth. Analyses of the organic materials on this same boundary in Spain have shown an abundance of hydrocarbons formed during fires – another probable effect of a major impact on land.

A very large, buried impact structure, called Chicxulub, was discovered during oil drilling on the Yucatan Peninsula in northern Mexico. This structure sits on the

Cretaceous–Tertiary boundary. Deep seismic profiling across the structure has shown that the impact affected the whole thickness of the Earth's crust. The deformed sedimentary sequences and normal faulting along the outer rings of the Chicxulub structure can be connected to zones of downthrown Moho, with a crater radius of 35–55 km.

Once the impact event was identified, the next job was to see if there was evidence of this event in the deep-sea sediments. ODP drilling of the seabed on the Blake Plateau southeast of Charleston (USA) in 1997 recovered cores with sections across the Cretaceous–Tertiary boundary, in water depths of 1345–2686 m. This site is 1500 km from the impact site, so scientists were interested to see the extent of the impact's effects.

The cores show a very sharp boundary with undisturbed or slumped Cretaceous material below the contact. Maybe the slump was caused by the impact. A 100 mm thick graded layer of green spherules occurs precisely at the Cretaceous–Tertiary time boundary. The grading from coarse at the base to fine at the top is evidence of gradual settling, with the heavier material reaching the seabed first. This layer contains abundant rock fragments, including limestones, chert, metamorphic rocks and mica fragments. These are rocks typical of the Mexican impact site. The layer also has a range of Cretaceous microfossils and, more importantly, altered tektite glass spherules. Above this is a thin layer with the iridium anomaly, interpreted as fallout from the fireball.

There is little doubt that this layer is the result of an impact, with the melted glass spherules, the iridium anomaly, the foreign rock fragments from 1500 km away and the mixed microfossil fauna.

Overlying the layer is grey ooze containing both Cretaceous and Tertiary microfossils, and anomalously high values of iridium. This passes upwards into undisturbed Tertiary deep-sea sediments.

A recent report has documented another major impact of a very similar age (60 Ma) buried in rocks below the North Sea. The Silverpit structure is 20 km across and has a multiringed pattern, with a central peak inside a 3 km wide crater. The sedimentary rocks of the impacted layer indicate that they formed in water depths of 50–300 m, so this meteorite also landed in the ocean. This report opens up the possibility that the major extinction at the end of the Cretaceous may be due to not just one, but several impacts.

Past and future of Earth in the Solar System

The origin of the Solar System is unknown. It is thought to have developed from a large nebula of dust and gas, from which the main planets gradually aggregated in orbits around the Sun. The denser parts moved towards the centre and the more volatile gases remained in the outer regions of the system. Thus the rocky planets are near the Sun and the gaseous giants are in the outer part of the Solar System.

Most of the rocky planets and moons, apart from Earth, exhibit massively cratered surfaces, and the dating of our Moon indicated that this happened some

4 billion years ago. So very early in the history of the Solar System there was a massive meteorite bombardment, presumably by debris left after the accretion of the main planets.

On Earth there is abundant evidence of meteorite impacts, even until the present day, but the Earth's surface is clearly not a cratered landscape like the Moon. Erosion by water and wind has worn most of the evidence away, and some craters have been buried by younger sediments. On the Moon the highlands preserve the craters, but the maria have been flooded by basalt filling in all but the most recent impact sites. On Venus it appears much of the evidence has also been covered by massive volcanic outpourings that have resurfaced the planet.

Some of the inner planets and moons appear almost dead, and their lack of a magnetic field indicates that there is no motion in the core. Earth still has a well-developed magnetic field, albeit dwarfed by the strength of the fields generated by the Sun and the gaseous planets such as Jupiter and Saturn. Mantle convection in the Earth is still evinced by plate tectonics, while for the inner rocky planets this has also apparently stopped.

Can we see in the other planets evidence of what Earth was like earlier in its history and perhaps how it will develop over the next few billion years? Eventually the heat will escape from Earth, and the internal motion will cease. At this point the Earth will become like the Moon and Mercury: tectonically dead, no plate tectonics and no magnetic field.

Our nearby neighbours, Mars and Venus, also have atmospheres, though they are dominated by carbon dioxide (over 95%) rather than oxygen, with minor nitrogen and some atmospheric water and sulphuric acid on Venus, and frozen water at the northern polar ice cap of Mars. Earth is unique in having an atmosphere dominated by nitrogen and oxygen, and with abundant water in solid (ice), liquid and gaseous forms. Two-thirds of the Earth's surface is covered in ocean waters, and at any one time about a third of the atmosphere is cloudy. This hydrological cycle is fundamental to maintaining the Earth's climate, with its equable temperatures, and together with the atmospheric oxygen is essential for the evolution and maintenance of life as we know it on this planet.

Earth sits at a remarkable position, just far enough from the Sun not to be baked and have its life-giving water and atmosphere evaporated, yet not too far out that it is frozen solid. Without its atmosphere Earth would suffer the same large swings in temperature experienced by the Moon, Mars and Mercury, depending on whether the surface faced towards or away from the Sun. If eventually the Sun's intensity increases, the oceans evaporate and the atmosphere disperses into space, Earth too will become like one of these planets.

Whether or not the Sun continues to radiate energy so fiercely, we can anticipate that the Earth will gradually lose the internal heat generated by radioactive decay, and the planet's interior will cool down. The mantle will stop churning and plate tectonic motions will cease. This will provide, at long last, relief from earthquakes and volcanic disasters. The ocean basins will level out, and mountain ranges cease to elevate. Eventually the core too will cease to convect and the Earth's magnetic field will diminish: no more spectacular auroras, and no more protection from ultraviolet radiation. How will we find our way around without compasses?

The other geological certainty is that another large meteorite will hit the Earth. While the massive bombardment occurred billions of years ago, there is plenty of evidence for continuing impacts and near misses.

Based on what is accepted as incomplete evidence, it was estimated by Eugene Shoemaker that the present rate of extraterrestrial bombardment produces a crater 20 km or greater in diameter on average every 400 000 years on Earth. Smaller iron meteorites up to 1 km size are estimated to impact about every 50 000 years, and the small 50 m wide stony meteorites no more often than about once every 200 years. However, we should note that the Tunguska devastation was probably caused by a 50 m object, and it has been calculated that the damage from such an object is equivalent to a 10 megatonne nuclear explosion.

The next large impact may be in a thousand years, or it may be next year. Recent reports indicate that the XF11 asteroid could pass within 50 000 km of Earth in October 2028. If it does hit the Earth the most likely impact site would be in the oceans, although the ensuing tsunami wave could dwarf those presently experienced from earthquakes. If it strikes on land, we can only hope it will be in a remote area.

Summary

Earth is one of four terrestrial planets in this Solar System. There is abundant evidence on Earth's Moon and on the other terrestrial planets for a meteor bombardment early in the formation of the Solar System. The craters formed during this bombardment have mostly been removed by tectonic cycling and water erosion on Earth. But meteorite strikes have continued, as evidenced by the impact structures marking the maria on the Moon and preserved on Earth. Many impact structures are well preserved in Australia. Some have surrounding fields of glassy tektites, formed by the melting of Precambrian rocks.

Major impacts in the ocean would have caused huge tsunamis, and consequential devastation on land. Impacts at 251 Ma and 65 Ma are thought to have been responsible for massive extinctions on Earth.

Plate tectonics on Earth is a symptom of heat loss: convection in the mantle moves the surficial lithosphere plates. The Moon and other rocky planets have generally cooled further and no longer have this movement. So they have no tectonic recycling, active volcanism or earthquakes, nor strong magnetic fields.

SOURCES AND REFERENCES

Ampol Exploration Ltd. 1991. *Yallallie No.1 Well Completion Report*. Open file report, Geological Survey of Western Australia.
'Australian Impact Structures', 1996. *AGSO Journal of Australian Geology and Geophysics*, 16. 625 pp.
Baldwin, S.L., McDougall, I. & Williams, G.E., 1991. 'K/Ar and 40Ar/39Ar analyses of meltrock from the Acraman impact structure, Gawler Ranges, South Australia'. *Australian Journal of Earth Sciences* 38: 291–298.
Bevan, A.W.R., 1996. 'Meteorites recovered from Australia'. *Journal of the Royal Society of Western Australia*, 79: 33–42.
Bevan, A.W.R., 1996. 'Australian crater-forming meteorites'. *AGSO Journal of Australian Geology and Geophysics*, 16: 421–429.
Bevan, A. & McNamara, K., 1993. *Australia's Meteorite Impact Craters*. Western Australian Museum. 27 pp.
Canup, R.M. & Asphaug, E., 2001. 'Origin of the Moon near the end of the Earth's formation'. *Nature*, 412: 708–712.
Cattermole, P., 2001. *Mars The Mystery Unfolds*. Terra Publishing. 186 pp.
Dentith, M.C. *et al.*, 1999. 'Yallalie: a buried structure of possible impact origin in the Perth Basin, Western Australia'. *Geological Magazine*, 136: 619-632.
Frazier, Ken, 1985. *Solar System*. Time Life Books Inc. 176 pp.
Gersonde, R. *et al.*, 1997. 'Geological record and reconstruction of the late Pliocene impact of the Eltanin asteroid in the Southern Ocean'. *Nature*, 390: 357–363.
Giovanelli, R.G., 1984. *Secrets of the Sun*. Cambridge University Press. 116 pp.
Glass, B.P., 1982. *Introduction to Planetary Geology*. Cambridge University Press. 469 pp.
Glikson, A.Y., 1996. 'Mega-impacts and mantle-melting episodes: tests of possible correlations'. *AGSO Journal of Australian Geology and Geophysics* 16: 587–607.
Haines, P.W. & Rawlings, D.J., 2002. 'The Foelsche structure, Northern Territory, Australia: An impact crater of probable Neoproterozoic age'. *Meteoritics and Planetary Science*, 37: 269–280.
Heiken, G., Vaniman, D. & French, B.M., 1991. *Lunar Sourcebook. A user's guide to the Moon*. Cambridge University Press. 736 pp.
Hill, D.H., Boynton, W.V. & Haag, R.A., 1991. 'A lunar meteorite outside the Antarctic'. *Nature* 352: 614–617.
Hutchison, R. & Graham, A. *Meteorites – The key to our existence*. Natural History Museum, London.
Kaufman, W.J. & Comins, N.F., 1997. *Discovering the Universe*. 4th edn. W.H. Freeman & Co. 436 pp.
King, Elbert A., 1976. *Space Geology*. John Wiley & Sons. 349 pp.
Kippenhahn, R., 1994. *Discovering the Secrets of the Sun*. John Wiley & Sons. 262 pp.
McNamara, K. & Bevan, A., 2001. *Tektites*. Western Australian Museum. 38 pp.
Malin, M.C. & Edgett, K.S., 2000. 'Evidence for groundwater seepage and surface runoff on Mars'. *Science*, 288: 2330–2335.
Mitton, S. (ed.) 1977. *The Cambridge Encyclopaedia of Astronomy*. Jonathan Cape. 481 pp.
Morgan, J. & Warner, M., 1999. 'Chicxulub: The third dimension of a multi-ring impact basin'. *Geology*, 27: 407–410.
Narlikar, J.V., 1999. *Seven Wonders of the Cosmos*. Cambridge University Press. 324 pp.
Price, M. & Suppe, J., 1994. 'Mean age of rifting and volcanism on Venus deduced from impact crater densities'. *Nature*, 372: 756–759.
Stewart, A. & Mitchell, K., 1987. 'Shatter cones at the Lawn Hill circular structure, northwestern Queensland: Presumed astrobleme'. *Australian Journal of Earth Sciences*, 34: 477–485.
Stewart, S.A. & Allen, P.J., 2002. 'A 20km-diameter multi-ringed impact structure in the North Sea'. *Nature* 418: 520–523.
Taylor, S.R., 1975. *Lunar Science: A Post-Apollo View*. Pergamon 372 pp.
Taylor, S.R., 1997. 'The origin of the Earth'. *AGSO Journal of Australian Geology and Geophysics*, 17: 27–31.
Veverka, P.C. *et al.*, 2001. 'Imaging of small-scale features on 433 Eros from NEAR: Evidence for a complex regolith'. *Science*, 292: 484–487.
Woodhead, J.D., Hergt, J.M. & Simonson, B.M., 1998. 'Isotopic dating of an Archaean bolide impact horizon, Hamersley basin, Western Australia'. *Geology*, 26: 47–50.
Zeilik, M., 1988. *Astronomy. The Evolving Universe*, Fifth edition. John Wiley & Sons. 544 pp.

WEBSITES

ABC Science Gateway – Space
www.abc.net.au/science/space/default.htm

Australian Astronomy
www.astronomy.org.au

Astronomical Society of Australia
[has information sheets and links to many other site]
www.atnf.csiro.au/asa_www/

NASA
http://spacescience.nasa.gov/

Photos and facts on all planets
www.pds.jpl.nasa.gov/planets/welcome.htm

NASA continuously releases images from the Cassini spacecraft, now of Jupiter and eventually of Saturn where it is due in 2004
www.ciclops.lpl.arizona.edu

Earth's Moon – 675 excellent prints from the 1971 Lunar Orbiter Photographic Atlas, now digitised.
www.lpi.usra.edu/research/cla

Hubble Space Telescope
www.hubblesite.org

Mars exploration
www.mars.jpl.nasa.gov

The Meteoritical Society
www.uark.edu/metsoc

60 Canadian meteorites including remnants of the January 2000 Yukon fireball
www.geo.ucalgary.ca/cdnmeteorites

The Interactive NGC Catalog Online
www.seds.org/~spider/ngc/ngc.html

Chicxulub impact core data
www.nmnh.si.edu/paleo/blast/K-t-boundary.htm

CHAPTER 12

CYCLES IN A CONTINENTAL JOURNEY

Global wandering

The story of Australia is the journey of a land mass wandering across the globe through geological time, at times locked with other land masses, at times being separate. And during all this time, being buried and then uplifted, accreting and eroding, alternately desiccating and being washed away, or iced over. Our present understanding of the global picture and Australia's history has been built on the careful work of thousands of geologists. It has involved both painstaking research on particular rocks, and also bursts of broader thinking to integrate all the data.

The idea of Gondwana was a great clue, though it was years later that the presence of an even earlier supercontinent Rodinia was also recognised.

The term Gondwana was first applied by Austrian geologist Eduard Suess in 1885, when he recognised that many of the rocks in India, southern Africa and Madagascar contained very similar plant fossils (*Glossopteris*), and hence perhaps were once part of the same land mass. Alfred Wegener in 1912 published the theory of continental drift – that it was possible whole continents had moved, an extraordinary concept at the time. The South African geologist Alex du Toit extended Gondwana to include northern Africa, Antarctica, Australia and South America in 1937. In fact it turns out that the distinctive plant fossils evolved only towards the end of Gondwana's existence, but they were still the critical clues in the overall picture.

The present continent of Australia has had four distinct stages in its geological development:

- before Rodinia 4600–1100 Ma
- part of Rodinia 1100–550 Ma
- part of Gondwana 520–100 Ma
- Australia 100 Ma to now.

The modern Australian lithosphere is a result of this long and complicated evolution. It has taken over 4400 million years of hard work by the Earth!

Before and during Rodinia

We know very little about the origin and arrangement of the Archaean cratons – that is, most of Earth history. But we do know a lot about the geology of individual parts, such as the Pilbara. There is plenty of evidence for the major events, such as the increasing oxygen content of the atmosphere and the rise of the eukaryotes after 2700 Ma ago. This was followed by the huge deposition of iron in the banded iron formations, both in Australia and elsewhere in the world.

We can date the major deformation events. These are very old events, such as those in the range 3700–2600 Ma in the Pilbara and Yilgarn regions, and 2500 Ma at Mount Isa and Broken Hill. Recent work indicates that these last two regions may originally have been joined and then separated before 2000 Ma ago.

Australia is an old and complex continent that is still changing. Our history is written in the rocks. What major themes do the rocks tell us about the history of this planet? Certainly that:

- *the Earth is dynamic, with the continents wandering over time and the oceans forming and closing,*

- *Australia has been subjected to repeated compressional events as the supercontinents were assembled, and repeated tensional events as the supercontinents disintegrated,*

- *the climates have fluctuated wildly, with icy poles the abnormal situation, and*

- *there have been five massive extinctions of life on Earth.*

Figure 12.1 *Aerial picture of the George Gill Range, central Australia (approximately 200 km west-southwest of Alice Springs). These sedimentary rocks started as layers of sand and mud deposited in rivers and on shorelines. Where did the sand and mud come from? It was eroded from older rocks, from landscapes long destroyed. The sediments have been buried, cemented, folded, uplifted, and are now being eroded to provide source materials for the next geological cycle. The George Gill Range is only a temporary feature, one of many ranges that have previously occupied this region. While there is heat within the planet this recycling on a geological scale is inevitable.*
PHOTO: RUA JOHNSON

In Australia, the three Archaean cratons – Pilbara, Yilgarn and Gawler – amalgamated to form the framework for southern and western Australia 2200–1600 Ma ago. Major volcanism and deformation occurred in the period 1800–1540 Ma, enlarging the cores of these Archaean cratons. Corresponding tension in the crust in northern Australia is evident between 1800 and 1670 Ma ago, forming sedimentary basins that host the massive lead and zinc ore bodies at Mount Isa and McArthur River in northern Australia.

Rodinia began to form 1450–1100 Ma ago when continental rearrangements joined the Australian cratons to India, Antarctica, Laurentia (the core of North America plus Greenland), and recent work indicates south China. The Grenville deformations 1300–1100 Ma ago are known in Australia, Canada and Antarctica, and may relate to the pushing together of separate blocks in the final assembly of the Rodinian supercontinent.

After 1100 Ma ago the full Rodinia was assembled. The main climatic and biological events were the Marinoan and Sturtian glaciations 780–575 Ma ago and the first signs of metazoan life – the Ediacaran faunas – some 550 Ma ago.

Gondwana

On present evidence, the break-up of the eastern part of Australia's Rodinian margin had occurred around 760 Ma ago, forming the initial palaeo-Pacific Ocean. Around 750 Ma ago the western side of present North America broke away from Australia. This earliest opening would later become the Atlantic Ocean.

This left a core of Australia, India and East Antarctica. During plate movements between 750 and 520 Ma ago, the fragments of later Africa and South America drifted around the globe towards India and Australia. By 520 Ma ago the cores of Africa and South America had locked together with Australia, India and East Antarctica to form the new supercontinent of Gondwana.

So it took around 230 Ma for the whole of Gondwana to be finally assembled after the disintegration of Rodinia. Further deformations occurred in the late Precambrian, between 600 and 550 Ma, such as the initial movements on the Darling Fault and the deformation in the Kimberley and in the Petermann Ranges. These coincided with the pushing together of the blocks to form Gondwana.

Between 350 and 255 Ma the present-day Northern Hemisphere land masses – Europe, Russia, North America and Greenland, called Laurussia – were also assembled as a supercontinent north of the western section of Gondwana, i.e. Africa and South America. This collision is expressed as deformation in eastern and central Australia. This super-sized supercontinent is the land known as Pangea.

Laurussia started to separate from Pangea around 240 Ma ago, opening the North Atlantic Ocean and the Mediterranean Sea. Eastern Gondwana did not break up till much later, starting around 180 Ma ago.

Australia's time in Gondwana was marked by two types of tectonics and sedimentation, either side of the Tasman Line. To the west, within the main land mass of Gondwana in that part which is now central and western Australia, the tectonics were mainly extensional. Basins formed where parts of Gondwana subsided and allowed marine incursions through northwestern Australia and into central Australia during the Cambrian, Ordovician, Devonian and Cretaceous.

However, to the east there were successive orogenic cycles at the edge of the supercontinent of Gondwana where subduction of ocean crust created compression and volcanism. The eastern third of present-day Australia has undergone cycle after cycle of such tectonism since the Cambrian. Each of these tectonic episodes – orogenies – has been given a name to relate it to the rocks where the evidence is most compelling. Australia has not always been the stable place it is now.

Australia

The break-up of Gondwana released Australia to be an independent continent for the first time in over 1600 million years. The tectonics were extensional, with signs of break-up in eastern Gondwana from 180 Ma, rifting along northwestern Australia from 154 Ma ago, and culminating 99 Ma ago with the start of the final disintegration of Gondwana. The separation of Australia started on the northwest corner and progressed anticlockwise around to the Tasman and Coral Seas. All but one of these spreading zones have stopped moving. The spreading zone in the Southern Ocean is still active, increasing the distance between Antarctica and Australia. The main movement for Australia now is more or less northeasterly at 67 mm/yr away from Antarctica.

The final result is a continental margin with large-scale tensional faults, with pieces of basement dropped down along these faults and overlain by flat sedimentary fills. Basalts erupted along the eastern highlands.

At first, while Australia was close to Antarctica, the climate was warm and humid, and there was rainforest on both continents. As Australia drifted northwards the cold circumpolar current began to refrigerate Antarctica, from around 33 Ma ago. Ice sheets became evident 15 Ma ago. The locking of water in the thick ice-sheets and the movement of Australia into warmer waters led to the rapid drying out of the continental interior. The sand deserts developed after 15 Ma ago, along with large salt lakes. More recently, the Great Barrier Reef became established off the northeastern coast as the continent moved into the tropical zone.

Eurasia

Earth's history has involved a series of supercontinents building and disintegrating. Supercontinents act like large insulating lids on heat released from the mantle. Studies of the chemical composition of ocean-floor basalts indicate that the mantle generating these basalts cools by 0.5–1.0°C per kilometre away from continental margins. That is, the continents must be keeping the top of the mantle warmer. Eventually the heat build-up forces rifting, and subsequent continental breakup. In this way the initial amalgamation of separate continental masses to form a supercontinent is the seed for the later disintegration.

So what is happening now? India has already docked with Asia, a collision forming the Himalaya. With Australia, Papua New Guinea and New Zealand currently heading north to join with Asia, and Africa heading north to dock with Europe, the amalgamation of the next supercontinent of Eurasia is only about 50 million years away.

Cycles of deformation

For virtually all Australia's geological history the cycles of sedimentation, burial and deformation have been caused by tectonic events within the greater supercontinents, of which Australia was just a part. Compressional events that produced deformed and metamorphosed rocks occurred as the supercontinents were being assembled. Then tensional events opened sedimentary basins as the continents broke up. Only the last event, the extension and break-up of the Gondwanan crust, to form the present continental shelves and the eastern Australian basalt volcanism are intrinsically related to the outline of the present Australian land mass.

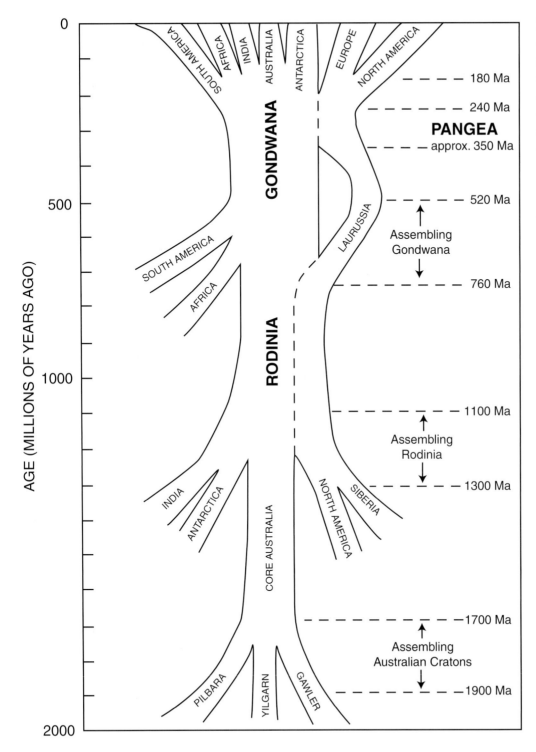

Figure 12.2 The present understanding of the movement of the major blocks during the assembly and disintegration of the supercontinents Rodinia, Gondwana and Pangea. Australia's core started with the assembly of the Pilbara, Yilgarn and Gawler cratons 1900–1700 Ma ago. This core was amalgamated with the Precambrian terranes of India, Antarctica, North America and Siberia 1300–1100 Ma ago to form Rodinia.

Part of Rodinia broke away after 750 Ma, and Laurussia drifted northwards across the globe.

Gondwana started assembling with the addition of South America and Africa between 750 and 520 Ma ago, and for a brief period was joined again by Laurussia about 350–255 Ma to form the giant supercontinent of Pangea. The break-up of Gondwana started after 180 Ma, with the final separation of Australia from Antarctica beginning 99 Ma ago.

Even though we can recognise several major orogenies, tectonism did not occur and cease everywhere at the same time. Tectonism tends to start in one place and then spread across a region or continent. The deformation is caused by a crustal compression somewhere, and that stress can take time to propagate through the entire continental mass.

During each orogeny, there is commonly a phase of burial and deformation that is followed by a phase of magmatism.

These volcanics and other material eroded off the previous edge of the continent accumulate in sedimentary basins. Such basins can either be shallow waters landwards of the volcanic arc, or deep waters seawards of the arc. Continued subsidence and compression deforms and metamorphoses these rocks, leading to the intrusion of granite magmas and consequent heating of the deformed and surrounding rocks. A new continental mass becomes attached or accreted to the previous continental mass. This process of continental accretion has happened repeatedly along eastern Australia, building the continent eastwards.

The six main deforming events or orogenies recognised along eastern Australia, with corresponding evidence elsewhere, are as follows.

Delamerian 520–490 Ma

During this time the western parts of New South Wales and Victoria and southeastern South Australia were deformed, as was western Tasmania and much of central western Queensland. Some tectonism extended to 480 Ma. Widespread granite and gabbro emplacement around 500 Ma ago was associated with regional uplift, erosion and the shedding of sediments at the start of the next orogenic cycle.

Benambran 460–435 Ma

In northern Victoria and parts of New South Wales, compression and uplift is evident at this time. The deformed rocks along the western edge of the Sydney Basin, and exposed on the lower parts of the Kanangra Walls, belong to this event. It was followed by intense magmatism in southeastern Australia between 421 and 390 Ma.

Tabberabberan 390–385 Ma

Crustal compression from the east produced major folding and deformation in Antarctica, Victoria and New South Wales. The fossiliferous marine limestones were folded, uplifted and eroded, and then overlain by quartzose sandstones across most of central and western New South Wales. Recent mapping has also shown that there was major movement of separate terranes in Victoria. Estimates of the lateral movement are up to 600 km. This seems large but is not extraordinary given the lateral displacements on a global scale due to plate tectonics.

The effects were felt across the continent. There was volcanism and deep water sedimentation in the New England area. Shallow basins were formed through central

Queensland, extending up to Charters Towers. Deformation this far north seems to be a bit later, up to 357 Ma ago. Parts of central Australia were uplifted by 5 km, with deposition in a basin now underlying Lake Amadeus. On the western coast there was also renewed basin development on both sides of the Kimberley Block.

Kanimblan 350–300 Ma

Compression from the east was most strongly felt in central New South Wales and Victoria. There was tight folding of former deep-water sediments and volcanics in the Hill End region, extending through the Snowy Mountains down to the coast of eastern Victoria. Further west the folding is more open. In northern Queensland the Hodgkinson region was also deformed.

Central Australian orogenies

The orogenies listed above are recognised mainly in southeastern Australia. There were also two major events in central Australia: the deforming of the Arunta region (northwest of Alice Springs) around 470 Ma and the long-lived Alice Springs Orogeny with thrust faulting between 450–300 Ma. Folding and faulting also occurred along the present western coast of Australia. How do these relate to the events in eastern Australia? We are not sure.

Hunter–Bowen 265–227 Ma

A series of gradually increasing compressional events thrust the eastern margins of the Sydney and Bowen Basins westward. The Late Permian (265–250 Ma) Hunter Orogeny deepened and compressed the coal basins in eastern Australia. Deformation was more intense to the east. During the Bowen Orogeny in the mid Triassic (233–227 Ma) the New England region pushed westwards over the Hunter–Mooki Thrust, and eastern Queensland from Rockhampton to Eungella pushed westwards, compressing the Bowen Basin. Some of the original normal faults that were part of the extension and formation of the basins were reversed. As a result, deeper parts of the basins were pushed back upwards along the same fault planes.

Early Cretaceous, around 120–105 Ma

The final orogeny and development of a volcanic arc down eastern Australia occurred just before continental break-up. There was widespread crustal uplift (indicated by fission track data) and granite intrusion, but not major deformation of the rocks as occurred in previous orogenies. Deformation is probably indistinguishable from much of the Late Triassic events. There was associated granite magmatism, such as the intrusions at Mount Dromedary in southern New South Wales and Urannah in northern Queensland.

The result of all these continental break-ups and reassemblings, of the cycles of erosion and sedimentation, and of the periodic downfaulting to form sedimentary basins, is that Australia is now a mosaic of overlapping geological pieces.

Cycles of climates

The climate changes on many scales. On a short-term scale of years or decades, we are aware of the cycles of droughts and rainy periods. The generally accepted theory is that variations in the strength of the ocean currents in the central Pacific (the El Niño effect) drive much of this variation in Australia.

There is also a scale of hundreds and thousands of years. There is plenty of historical evidence from the 1200s onwards of the Little Ice Age, which peaked in the 1600s. Arctic pack ice reached Scotland, the Thames River froze, and there were famines and many perished in Europe.

Recent work on deep-sea cores shows that there have been cycles of major iceberg movement every 1500 years or so, presumably as a result expansion of continental ice sheets. These iceberg armadas transport glacial debris much further south than normal, and the plankton preserved in the sediments show that the seawater can be 2°C colder than normal. The ocean cooling starts around 500 years before the iceberg armadas, which is evident from the increased debris dropped onto the seabed which is preserved as layers of gravel in the cores.

Two other discoveries provide evidence of these events. Firstly, the analysis of oxygen isotopes in ice cores from Greenland showed that atmospheric temperature drops of 2°C occurred at the same time. Secondly, these temperature excursions occur in both glacial and interglacial times. In other words, some different mechanism is providing regular periods of cooling on a millennial scale, irrespective of the main factors changing the global climate from glacial to non-glacial.

We are not sure what causes these cycles, but the evidence is pointing to changing alignments of the Earth, Moon and Sun that result in a greater tidal pull on the oceans. The tides may be helping to turn over the ocean water body, changing the balance between the colder deep water and the warmer shallow waters. This is an important factor because some of the present global warming may be due to the emergence of the planet from this latest mini-glacial cycle, as well as the effect of greenhouse gases.

It seems clear that the oceans play a key role in the emergence from glacial periods. The major surface ocean currents move heat from the tropics to the polar regions, especially in the Atlantic Ocean. This water cools and sinks and returns along the deep ocean floor towards the equator. Changes in the speed of this transfer may also be caused by injections of fresh water from the initial melting of the Antarctic ice sheet. This fresh, lower-density water disturbs salt balances and water circulations in the South Atlantic Ocean, causing increased movement in the North Atlantic.

Another view on the significance of the greenhouse gases has emerged from studies of deep-sea cores. Sediments from the Late Palaeocene show an extraordinary rise in ocean temperatures 55 Ma ago, with the temperatures of even deep ocean waters rising 5–7°C. It seems this happened in less than 10 000 years and that the normal state returned after about 140 000 years. The source of this temperature excursion is thought to be the release of methane from ocean sediments into the atmosphere.

Then there is the millions of years scale. Four major glaciations of the Earth –four icehouses – have been identified in the rock record. These are evident in Australian rocks and in the rocks of other major continental masses. That is, we are sure these represent major fluctuations in the total climate of the Earth, when the whole planet was plunged into lower temperatures for long periods, resulting in the growth of huge polar ice sheets. These icehouses alternated with periods of warmer Earth – the greenhouses – when average global temperatures were higher. In greenhouse times there were no polar ice sheets, and forests extended to polar regions. The icehouses have been:

Tertiary–Recent	15 Ma to present
Permo-Carboniferous	330–255 Ma ago
Late Proterozoic	780–575 Ma ago
Early Proterozoic	2300–2000 Ma ago

Two interesting perspectives arise from this summary. Firstly, each of the major icehouses of the geological past lasted 75 million years or more. The present icehouse is the culmination of a long climatic change, with evidence for cold conditions in Antarctica starting some 33 Ma ago, although there were also beech forests on the continent at that time. Expansion of ice to cover the entire land mass of Antarctica, and extending outwards to form the floating ice shelves, started only about 15 Ma ago. Intense glaciation in Europe and North America started around 2.5 Ma ago. In comparison with the duration of the older icehouses, we are still in the early part of this icehouse, unless the global warming due to human production of greenhouse gases propels us out of it earlier than the natural cycle would allow.

Secondly, the total time spent in icehouses adds up to about 600 million years. This is about 11% of Earth's history. Icehouses are the exception rather than the rule. The normal state for Earth is much warmer, with only one or even no polar ice caps, though probably with ice sheets and glaciers on high continental mountain ranges. Consequently the sea level would generally have been much higher because all the water locked up in the present ice caps would melt and flow into the oceans. In this situation the Antarctic continent would have been forested, and the Arctic Sea navigable by surface shipping.

It is so hard to imagine the world radically different from the one we know from books and television today. We are accustomed to the images of tropical jungles or dry grasslands near the equator with ice fields near the poles. We think of this as normal. We have adjusted our life patterns and economies to the normal, regular variations in climate, and we are made aware how limited is our concept of 'normal' only when droughts, severe winters or tropical cyclones upset our plans and comforts – as frequently they do.

Yet the geological record tells us that the present situation, in an icehouse, is abnormal. If we remark that there were times when tropical breadfruit trees grew in Greenland and lemurs swung in the trees, many people would be astonished, barely able to imagine it. Similarly, Antarctica has been a warm, sometimes dry, sometimes moist continent, with peat accumulating in lush vegetated landscapes until at least 30 Ma ago. The deserts and Lake Eyre in central Australia have been covered in wet forests.

What causes icehouses?

The factors that trigger and then maintain an icehouse are not precisely known. Possibilities include:

- changes in Solar activity
- variations in Earth's orbit
- volcanic eruptions
- mountain-building.

An obvious trigger would be variations in the intensity of solar radiation. Perhaps it was less intense at various times, before returning to the more typical state. Alternatively, the Earth might have experienced major changes in its orbit, at times being much further from the Sun, so that solar heating would be lessened. Volcanic eruptions can fill the atmosphere with dust and aerosols, screening out solar radiation. Mountain-building can thrust land masses upwards to intersect the colder air higher in the atmosphere.

Does the Sun's output change?

Our current understanding of the Sun's nuclear processes and of planetary motions suggests that this explanation is highly unlikely on a scale of millions of years. Certainly solar irradiance varies with the 11-year magnetic cycle, because of variations in sunspot and flare activity and eruptions of the Sun's corona. However, these cause only weak variations in the solar heating received by Earth. As we all know, there is no obvious 11-year cycle in the weather.

But solar activity does not oscillate within a constant range, and there are decades when the activity is higher than the long-term average, and decades when it is lower. There is an 80-year cycle that envelops these peaks and troughs of solar activity (see Figure 11.2), and the maxima and minima in this cycle correlate well with some major climatic fluctuations, especially the advance and retreat of Northern Hemisphere glaciers, over the last few hundred years. The minima in solar activity have been around 1670, 1740, 1810 and 1900, with recent maxima around 1630, 1700, 1780, 1850 and 1950. On this basis the Northern Hemisphere has been approaching a minimum since the 1990s, and will be coolest around 2030 and then warmer towards 2070. Will this happen, or will increasing greenhouse gases or other processes interrupt the trend?

Another theory notes that the Sun rotates around the centre of gravity (barycentre) of the Solar System, which is not always at the centre of the Sun. Just as the circling of the Moon around the Earth distorts the oceans forming the tides, the motion of the Sun around the barycentre may alter its internal circulation and turbulence, the number and intensity of solar storms and flares, and hence the luminosity of the Sun.

Variations in the Earth's orbit

There are well-known variations in the Earth's orbit that appear to be sufficient to alter the net solar radiation reaching the Earth's surface, and hence cause climatic variations. The link between these secular changes of the Earth's orbit and climate were proposed by English geologist James Croll in the 1860s and 1870s. In the 1920s Serbian mathematician Milutin Milankovitch calculated the time periods for these orbital changes, a painstaking job by hand in the days before electronic calculators. He calculated the variations in the tilt of the Earth's rotation axis relative to Earth's orbital plane around the Sun, and found that the tilt varies between 21.5° and 24.5° over a period of about 41 000 years. The change in tilt affects the strength of solar radiation reaching high latitudes, and thus climate. Also, the Earth's orbit around the Sun is elliptical rather than circular, and the size of this ellipse increases and decreases over time with a dominant period of 100 000 years.

Combining these facts with calculations of precession of the Earth's position in the orbit, Milankovitch confirmed recurring time intervals of 23 000, 41 000, 100 000 and 400 000 years. It turns out that these match many of the known glacial–interglacial cycles in the Pliocene–Pleistocene ice age over the last 2.5 million years. These climatic cycles due to orbital variations are well accepted and are now known as Milankovitch variations.

Scientific studies of greenhouse gases in ice cores have shown a strong correlation between warm climates and higher gas contents over the past 160 000 years. More recent modelling of orbital changes in relation to established variations in greenhouse gas (carbon dioxide, methane) contents in ice indicates that ice sheets should have appeared 6000–3500 years ago, and that the gas concentrations should have fallen steadily from 11 000 years ago. But the new ice did not appear, and unexpected increases in carbon dioxide and methane occurred 8000 and 5000 years ago, perhaps as a result of human activity. So human influence on the climate may have started millennia ago, though it has clearly accelerated at an alarming rate over the last 150 years.

However, it seems that the wobbles in the Earth's orbit should persist through both ice-house and greenhouse phases. The wobbles can account for variations in glacial advance and retreat, and consequent sea level changes, but we do not think they are the primary cause of icehouse or greenhouse states for the whole planet.

Volcanic eruptions

Large volcanic eruptions eject vast amounts of dust and sulphur dioxide into the atmosphere. The initial eruption columns may be up to 10 000 m high, and major eruptions extend 20 to 30 km up into the atmosphere. The jet streams swirl the dust around the globe, causing vivid sunsets and other phenomena, such as those that followed the eruption of Krakatau in 1883 for nearly three years.

The dust particles may be only 0.5 to 1.0 micrometres in diameter, so they take several months to settle into the lower levels of the atmosphere, where they can be removed by precipitation. Even after the dust settles, the sulphur dioxide and other gases ejected from the volcano remain. These dissolve in water droplets to form an aerosol layer of sulphuric acid that can remain for years at around 20 km altitude.

The eruption of Krakatau was studied by British and Dutch scientists, and it is interesting to note that the British report, by the Royal Society, is entitled *The Eruption of Krakatau and Subsequent Phenomena*. The subsequent phenomena included blue and green Suns, then a peculiar haze, and extraordinary twilight glows as far away as the British Isles. These effects have been studied many times after the volcanic eruptions this century, especially Mount Agung in Bali (1963), Mount St Helens in the USA (1980), El Chichon in Mexico (1982), and Mount Pinatubo in the Philippines (1991).

The volume of material ejected by volcanoes can be enormous. The spectacular eruption of Mount St Helens was small by geological standards. The largest historical eruption in the Andes, was the Huaynaputina volcano in southern Peru, which exploded between 19 February and 6 March 1600. Some 10–13 cubic kilometres of volcanic debris was erupted at an average rate of over 70 000 cubic metres per second, and spread across more than 115 000 square kilometres.

The Pinatubo eruption plume reached 30 km into the atmosphere and injected probably the largest amount of aerosols into the atmosphere last century, estimated at 30 billion tonnes. A particular characteristic of a volcanic dust cloud is that it causes a

Figure 12.3 *Vertical scale of features on Earth. The ocean basins are up to 5 km deep (except for the narrow trenches at subduction zones, which reach 11 km deep and are not shown). Hawaii is a very large object, and Mt Everest is the tallest land feature, some 8 km above sea level. Modern jet passenger planes travel at 10–11 km above sea level at the boundary between the troposphere and the stratosphere. Most rain comes from slowly moving clouds at 5–8 km altitude, whereas the jet stream winds at 10–11 km altitude encircle the globe at high speeds but cause no rain.*

Volcanic eruptions form thermal convection columns from which most ash is dispersed as ash clouds in the troposphere. Aerosols are dispersed at much higher altitudes in the stratosphere, where they can reflect sunlight and lead to measurable global cooling.

Note the temperature profile. Most ocean water has a temperature less than 10°C, and much of it is at less than 4°C. The atmosphere is below freezing point for most of its height. Human life depends on a very thin layer on Earth, near sea level where temperatures enable survival.

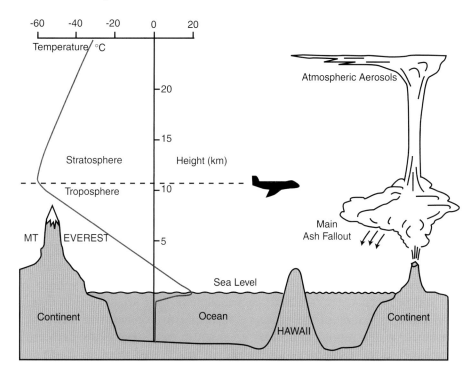

brilliant glow lasting some 45 minutes to an hour after dusk, when the Sun has left the Earth's surface in darkness but its rays are scattered and reflected off this layer much higher in the atmosphere. Following the Pinatubo eruption surface air temperatures in the northern hemisphere were up to 2°C cooler than normal in 1992, though the effect is complex as some regions were warmer in the following winters.

Significant temperature drops do occur as dust particles and droplets reflect Solar radiation back into space, causing a cooling of the Earth and atmosphere. This effect was proposed as long ago as 1784, by Benjamin Franklin, who noted in a paper read to the Philosophical Society of Manchester on 22 December 1784:

> During several of the summer months of the year 1783, when the effects of the Sun's rays to heat the earth in these northern regions should have been greatest, there existed a constant fog all over Europe, and great part of North America...and the rays of the Sun seemed to have little effect towards dissipating it.
>
> ...Hence the surface was early frozen. Hence the first snows remained on it unmelted, and received continual additions. Hence the air was more chilled, and the winds more severely cold. Hence perhaps the winter of 1783–4 was more severe, than any that had happened for many years.
>
> ...or whether it was the vast quantity of smoke, long continuing to issue during the summer from Hecla in Iceland, and that other volcano which arose out of the sea near that island, which smoke may be spread by various winds, over the northern part of the world...

One of the best known examples of the chilling effect of volcanism occurred in 1816, 'the year without a summer', following the eruption of Tambora in Indonesia in late 1815. Tambora was a stupendous eruption, much larger than Krakatau, and produced more than 100 cubic kilometres of ash.

In the early 1980s scientists obtained long cores of ice from the Greenland and Antarctic ice caps. These cores were accurately dated, and the acidity of the ice was measured by dragging an electrode down a cleanly cut face of the core. Peaks in acidity in the Greenland core matched the times of known volcanic eruptions in the Northern Hemisphere. The acidity is caused by volcanic sulphur dioxide that has been dissolved in moisture to form sulphuric acid in the upper atmosphere, and which has then been precipitated as snow.

The ice core data confirm suspicions going back to the time of Franklin. We know now that periods of intense volcanism match both glacial advances and also the precipitation of sulphuric acid in snow. The link between volcanism and climatic fluctuations is well established.

The atmosphere is the central component in the regulation of the global climate, but the ocean waters and the vegetation cover are also important. Furthermore, it is the processes of interaction between the atmosphere and the ocean, and between vegetation and the atmosphere, that are critical in determining weather patterns. Many of the processes form positive or negative feedback loops. For instance, water vapour and other greenhouse gases (especially carbon dioxide and methane) in the atmosphere absorb radiation reflected from the Earth's surface, and re-radiate it as infrared heat, so contributing to warming of the atmosphere. This warming increases evaporation, further raising atmospheric humidity and leading to more warming – a positive feedback. Since they cover 70% of the globe, the oceans receive most of the Earth's Solar radiation, and the evaporation from the ocean to the atmosphere is the overwhelming source of atmospheric moisture that falls as rain and snow.

In contrast, a cooler atmosphere restricts vegetation and leads to greater snow cover in polar regions. The snow cover tends to reflect more solar radiation back into space, decreasing warmth near the surface, and hence tending to lead to an even cooler atmosphere. Again this is a case of positive feedback, though in the direction of cooling rather than of warming. However, the reduced vegetation cover increases carbon dioxide levels because less is used in photosynthesis. Carbon dioxide absorbs infrared thermal radiation, and so tends to assist atmospheric warming. So if vegetation cover is less, carbon dioxide builds up and counteracts the cooling atmosphere – a negative feedback.

Hence climatic change can be viewed as the result of adjustments between a very wide range of processes affecting the heat balance of the atmosphere and oceans. The basic heat input is the Sun, and major changes in Earth's climate probably involve changes in solar radiation reaching the Earth.

We are only coming to grips with these effects now, but they were predicted over 100 years ago. The Swedish chemist Svante Arrhenius was a great scientist. Just one of his efforts followed the collapse of his marriage on Christmas Eve 1894, when he focused his energy on a problem which puzzled many scientists: how did the Earth cool during the ice ages? Without modern computers, and working long hours every day for a year, he divided the Earth's surface into small squares and calculated for each square the incoming radiation and the outgoing heat, which depended on whether that part of the Earth was covered in ocean, forest, grassland or ice.

Arrhenius calculated the effects of lower atmospheric carbon dioxide levels and showed that a reduction in CO_2 by a third to a half would cool the planet by 4–5°C – sufficient to glaciate all Europe. Then he tackled the opposite problem, and calculated that a doubling of carbon dioxide and other gases would result in a global warming of 5–6°C. The same result has been predicted by recent reports of the Intergovernmental Panel on Climate Change. Arrhenius had also predicted the main patterns: greater warming in the higher latitudes than the tropics, with more warming at night, over land and in winter.

It was noted in Box 7.2 that the correlation between global temperature and greenhouse gas levels has been established from ice cores, and it seems that the present communities on Earth may be starting to feel its effects now.

Plate movements

We have identified several causes for fluctuations of the climate on scales of decades to millennia, but we have not yet solved the puzzle of why the whole globe is tipped from a greenhouse into an icehouse.

A primary requirement of an icehouse is accumulation of major ice sheets. These may be either mountain ice sheets covering extensive mountain ranges (such as the Greenland ice sheet today and the ice sheets that covered the Rocky Mountains along western North America and Canada until about 5000 years ago) or continental ice sheets such as that covering Antarctica. Ice accumulates because of perennial temperatures below zero, and major ice sheets form when elevated land masses intersect the cold air of the atmosphere higher than 5000 m above the Earth's surface. The formation of high altitude ice masses is thought to set in train a series of climatic feedbacks that can lead to greater global cooling. This all requires mountain-building – a fundamental geological process.

The most recent icehouse closely follows the collision of India with Asia and the uplift of the Tibetan Plateau and the Himalaya. This collision occurred around 54–49 Ma ago, and was followed by a period of intense deformation as the crust was thrust upwards some 5–8 km, forming the mountain ranges. The Late Proterozoic glaciations (780–575 Ma) coincide with plate collisions and mountain-building associated with amalgamation of Gondwana between 750–575 Ma.

It turns out that the geological record shows clear evidence of major mountain-building episodes (from plate collisions) at each of the icehouse periods listed above. Not all are evident in Australia; for instance, the Permo-Carboniferous glaciation seems to follow mountain-building around 350 Ma ago in South America.

Are the periodic surges of mountain-building in Earth's history simply random events because every so often two large continents end up colliding, or are these events controlled by fundamental processes that determine periods of slower or faster plate movements?

Solar activity, volcanism, plate tectonics and climate

There is also evidence that eruptions in New Zealand, Japan and South America have occurred at the same time. If several large volcanic eruptions occurred in a short space of time, propelling even more volcanic dust into the atmosphere, it is possible that a significant change in global climate could follow. Why do the volcanoes show periodic increases in eruptions? And more importantly, how do these short-term climatic variations relate to the major icehouse–greenhouse cycles on a time scale of millions of years? Here we are really pushing the limits of our current understanding, but let us speculate.

There is a strong interaction between the solar activity and the Earth's magnetism. The velocity of the solar wind is dependent on the Sun's activity, with higher winds during periods of increased activity. The Earth's magnetic field is probably produced by the rotation of the fluid, iron-rich core inside the Earth. Charged particles in the solar wind interact with the Earth's magnetic field, producing the auroras seen in the night skies in high latitudes. But the effect goes deeper than this display of bright lights. There is a magnetic braking effect: after very strong solar flares the speed of the Earth's rotation slows, although only by about 2 milliseconds per day.

How does this relate to volcanism? Most active volcanoes lie along plate boundaries or over hotspots where mantle plumes are pushing basalt through the overlying plate. As in any spinning body, accelerations and decelerations will influence the behaviour of rigid bodies (such as the lithospheric plates) floating on a rotating fluid (such as the mantle). Thus any coupling between the solar wind and the Earth's magnetic field has the potential to alter the rates of convergence of the plates and hence the amounts of volcanic activity.

There is other evidence that the Earth's magnetic field and volcanism are linked. We know that the Earth's magnetic field has reversed its polarity repeatedly in the past. There is a strong correlation between normal epochs and hotspot volcanism, whereas reversed epochs are characterised by flood basalts. Flood basalts are thought to be due to the arrival of major mantle plumes at the Earth's surface, while the hotspots represent minor plumes.

It seems there is a fundamental process occurring at the mantle–core boundary in the Earth. Heat transfer across this boundary could initiate a major plume, causing surface volcanism, and at the same time alter convective flow in the outer core and hence change the Earth's magnetic field.

There is a wonderfully challenging job here for some young scientists – to relate the motion and behaviour of the Sun to volcanism and the climate on Earth. We are starting to gather enough data to show some very curious connections. So far we do not have a satisfactory theory of solar activity that can predict the Sun's behaviour from its motion around the centre of mass of the Solar System and other factors. Nor can we compute accurately the braking effect on the Earth, or the influences on the mantle convection cells and hence on plate motion.

If, as has been proposed, the major icehouse–greenhouse changes are caused by renewed plate activity, when continental collisions force mountain chains upwards to intercept the cold air of the jet streams, these too are caused by changes in mantle convection that propel the plates.

In summary, there are variations in climate on many scales. There are seasonal changes, and longer-term alternations of drought years and wet years, and at the longest geological time scale of millions of years to times when the Earth is in icehouse or greenhouse conditions. We know many of the processes that affect these changes but probably not all of them, and we certainly do not understand enough even of the ones we know about to be able to predict the future climate in detail.

Somehow one senses we are on the verge of a major new understanding that will take some excellent data-gathering and incisive thinking to unravel.

Evolution and extinctions

It is clear from the fossil record that evolution has operated at two major levels:

- Macro-evolution, in which major new groups of organisms evolve. Examples include the hard-shelled faunas formed in the Cambrian, the first land plants in the late Silurian, the start of the ammonites in the Devonian, and the rise of the primates in the Tertiary. The extinction of some species may be related to the rise of new predator groups. This is thought to be the fate of trilobites with the evolution of nautiloids, and a similar case can be made for the mammal megafaunas with the introduction of humans.

- Micro-evolution, in which there is continuing modification of existing lines. Some species die out, and others take their places. Examples are the changes in the graptolites, and the evolution of horses from small animals less than a metre high around 60 Ma ago to the modern forms.

In some cases new species evolve but become extinct after a short time as they have not developed any new advantageous traits. In contrast, some organisms got it right very early on – the bivalves and gastropods, the sharks and the algae have all been around with little change for hundreds of millions of years.

Along the way there have been times of enormous ecological change that have caused major extinctions of life on Earth. Five stand out:

end of the Ordovician	440 Ma	60% of marine genera lost
late Devonian	370 Ma	57% of marine genera lost
end of the Permian	251 Ma	82% of marine genera lost
end of the Triassic	210 Ma	53% of marine genera lost
end of the Cretaceous	65 Ma	47% of marine genera lost

Losses of 50–80% of all marine life represent an enormous change. Even the rapacious activities of humans have not done that so far. Luckily after each event there has been a major radiation of new life and a host of new species have evolved. The pattern of recovery and survival of life seems to be a global order as well as an individual striving.

Four possible reasons are normally envisaged:

- Changes in sea level or major chemical changes of the ocean waters, which alter the available space and conditions for life, especially in shallow marine areas. The Cambrian explosive radiation, for example, may have been triggered by an initial sea level rise that flooded shallow areas, providing a new range of habitats. Conversely, a sea level fall may wipe out a large range of ecological niches, as is thought to have occurred in the extinction at the end of the Ordovician.

- Global climate change, such as the oscillation between greenhouse and icehouse conditions that drastically alters the ecology, may provide opportunities or wipe out species. There is evidence of evolution happening, such as the development of the cold-climate Gondwanan flora in the Permo-Carboniferous icehouse. Certainly such climate changes cause regional annihilation of faunas and floras, such as the loss of rainforests in Antarctica, but there is no clear evidence yet for such changes causing extinctions on a global scale.

- Massive volcanism that blankets the Earth's atmosphere in volcanic dust from eruptions. Such eruptions can also generate sulphurous gases. The mass extinction at the Triassic–Jurassic boundary 210 Ma ago has been attributed to volcanic eruptions.

- Meteorite impacts that cause global atmospheric blanketing by dust, extensive wildfires, and geochemical changes. Recent evidence for meteorite impacts at the time of the Permo-Triassic (251 Ma) and the Cretaceous–Tertiary (65 Ma) extinctions, which were also accompanied by massive volcanism and continental fracturing, has raised the possibility that the impacts themselves are capable of disturbing the entire thickness of the crust, causing the rifts and ensuing magmatic activity.

Understanding these changes, and especially the ecological and climatic effects of the human take-over of the planet, will require excellent scientific work.

Epilogue – lessons of geological perspective

A geological perspective provides a different idea than most people could gain from their daily rounds on the planet. I want to outline three examples now.

The geological video

One of the comforting though quite naive attitudes is our general assumption that the present state of the world is the normal state, and that other states are abnormal. That is, the globe has a tropical region with Asian and Amazonian jungles, increasingly cold regions at higher latitudes, and polar ice caps. Uluru is much the same now as when it was named Ayers Rock by Gosse in July 1873. The seasons come and go, rainfall varies from year to year with droughts and floods, but the normal pattern is well established.

Similarly, it is easy to imagine in reading about the ancient volcanic chains that erupted down eastern Australia, or about the immense icy peatlands that were the start of our black coal deposits, that these happened in a land somewhere else. It is commonly hard to fully accept they actually happened right here, just the other side of the Blue Mountains or wherever. It is hard to imagine that Australia was not always the one we know now.

Certainly, the Earth and the landscapes in Australia have been very different in the past, and will be different in the future. Every view we see of the Australian landscape now is like a still image from a video; it is only one of millions that could have been taken on the same spot since the Earth was formed. Our geological video runs at a slow speed, but you can be certain that in a million years it will look very different.

One day there will be no Uluru, no Blue Mountains, no Kosciusko, no Dandenong Ranges, no Flinders Ranges. They will all be washed away, or submerged under the sea and sediments.

Becoming green

The planet has been green only since the Silurian, some 400 Ma ago. Before that there were no land plants as we know them. Perhaps there were algae and fungi in suitably wet niches, but the only plants were primitive vascular plants growing in shallow waters, without the strong cellulose fibres to hold them upright. One of the earliest plants known comes from Victoria – *Baragwanathia* – and Australia holds an almost continuous record of plant evolution from then on.

The time from the Silurian to the present represents less than 10% of Earth's history. So for most of its history the Earth has been a brown planet. Viewed from afar in space it has probably always appeared as it is now – blue and white from oceans and clouds – but up close it was brown. So being green is something pretty recent.

Facing burial

And what of the future? Some will recognise the lines from Percy Shelley's 'Ozymandias':

> *And on the pedestal these words appear:*
> *'My name is Ozymandias, king of kings:*
> *Look on my works, ye Mighty, and despair!'*
> *Nothing beside remains. Round the decay*
> *Of that colossal wreck, boundless and bare*
> *The lone and level sands stretch far away.*

These lines were not dreamt up, but modelled after the inscriptions on a funerary temple to the ancient Egyptian king Rameses II. The original words on the monument were:

> *I am Ozymandias, king of kings. If anyone wishes to know how mighty*
> *I am and where I lie, let him surpass my works.*

Shelley has used a little poetic licence to dramatise the effect.

The supreme irony is that the monument to Rameses II now stands, as Shelley says, alone in the middle of the sands. The magnificent gardens or the palaces or whatever were Ozymandias' great works have long gone. The legacy of this man is now the barrenness of the desert sand – a small yet very neat lesson of geological perspective!

What about us? What about the Australian landscape? There are only two alternatives: either there will be further tectonic uplift, or there will be continued subsidence.

If there is uplift in northern Australia, the present northern plains south of Darwin and the Gulf of Carpentaria will become a plateau or mountain range. This could happen as Australia moves northwards and, in colliding with Asia, is jacked upwards by the compression, forming a mountain belt like the New Guinea

Highlands. All the rocks and soils will end up like all other mountains, eventually eroded away. Features can remain for a few million years, but eventually all goes downhill as sediment to be buried in the oceans.

If there is subsidence, the present land surface will gradually come down to sea level and then sink under the ocean. This has happened thousands of times in the past. Plenty of drillholes in the continental shelf have penetrated more than 1000 m of sediments, overlying granites that were once high enough to have been weathered and eroded by streams. The rate of continental subsidence is slow, perhaps 20 mm per century, but these eroded granites now at depth show that the process is inevitable.

And what will happen to tall city buildings? They will all end up down there too, with their foundations, or what is left of them, a kilometre or more underground, covered by layers of sand and mud. Eventually they may even get taken deep enough in the Earth to be melted down into granite again.

It seems hard to imagine, and perhaps a bit scary for some. But it is as true as the pie you eat for lunch, although granite just takes a little longer to bake.

Links to the landscape

So why worry about preserving anything? The Earth, its climate, and its ability to sustain life constitute a very complex web. The time frames for change are longer than human lifetimes: it normally takes a few decades to severely pollute a water body, erode a coastline or cause salination of arable land.

The geological scale of Earth movements, reversal of the magnetic field, and climate change is much longer. Over time the Earth can recover from devastating meteorite impacts and volcanism, even where 80% of species are extinguished. In geological terms all will be well, but that will be too long for our human survival.

We evolved within this web and are dependent upon it. It is in our interests to minimise our alteration of this complex web; we have little idea of the effects of our actions, and almost no ability to restore the system when it has been damaged.

More fundamentally, we have a spiritual relationship more important than a physical dependence on the Earth and its landscape, as anyone who has been locked away in a closed prison for long periods will tell you. The landscape has a power to evoke wonder and humility, and an ability to restore the soul. Many others after us will need that.

The Aboriginal people testify to the truth of Charles Laseron's words in the 1950s:

> Let us therefore not only love our country, but let us consider this natural wonderland of Australia as a sacred heritage, something to be preserved unsullied and intact for the benefit of generations to come.

SOURCES AND REFERENCES

Betts, P.G., Giles, D., Lister, G.S. & Frick, L.R., 2002. 'Evolution of the Australian lithosphere'. *Australian Journal of Earth Sciences*, 49: 661–695.

BMR Palaeogeographic Group, 1990. *Australia – Evolution of a Continent*. Australian Government Publishing Service. 96 pp.

Bok, B., & Bok, P.F., 1974. *The Milky Way*. Harvard University Press. 273 pp.

Bray, J.R., 1974. 'Volcanism and glaciation during the past 40 millennia'. *Nature* 252: 679–680.

Eyles, N., 1993. 'Earth's glacial record and its tectonic setting'. *Earth Science Reviews*, 35: 1–248.

Hammer, C.U., Clausen, H.B. & Dansgaard, W., 1980. 'Greenland ice sheet evidence of post-glacial volcanism and its climatic impact'. *Nature*, 288: 230–235.

Humler, E. & Besse, J., 2002. 'A correlation between mid-ocean-ridge basalt chemistry and distance to continents'. *Nature*, 419: 607–609.

Jenkins, R.B., Landenberger, B. & Collins, W.J., 2002. 'Late Palaeozoic retreating and advancing subduction boundary in the New England Fold Belt, New South Wales'. *Australian Journal of Earth Sciences*, 49: 467–489.

Landscheidt, T., 1987. 'Long-range forecasts of Solar cycles and climate change' in M.R. Rampino, J.E. Sanders, W.S. Newman & L.K. Konigsson, *Climate: History, Periodicity and Predictability*, Van Nostrand, New York. p. 421.

Larson, R.L., 1991. 'Geological consequences of superplumes'. *Geology*, 19: 963–966.

Laseron, C.F., 1954. *The Face of Australia*. Angus & Robertson. Second Edition. 244 pp.

Lean, J. & Rind, D., 2001. 'Earth's response to a variable Sun'. *Science*, 292: 234–236.

Li, Z.X. & Powell, C. McA., 2001. 'An outline of the palaeogeographic evolution of the Australasian region since the beginning of the Neoproterozoic'. *Earth-Science Reviews*, 53: 237–277.

Mackness, B., 1987. *Prehistoric Australia*. Golden Press. 191 pp.

McCormick, M.P., Thomason, L.W. & Trepte, C.R., 1995. 'Atmospheric effects of the Mt Pinatubo eruption'. *Nature*, 373: 399–404.

Meinel, A. & M., 1983. *Sunsets, twilights and evening skies*. Cambridge University Press. 163 pp.

Rampino, R., Sanders, J.E., Newman, W.S. & Konigsson, L.K., 1987. *Climate – History, Periodicity, and predictability*. Van Nostrand Reinhold. 588 pp.

Richards, M.A., Duncan, R.A. & Courtillot, V.E., 1989. 'Flood basalts, and hot-spot tracks: Plume heads and tails'. *Science*, 246: 103–107.

Ruddiman, W.F., 2003. 'Orbital insolation, ice volume and greenhouse gases'. *Quaternary Science Reviews*, 22: 1597–1629.

Sepkoski, J.J., 1996. 'Patterns of Phanerozoic extinction: a perspective from global data bases', in O.H. Walliser, (ed.), *Global Events and Event Stratigraphy in the Phanerozoic*. Springer. p. 37.

Stommel, H. & Stommel, E. 1979. 'The year without a summer'. *Scientific American*, 240: 176–186.

Thouret, J-C., Davila, J. & Eissen, J-P., 1999. 'Largest explosive eruption in historical times in the Andes at Huanaputina volcano, A.D. 1600, southern Peru'. *Geology*, 27: 435–438.

Veevers, J.J., (ed.), 1984. *Phanerozoic Earth History of Australia*. Oxford Monographs on Geology and Geophysics No.2. Clarendon Press. 418 pp.

Willman, C.E., VandenBerg, A.H.M. & Morand, V., 2002. 'Evolution of the southeastern Lachlan Fold Belt in Victoria'. *Australian Journal of Earth Sciences*, 49: 271–289.

WEBSITES

Cycles of global and climate change
www.scotese.com/
www.geochange.er.usgs.gov/

World Data Center for Paleoclimate Change
www.ngdc.noaa.gov/paleo/climap.html

Gateway to global change data
www.gcdis.usgcrp.gov

Index

Page numbers in *italic* refer to illustrations.

aeolian deposits, 45, 93, 202
Albany, 61, 147, 201
Alice Springs, 156
alluvial fan, 111-12, *112*
ammonite, 83, *120*, 136, *136*
amphibians, 123–5, *124*
amphibole, 31, 35–6
amygdule, 35
Anakie, 170, 173–4
andesite, 34, 90, 93
Antarctica, *28*, 73–6, *74*, *75*, 99, 103, *140*, 141, 151-2, *154*, 155–6, 253–5, *256*
apatite, 25, 32
aplite, 36
Arc, Baldwin, 94
Arc, Calliope, 93
Arc, Connors, 94
Arc, East Australian, 132–3, 141
Arc, Eungella, 107
Arc, Molong, 90, 93
Arc, Stavely, 88
Archaean, *50*, *53*, 55–61, *56*, *57*, 68–71, *144*, 253
archaeocyathids, 73, 81, *81*
Arckaringa, 142, *143*
Arnhem Land, 60–1, 90, 144
asthenosphere, *20*, 20
Atherton Tableland, 107, 164, *165*, 167
atmosphere, *20*, 20–1, 52, 235–6, 249, 262–5, *263*, 269
Australia: age, 3, 255; climate, *10*, 10–11; drainage and relief, 11, *12*

banded iron formations, *62*, 62–4, 74, 104, 252
Baragwanathia longifolia, 85, 94
Barkly Tableland, 80
basalt, *27*, 33, 35, 63–4, *107*, 107, 126, *128*, 131, 145, 164–80, *165*, 267
Bass Strait, 64, 139
Batemans Bay, 100, *101*, 128
batholith, 35
Bathurst, 107
Beachport, 8
bedding, *36*, 36, *192*; graded, 48, *90*
Bega, 95, 178
Benioff zone, 27
BIFs, *see* banded iron formations
biotite, 32, 40
birds, 134, 150
bivalves, 83, *85*, *120*
Blue Mountains, *118*, *128*, 269
Bourke, *119*, 119
brachiopods, 83–4, *120*
breccia, *29*, 30, 242
Britomart Reef, 228, *229*

Broken Hill, 30, 62, *65*, 65, 79, 89
Brunhes-Matuyama reversal, *23*, 23
Bunbury, 131
Bungle Bungle Range, 94
Burdekin, *13*, 94, 178, *195*, 227

Cairns, 8, 93, 197, 203, 217, 225–6
calcite, 32, 37
calcrete, 42, 214
calderas, 35, 107, 209, *210*
Callide, 123
Cambrian, *50*, 79, 80, 88–9, *89*
Canberra, 90, 95, 173
Canowindra, 85–6
Cape Byron, 204
Cape Flattery, 203
Cape Hillsborough, 166, 170, 204
Cape Keerweer, 1
Cape Naturaliste, 201, *201*
Cape York, 107–8, 178
Carboniferous, *50*, 99, 104, 110, 111, 114–15
cement, 37
cephalopods, 83, *135*
channels, *45*, 45
chenier, 203, *200*
chert, 38
Chillagoe, 93
chlorite, 32, 40
chron, *23*, 23
clay, 36
cleavage, mineral, 30
cleavage, rock, *39*, 39
Clermont (Qld), 93, 166, 170
climate change, causes, 259, 261–6
Cloncurry, 10
coal, 38, *38*, 108–11, 123, 125, 128, 130, 133, 147–9, *148*
Collie, 104
Collinsville, 95, 100, 107, 114, 128, 137
collision zone, 28, *40*, 41
Comboyne Plateau, 170, 177
conglomerate, 37, 67
continental arc, 28, *28*, *40*, 41
continental crust, 19, *20*
continental drift, 3, 6, 21, 73, 252
continental shelf, *47*, 47, 182–6, *183*, *185*, 225–6
continental slope, *47*, 47, 182–6, *185*, 229–30
convergent margins, 6, 27
Cooktown, 106, 213
Cooma, 30, 96
Coorong, 147, 196
coral, 83, *83*, 106, *120*, 188, 213, 217–22, *222*, 225, 227–8, *228*
Coral Sea, 133, 139, *165*, 178, 224
core, Earth's, 19, *20*

273

corestone, 41
corundum, 30
Cowra, 93
craton, 49, 56, 253
Cretaceous, 50, 131–7, *132*, *135*, *144*, 144, 203–4, 258
cross-beds, *122*, 122, *123*
Curie point, 23
Currabubula, 107
cyanobacteria, 68–9
cyclones, 214–16, *215*, *216*, 220

Darling Fault and Range, 55, 89, 201
dating: biostratigraphic, 24: magnetic signature, 22–3; radiometric, 3–5, 49; thermoluminescence (TL), 24, 158; zircon, 5, *5*, 53
debris flow, 44
Deep-Sea Drilling Program, *see* ODP
deep-sea fans, 47
deformation, 29–30, 67, 96, 255, 257
deltas, 46, *195*, 195–6
denudation rate, 180
Derwent River, 175, *190*
deserts, 45, 145, 151–2, 155–9, *157*
Devonian, 50, 93–4, 257
diamond, 30, 131, 173–4
diatomite, 172–3
diatreme, 34, 126, 167
dinosaurs, 125, *127*, 127, 133–4, *134*
Diprotodon, 152–3, *153*
divergent margins, 6, 26
dolerite, 35, 125
dolomite, 37
drainage basin, 44
dropstone, 64, 100, *101*, 104, *106*, 132
Dubbo, 90
dunes, 46, 93, *157*, 196, 202, 203, *205*
duricrust, 42
dykes, *34*, 35, 49

Earth: age, 3, *5*, 49, 52; crust, 19, *20*; formation, 6; heat loss, 54–5, 249–50, 255; internal structure, 19–20, *20*; internal temperature, 54; magnetic field, 19, 249–50, 266–7; orbital variations, 262; origin, 52; oxygen content of atmosphere 64, 70–1
earthquakes, 6–8, 9, 27, 194; Australian, 8–9, *9*; magnitude, 7–8
echinoderm, 84–5
Ediacara, 71–2, *72*, 80, 253
EEZ, *see* Exclusive Economic Zone
Eighty Mile Beach, 203
Einasleigh, 89, 168, *179*

erosion, coastal, 190–3, *191*, *192*, *193*, 197–9
erratic, 46, 101, 103
Esperance, 55
evaporite, 38, 88
evolution, 70–1, *80*, 80–2, *83*, 91, *92*, 94, 133, 151–6, 267–8
Exclusive Economic Zone, 16, 186–7, *187*
extinction, viii, *80*, 83, 92, *120*, 153, 268; 251Ma, *80*, 119–20, *120*, 269; 65Ma, *80*, *120*, 247–8, 269; causes of, 119–20, 153, 247–8, 268–9
Eyre Peninsula, 55, *57*, 158

Fantome Island, *218*, 218
faults, 29, 67, 130, *185*, *208*, 209
feldspar, 32, 35–6, 60, 67–8
Ferrar dolerites, 125–6
ferricrete, 42
Fingal, 123
fish, 85–6, *86*, 125, *135*
fission-track dating, 25–6
Fitzroy Crossing, 93
Fleurieu Peninsula, 101–2
Flinders Ranges, 64, *65*, 65, 66, 71, 81, 88
Flinders, Matthew, 1, 213
floodplains, 45 112–13
fluorite, *31*, 32
folds, 29, *29*, 67
foliation, 39
fossils, 37, 60, 81–7, *85*, 87, 94, *121*, 150; preservation, 86–7, *87*, 91, 123–5
Fraser Island, 182, 204

gabbro, 35, 166, 172
galena, 31
garnet, 31, 173, 40
gastropod, 83, *85*, 120
Gawler craton, 56, *57*, 57–8, 61, 244, 253
gemstones, 173–4
geochemistry, 13
geological time scale, 49, *50*
geomorphology, 13
geophysics, 13, 46, 91
Georgetown, 56, 61
Geraldton, 93, 100, 104
Gerringong, 131
Gippsland, *206*, 206
glaciation and glacier, viii, 46, 65, 99, 99–106, 151–2; Huronian, 64; Ordovician, 91; Permo–Carboniferous, 99, 99–106, 260, 266; Proterozoic, 64–7, *65*, 253, 260, 266; Tertiary-Recent, 151–2, 260, 266
Glasshouse Mountains, *165*, 166

Glossopteris, 73, 110–11, *111*, 252
gneiss, 39, 97
Gondwana, 3, 73, *75*, 75–6, 79–80, 132, 252, 254, 256; break-up of, 118, 131–2, 139–41, *140*, 255, 256
Gosford, 106, 124
Gosse Bluff, *15*, 85, 244
graben, 29, *201*, 201
Grafton, 93
granite, 35, 49, 56–7, *57*, 59, 95, *95*, 97, 107; emplacements, 57, 59–60, 96
graptolite, 91, *92*
gravel, 36, 47
Great Australian Bight, 182
Great Barrier Reef, 50, 213–30, *214*, *215*, *216*, *217*; age, 224–6, 228–9, *229*; formation, 219, 223–7, *224*; growth, 220–1, *219*, *224*
Great Divide, *12*, 145, *165*, 170, 173, 177–9, *179*
Great Escarpment, *171*, 173, *177*, 177–8
Great Sandy Desert, 90, 93, 100, 104, 156, *157*, 202–3
greenstone, 33, 58
Grenville series, *74*, 74
Gulf of Carpentaria, 136, 145, 182, 203
Gulgong, 125
gypsum, 32, 38

Hadean, 50, 54
half-life, 4
halite, 32, 160
Hamersley, *62*, 62
hardness, *see* Moh's hardness scale
Heathcote, 103
hematite, 30, 61, 87
Herberton, 167
Heron Island, 220
Hill End, 93
hornfels, 38
hotspots, 26–7
Houtman Abrolhos, 202, *214*, 214
Hughenden, 122, 168
Hunter Valley area, 109, 113, 130, 205

icehouse, 99, 260, 261
icthyosaurs, 134
ilmenite, 30
impact craters, *15*, 237, 238, 241, 241–5, *242*
Indo-Australian plate, 6, *6*, 141, 166
Inman Valley, 101–3
inverted topography, 142, *143*, 168
island arc, 28, *40*
isoclinal folds, 29

274 INDEX

isostasy, 29
isotopes: radioactive, 4–5, *5*; stable, 106, 154–5, *159*, 159, 188, 227

Julia Creek, 134, 136
Jurassic, *50*, *122*, 125–7, *126*, 131

Kalbarri, 93
Kalgoorlie, 56, *142*
Kambalda, 56
Kangaroo Island, 79, 102
kaolinite, 32
karst, 44
Kata-Tjuta, 67, *67*–8, 80
Kiama, 106–7, *107*, *129*
Kiandra, 90
Kimberley, 56, 61, *65*, 65, 88–9, 93, 131, 190, 202
Kimberley craton, 61, 202
King Island, 64, *65*
komatiite, 58–9, *58*, 74

Lake Alexandrina, 11
Lake Eyre, 10–11, 147, 156, 158–9
laminae, 36
Larapinta Seaway , 89, 90
large igneous province (LIP), 26–7
laterite, 42, 144–45, 180
Latrobe Valley, 131, 147
lava, 33, 59, 169,
lava tubes, 33, *169*, 169, 175
Leigh Creek, 123
levees, 45
Lightning Ridge, 134, 150
limestone, *37*, 37, 44, 88, 90
Lithgow, 110, *128*, 128, 130
lithosphere, 20
Lord Howe, *12*, 132, 141, *165*, 176–7

maar, 167
MacDonnell Ranges, 88
Mackay, 107, 204
magma, 33, *34*, 35, 59
magnetic reversal, 19, 21–4, *23*, 66, 145
magnetite, 23, 30, 61
mammals, 125, 150
mantle, Earth's, 19, *20*, 139
mantle convection, 22, 54, 249, 267
mantle plume, 26, 166, *176*, 176–7
marble, 39
Marble Bar, 10
Maryborough, 133
matrix, 37
Meckering, 8–9

Meekatharra, 8
megafauna, 152–3, *153*
mesa, 42, 142, *143*, *168*, 168
Mesozoic, *50*, 118–37, 139–46
metamorphic grade, 40, 96, 97
metamorphism, 38, 39–41, *40*, 59, 61–2, *95* 96, 114; P–T–t, *40*, 41, 96, 97
meteorites, 53, *240*, 240–1; impact, 120, 241–8, *247*, 250, 269
mica, *31*, 31–2, 39–40, 97
mid-oceanic ridges, 22, 26, *28*
minerals, 30–2, *31*
Mirackina Palaeochannel, 142, *143*
Moho, 19, *55*
Moh's hardness scale, 30
molluscs, 83
Molong, 93
Monaro Plateau, 173
monazite, 32
montmorillonite, 32
Moon, 53, 236–40, *237*; age, 238; composition, 237–8
moonquakes, 239
moraine, 46, 99
Moruya, 178, 193
Mount Bellenden Ker, 10
Mount Canobolas, *128*, *165*, 173
Mount Desolation, 168
Mount Dromedary, 258
Mount Gambier, 175
Mount Isa, 30, 61, 70, 122, 144
Mount Koscuisko, 11, *184*, 184
Mount Morgan, 127
Mount Mulligan, *42*, 123
Mount Schank, *174*, 175
Mount Warning, 170, *172*
Mount Wingen, 113–14
mud, 36, 47, 200
mud flow, 44
mudstone, 37
Murchison River, 55, 93
Murgon, 150, *168*
Murray Basin, *95*, 145
Murray River, *45*, 146
muscovite, 25, 32
Musgrave Ranges, *56*, 61

Nandewar Range, *165*, 166, 171
Naracoorte, 209, *210*
New England (NSW), 130, 145
Newcastle, 8–9, 104–6, *105*, 116
Normanby River, 89
Norseman, 56, *142*
North Pole (WA), 68
nuees ardentes, 34

Nullarbor Plain, 132, 141, 147, 151, 160–1, *161*, 200, *240*, 242

obsidian, 35
Ocean Drilling Program, *14*, 14–15, 224
oceanic crust, 19–20, *20*, 28
ODP, *see* Ocean Drilling Program
oil shale, 110, 136, 149
olivine, 31, *58*, 59, 173, *174*
opal, 30, 87, *135*, 150
optically-stimulated luminescence (OSL), 24
Ordovician, *50*, 89, 89–92
orebodies, 61–2, 253
origin of life, 68–71
orogenesis, 48–9, 61, 67, 88, 95, *254*, 257–8
Orogeny, Alice Springs, 67, 258
Orogeny, Benambran, 91, 92, 257
Orogeny, Delamerian, 89, 92, 257
Orogeny, Early Cretaceous, 132–3,137, 258
Orogeny, Hunter-Bowen, 123, 258
Orogeny, Kanimblan, 88, 258
Orogeny, Tabberabberan, 94, 257
orthoclase, 32
Otway, 133, *192*, 207

palaeoclimate, 43, 55, 64–7, 88, 99, 110, 118–19, 122, 125, 146–7, *149*, 149, 151–2, *154*, 154–5, *157*, 188, 233, 259–66
palaeomagnetism, 66, *119*, 119
palaeontology, 13
Palaeozoic, *50*, 79–97, 99–16
Palm Valley, 93, 156
Pangea, 73, *254*, *256*
Parkes, 93, 95
partial melting, 39
pelagic sediment, 36, 48, 229
Permian, *50*, 99–116, *116*, 136–7, 258
Petermann Ranges, 67, 80
petrified wood, *87*, 87
petroleum, 38
Pilbara, 53, 56, 57, 60, 68, 70, 89, 93, 104, 202, 252
Pilbara craton, 56, 57, 61–4, 68, 253
pillow basalt, 58, *63*
Pittwater, 190, 207, *208*
plagioclase, 32, 35
plains: abyssal, 48; coastal, 113
planets, 233–6, 248–50
plants, 94, 108–12, *111*, *121*, 122, 125–7, *126*, *127*, 133, 145–7, *148*, 149
plate tectonics, 6, *6*, 15, 21–2, *28*, 48, 139, 249, 265–6

plateaus, *42*, 42, 47, 182–4, *187*
plugs, volcanic, 35, *170*, 170, 172–3
Point Quobba, 202
Port Jackson, 190, 207
Port Philip Bay, *208*, 208–9
Port Stephens, 205
Portland, 175, 209, *210*
Precambrian, *50*, 52–76, *56*, 79
Princess Charlotte Bay, 79, 214
Proserpine, 149
Proterozoic, *50*, 54, *56*, 60–76, *144*
pumice, 189
pyrite, 31, 87, 91, 113, 130
pyroxene, 31, 35, *58*, 59

quartz, 30, *31*, 35–6, 60

radioactive decay, 3–5, *5*
radiocarbon, *5*, *218*, 218, *229*
reef environments, 221–2
reefs: fringing, 217–18, *218*; shelf, 218–20, *219*, 221–2, 228, *229*
regolith, 43
rhyolite, 33, 35–6, 63, 93, 107, 132, 164, 166
Richter Scale, *see* earthquakes, magnitude
river deposits, *45*, 94, *105*, *112*, 112–13, *118*, 122–3, 131
Riversleigh, 150, 159
Rockhampton, 93
rocks: extrusive, 33; igneous, 32–6; intrusive, 35; metamorphic, 38, *53*, *97*; sedimentary, *36*, 36, 60, *106*, *135*, *253*; *see also under individual names*
Rodinia, 3, 73–6, *74*, 252–3, *256*
rutile, 30

salt lakes, 141, *142*, 147, 156, *159*, 159–60
sand, 36, 47, 196–9, 201–2
sandstone, 37, 42, 49, 67, 90, 93, *118*, *122*, *129*
sapphire, 30, 173–4
schist, 39
Scone, 164
scoria, 34, *166*, 167, 175
sea-level change, 152, *157*, 158, *185*, 188–90, *190*, 208, 219, 220, 224–6
seamount, 12, *165*, *176*, 176–7
sedimentary basin, 36, *56*, 100, 130
sedimentation, coastal, 47, *112*, 195–200, *200*, *205*, 205–6, *206*, 226–7
seismic waves, 19

serpentine, 32
Shark Bay, *69*, 69, 202
Shoalhaven, 195, 205
SHRIMP, *5*, 5, 53, 96, 158, 199
siderite, 32, 87
silcrete, 42, 142
silicate structures, 31–2
sill, 35, 125
silt, 36
Silurian, *50*, *89*, 92–3
Simpson Desert, 88, 90, 156, *157*
sinkhole, 44
slab-pull, 22
slate, 39
Snowy Mountains, 93, 95, *165*, 173
soil creep, 44
soil profile, 41
soils, Australian, 43–4, 164, *166*
Solar System, 52, 232–6, 248–50
sphalerite, *31*, 31
spinel, 30
spinifex texture, *58*, 58
stalactites, 44, 161
stalagmites, 44, 161
striated pavement, *46*, 65, 100, *102*, 102–4
stromatolites, 68–9, *69*, 70
stromatoporoids, 84
subduction zone, 27, *28*, 34, *40*
subsidence, 29, 48, 108, 130, 137, 182, 184, 223, 271
Sun, 232–3, 261–2, 267
Superchron, 23
Sydney Basin, 95, 128–31, *128*, *129*, 136–7

Taemas–Buchan area, 85, 94
talc, 32
Tamworth, 104
Tasman Line, *79*, 79–80, 88
Tasman Sea, 133, 137, 139, *140*, *165*, 176, 178
tectonics, 21, 80, 88–90, 94–5, 136–7
tektite, 25, *240*, 245
Tennant Creek, 8, 61, 80
terrane, 49
Tertiary, *50*, 139–61, 149, *154*, *161*, 164–80
thylacines, 161
Tibooburra, 93
till, *46*, 46, 64–5, *66*, 100–2, 132, 151
tillite, *see* till
Torrens Creek, 145, 168
Torres Strait, 1, 164
Tower Hill, 175

transform margin, 6, 28
trench, 27, *28*
Triassic, *50*, 119–25, 137, 258
trilobites, 84, *85*, *120*
tsunami, 193–5
tuff, 34, 100, 126, 174–5
Tumut, 93
turbidite, 48, *90*, 90–1
turbidity current, 47
Tweed River, 198, *199*

Uluru, 67, 67–8, 80
unconformity, 49, *67*, 81, *128*, *129*, 130–1, *144*
Undara, *169*, 169
uplift, 29, 49, 80, 89, 93, 104, 178–9, *210*, 211

varve, 104, *105*
vesicle, 35
Victor Harbour, 101
volcanic arc, *28*, 28, 88, *89*, 90, 94, 100, 106–7, 116, *132*
volcanic glass, 35
volcanism, viii, 27–8, *28*, 33, 33–5, 48, 58–9, 63–4, 88–93, 106–7, *129*, 131, 164–80, *165*, *166*, *172*, 209, 238, 262–4, *263*
volcanoes: central, 164–6, *165*, *172*; shield, 164, *165*

Warrumbungle, *165*, 166, 172
weathering, 41, 43, 68, 144, 164, 167, 200, 225
well sorted, 45
Wellington, 44, 95
Whitsunday Group, 133, 217
Wilkinson Lakes, 57
Wilpena, 71, *72*, 80
Wilsons Promontory, 206
Wolfe Creek crater, *242*, 244
Wollemi pine, *126*, 126–7
Wollongong, *129*, 130, 177, *192*, *193*
Wynyard (Tas), 103, 175

Yallalie, *243*, 243
Yass, *37*, 93
Yea, 94
Yilgarn craton, 55–7, *56*, *57*, 61, 253
Yorke Peninsula, 102, 158

zircon, 25, 31